K. Ramesh, Digital Photoelasticity

Springer-Verlag Berlin Heidelberg GmbH

K. Ramesh

Digital Photoelasticity

Advanced Techniques and Applications

With 211 Figures and 39 Tabels

Springer

Professor K. Ramesh

Indian Institute of Technology
Department of Mechanical Engineering
208016 Kanpur
India

e-mail: kramesh@iitk.ac.in

Additional material to this book can be downloaded from http://extras.springer.com.

ISBN 978-3-642-64099-5 ISBN 978-3-642-59723-7 (eBook)
DOI 10.1007/ 978-3-642-59723-7
CIP data applied for

Die Deutsche Bibliothek - CIP-Einheitsaufnahme
Digital photoelasticity : advanced techniques and applications / K. Ramesh. -
Berlin ; Heidelberg ; New York ; Barcelona ; Hong Kong ; London ; Milan ; Paris ;
Singapore ; Tokyo ; Springer, 2000

Typesetting: Camera ready by author
Cover design: de'blik, Berlin
SPIN: 10752625 62/3020 hu - 5 4 3 2 1 0 -

Dedicated to

His Holiness Jagadguru Śrī Candraśekharendra Saraswathi Swāmi
Śankarācārya of Kāñci Kāmakoti Pīṭham
Kāñcīpuram

Preface

Rapid strides have been made in the use of digital image processing techniques for data acquisition in photoelasticity in the last two decades. Techniques such as fringe thinning, fringe clustering, fringe tracing, phase shifting, polarization stepping and Fourier transform methods have significantly contributed to the automation of data acquisition. The recent developments in colour image processing and development of tricolour light source have added a new dimension. The use of time delay and integration (TDI) camera techniques has extended digital photoelasticity for dynamic analysis. Now the field of *Digital Photoelasticity* has matured to a level where it could be used to solve problems in industries. Apart from developments in data acquisition techniques, several methods have also come into existence for efficient processing of experimental data. Extensive use of computer graphics has found a unique place in presenting the experimental results in a meaningful way.

Though there has been significant developments in data processing and data acquisition in the last two decades, there is no book available yet to present these developments in a comprehensive way. The motivation for this book is based on the experience of teaching the course on *Experimental Methods in Stress Analysis* at IIT Kanpur for the last 10 years. I have always felt a need for introducing Digital Image Processing in an appropriate way, which will be useful for an experimentalist. Further, a need was also felt in improving the programming skills of Mechanical, Aerospace and Civil engineering students on the use of 'C' language in writing codes for experimental mechanics.

The principal objectives of the book are to provide an introduction to basic concepts and methodologies for digital photoelasticity and to develop a foundation on which the future researcher and student can develop his own new and fruitful ideas and promote research in formulation of problems in digital photoelasticity and application of these techniques to industries. Particular care has been taken to give a clear, straightforward, and simple presentation of the various topics.

In one volume it provides, the basics of photoelasticity, data acquisition by DIP techniques, data analysis by statistical techniques, data presentation by computer graphics and the use of Rapid Prototyping technologies to enhance the speed of the entire process. The book not only presents the various techniques but also provides time-tested software codes for important techniques. Exercises designed to support and extend the treatment have been placed at the end of each chapter.

The book is divided into eleven chapters. Jones calculus has been used extensively to explain the various optical arrangements used in this book. The first three chapters provide the necessary background on photoelasticity, reflection photoelasticity and digital image processing. Chapter 1 not only reviews the basic principle of photoelasticity but also provides a discussion on the use of modern methods for data reduction and the use of computer graphics for theoretical reconstruction of fringe patterns. A comprehensive discussion has also been provided on fringe ordering in photoelasticity. Chapter 2 briefly reviews the methodology of reflection photoelasticity and emphasises the need of correction factors for an engineering analysis. Chapter 3 gives a comprehensive discussion on digital image processing. The techniques that will have potential application for digital photoelasticity are highlighted. As image acquisition is a very crucial step in digital photoelasticity, a discussion on the design and principle of operation of CCD cameras is also provided in this chapter.

Chapters 4–7 give the various methodologies developed for data acquisition in digital photoelasticity exhaustively. At the beginning of each of these chapters, the relevant conventional techniques are reviewed to appreciate the connectivity of these with the modern developments. Chapter 4 deals with those techniques that mimic the conventional way of data acquisition using modern gadgets. Techniques related to fringe multiplication, fringe thinning and fringe tracing are discussed in this chapter. Chapter 5 deals with the whole-field extraction of photoelastic data based on direct processing of the intensity recorded. Techniques such as phase shifting, polarization stepping and Fourier transform techniques are discussed in this chapter. These techniques represent a fundamental shift in the methodology of acquiring data in photoelasticity and have provided hope for minimal operator intervention in interpreting photoelastic data. To get a comprehensive understanding of the various techniques and also to provide an easy comparison between them, the various optical arrangements and the relevant intensity equations are presented in tabular form. Chapter 6 deals with the algorithms meant for unwrapping the phase determined by various techniques in Chap. 5. Issues related to enhancing data extraction in stress concentration zones are also discussed in this chapter. Chapter 7 deals with the use of colour image processing techniques for data acquisition in digital photoelasticity. Methodologies of three-fringe photoelasticity, phase shifting in colour domain, polarization stepping in colour domain, spectral content analysis and tricolour photoelastic method are discussed in this chapter.

Chapters 8–11 deal with the application of these methodologies and summarise the recent trends and future possibilities of digital photoelasticity. Chapter 8 presents the application of digital techniques for data extraction in high stress gradient problems such as contact stress and evaluation of stress field in fracture mechanics. This chapter provides a synthesis of data acquisition by digital techniques, statistical methods for data interpretation and finally verification of the experimental result through computer graphics. The main emphasis of the chapter is to show that the experimental analysis could be self-consistent. Chapter 9 deals with a comprehensive review of the techniques for stress separation. Chapter 10 deals with the early initiatives on the stress analysis of RP models by photoelasticity. Chapter 11 initially deals with the extension of digital techniques developed for 2-D transmission photoelastic analysis to other branches of photoelasticity such as *integrated photoelasticity* and *scattered light photoelasticity*. The recent developments in DIP hardware and software, particularly the need for developing a device independent software for digital photoelasticity are then indicated. Developments in CCD camera technology and its role in digital dynamic recording are presented. Finally a comprehensive discussion of photo-orthotropic elasticity theories and the role of digital techniques for data acquisition is presented.

A new one-semester course on Digital Photoelasticity could be designed based on this book. Selected chapters of this book can form part of courses such as Experimental Stress Analysis (Chaps. 1–4, 8 and 9), Fracture Mechanics (Chaps. 1 and 8) and Whole Field Optical Measurements (Chaps. 1–7). The designers and consultants can adopt the techniques discussed in this book in practice as source code of computer software for most of the techniques are provided in the accompanying CD.

The credit for various methodologies of digital photoelasticity discussed in this book belongs to many investigators across the world who have contributed to the growth of the field. Many of the fringe patterns and graphs presented in this book stem from research in the Photomechanics Laboratory at IIT Kanpur. Credit for these accomplishments belongs to numerous former students whose publications are cited throughout the book. The book also has results of several other researchers. Figures 5.20 and 5.21 are courtesy of Professor Y. Morimoto (Wakayama University, Japan) and these were reproduced with permission from Society for Experimental Mechanics. Figures 7.14–7.16 are courtesy of Professor S. Yoneyama (Gaukin University, Japan) and Fig. 11.12 is courtesy of Professor A. Asundi (Nanyang Technological University, Singapore) and these were reproduced with permission from Elsevier Science Limited. Figures 9.24 and 10.12 and part of Table 2.1 are courtesy of Measurements Group, Inc., NC, USA. Abridged list of PIP commands in the enclosed CD is reproduced with permission from Matrox Electronics Systems, Canada.

Permission was extended by the Society for Experimental Mechanics, SPIE – The International Society for Optical Engineering, British Society for Strain Measurement (BSSM), John Wiley & sons Limited and Elsevier Science Limited to use the published material in which I was one of the authors. Research by the author on Digital Photoelasticity was supported in part by the Aeronautics Research and Development Board (Structures Panel), Government of India.

I am indebted to a number of academicians who have provided critical reviews to several chapters of this book. In particular I wish to extend my appreciation to Professor G. Subramanian (Aerospace Engineering, IIT Madras), Professor P. M. Dixit (Mechanical Engineering, IIT Kanpur),

Professor Sumana Gupta (Electrical Engineering, IIT Kanpur) and Professor B. Rathish Kumar (Department of Mathematics, IIT Kanpur). Although these reviewers provided valuable comments, the accuracy of the text is the sole responsibility of the author. I will feel grateful if readers bring to my attention any errors and omissions.

I am also indebted to a number of individuals who, have contributed in different, but important, ways to the preparation of this book. In particular I wish to extend my appreciation to Professor Om Prakash (Mechanical Engineering, IIT Kanpur), Dr. D. P. S. Chauhan (Mechanical Engineering, SGSITS, Indore) and the spirited team of students comprising of Mr. Sanjay Kumar Mangal, Mr. Dhiraj Kumar Tamrakar, Mr. Vinayak Gadre, Mr. Anant Gopal Kulkarni and Mr. Abhishek Goel. Thanks are due to a large number of graduate and undergraduate students of IIT Kanpur on whom most of the material presented in this book was tried and tested. Thanks are also due to Mr. S. S. Kushwah and Mr. G. K. Shukla for their meticulous work in preparing the figures and to Mr. Radhey Shyam and Mr. Lavendra Singh for assisting me in preparing the camera-ready manuscript.

I express my appreciation to my editor Dr. Dietrich Merkle at Springer-Verlag and to my production editors Ms. Gabirele Mass and Ms. S. Cuneus and also Ms. Nicole Grünewald Heller for their positive and accommodative approach during the entire period of this project.

I have received words of practical wisdom from Professor K. A. Padmanabhan (Director, IIT Kanpur) and Professor N. Sathyamurthy (Department of Chemistry, IIT Kanpur). My thanks are due to them.

No words of gratitude can fully explain the role played by my parents in bringing out this book. My father Sri. M. Krishnamurthi meticulously corrected the language of the book. Inquisitiveness showed by my daughter Sow. Mathangi, aged six, on the need of correction and re-correction of the manuscript almost on a daily basis was a source of constant encouragement to me. My son Chi. Sankaraprasad aged two also took an active part as he felt that it was his privilege to keep the multi-coloured pens and would issue it to me one by one to implement the corrections. Last but not the least my thanks are due to my wife Sow. Anuradha for all her indirect help in bringing out this book to its logical conclusion.

I had the unique and rare privilege of explaining portions of my PhD thesis to His Holiness Jagadguru Śrī Candraśekharendra Saraswathi Swāmi, Śankarācārya of Kāñci Kāmakoti Pīṭham in 1988. His Holiness's message of perfection and excellence in the chosen field of activity was a source of inspiration for me to write this book.

Kanpur
9th September 1999

K. Ramesh

Contents

1 Transmission Photoelasticity 1

1.1 Introduction 1
1.2 Physical Principle Used in Photoelasticity 1
1.3 Nature of Light 2
1.4 Polarization 3
1.5 Passage of Light Through Isotropic Media 4
1.6 Passage of Light Through a Crystalline Medium 5
1.7 Light Ellipse 6
1.8 Retardation Plates and Wave Plates 8
1.9 Stress-Optic Law 9
1.10 Plane Polariscope 10
1.10.1 Analysis by Trigonometric Resolution 13
1.11 Jones Calculus 14
1.11.1 Rotation Matrix 14
1.11.2 Retardation Matrix 15
1.11.3 Representation of a Retarder 16
1.11.4 Polarizer 17
1.11.5 Quarter-Wave Plate 17
1.12 Analysis of Plane Polariscope by Jones Calculus 17
1.13 Circular Polariscope 18
1.14 Use of White Light 20
1.15 Determination of Isoclinic and Isochromatic Fringe Order at a Point 21
1.15.1 Ordering of Isoclinics 21
1.15.2 Ordering of Isochromatics 22
1.16 Tardy's Method of Compensation 22
1.17 Calibration of Photoelastic Model Materials 24
1.17.1 Stress Field in a Circular Disc Under Diametral Compression 24
1.17.2 Conventional Method 25
1.17.3 Sampled Linear Least Squares Method 26
Need for a better methodology 26
Use of whole field data to evaluate material fringe value 26
1.17.4 Theoretical Reconstruction of Fringe Patterns 28
1.18 Further Comments on Fringe Ordering 29
1.18.1 Properties of Isochromatic Fringe Field 30

1.18.2 Properties of Isoclinic Fringe Field 31
1.18.3 Use of Fringe Field Properties to Identify Fringe Ordering 32
1.18.4 Role of Principles of Solid Mechanics in Fringe Ordering 33
1.19 Determination of the Sign of the Boundary Stresses 34
1.20 Resolving the Ambiguity on the Principal Stress Direction . . 34
1.21 Introduction to Three-Dimensional Photoelasticity and Integrated Photoelasticity 35
1.21.1 Conventional Three-Dimensional Photoelasticity . . . 36
1.21.2 Principle of Optical Equivalence 37
1.22 Model to Prototype Relations 39
1.23 Closure 42
Exercises 42
References 46

2 Reflection Photoelasticity 47

2.1 Introduction 47
2.2 Reflection Polariscope 48
2.3 Stress and Strain-Optic Relations for Coatings 49
2.4 Coating and Specimen Stresses 50
2.5 Correction Factors for Photoelastic Coatings 52
2.6 Poisson's Ratio Mismatch 56
2.7 Coating Materials 57
2.8 Bonding the Coating 59
2.9 Selection of the Coating Thickness 60
2.10 Calibration of the Coating Material 62
2.11 Data Collection and Analysis 64
2.12 Application of Photoelastic Coatings 65
2.13 Closure 65
Exercises 65
References 66

3 Digital Image Processing 67

3.1 Introduction 67
3.2 Image Sampling and Quantization 67
3.2.1 Pictures as Functions 67
3.2.2 Uniform Sampling and Quantization 68
3.3 Video Standards 69
3.4 Image Sensors 71
3.5 Image Display 72
3.6 Image Perception 73
3.7 Image Storage 74

3.8 Some Basic Relationships and Mathematical Operations Between Pixels . . . 74
3.8.1 Neighbours of a Pixel . . . 74
3.8.2 Arithmetic and Logic Operations . . . 75
3.8.3 Neighbourhood Oriented Operations . . . 75
3.9 Basic Steps in Image Processing . . . 77
3.10 Typical Image Processing Systems for Digital Photoelasticity 77
3.11 Software Structure and Design . . . 79
3.12 Image Acquisition . . . 80
3.13 Tools for Image Understanding . . . 82
3.13.1 Pseudo Colouring . . . 82
3.13.2 Histogram . . . 83
3.13.3 Two-Dimensional and Three-Dimensional Intensity Plots . . . 84
3.14 Filtering in Spatial Domain . . . 86
3.14.1 Low Pass Spatial Filtering . . . 88
3.14.2 Median Filtering . . . 88
3.15 Image Enhancement . . . 88
3.15.1 Contrast Stretching . . . 89
3.15.2 Histogram Equalisation . . . 90
3.16 Image Segmentation . . . 91
3.16.1 Thresholding . . . 91
Global thresholding . . . 92
Semi thresholding . . . 92
Dynamic thresholding . . . 92
3.16.2 Edge Detection . . . 94
Edge detection by convolution filters . . . 94
Edge detection by non-convolution filters . . . 95
Edge detection by thresholding . . . 96
3.17 Morphological Filters . . . 98
3.18 Further Discussions on Image Sensors . . . 98
3.18.1 Operation of CCD Arrays . . . 98
3.18.2 Interline Transfer CCD . . . 100
3.18.3 Linearity and Dynamic Range . . . 101
3.18.4 Sources of Noise . . . 102
3.19 Digitisation of the Camera Video Signal . . . 103
3.20 Resolution of an Image Processing System . . . 103
3.21 Gamma Compensation . . . 104
Exercises . . . 104
References . . . 105

4 Fringe Multiplication, Fringe Thinning and Fringe Clustering . . . 107

4.1 Introduction . . . 107
4.2 Fringe Multiplication . . . 108

4.3 Half Fringe Photoelasticity (HFP) 110
4.4 DIP Methods for Fringe Thinning 111
4.5 Algorithms Based on Considering the Fringe Patterns as a Binary Image 112
4.6 Mask-Based Algorithms for Skeleton Extraction Using Intensity Variation within a Fringe 113
4.7 Global Identification of Fringe Skeletons Based on Intensity Variation 115
4.7.1 Edge Detection 115
4.7.2 Fringe Skeletonization 116
Row-Wise scanning algorithm 116
Algorithm for fringe skeleton extraction for arbitrarily shaped fringes 117
4.7.3 Applications of the Algorithm to Actual Experimental Conditions 120
4.8 Further Improvements on the Global Thinning Algorithm . . 123
4.9 Performance Evaluation of Various Fringe Thinning Algorithms 126
4.9.1 Comparison of the Skeleton Extraction 127
Computer generated test images 127
Images recorded from actual experimental situations . 129
4.9.2 Comparison of the Computational Effort 131
4.10 Use of Tiling to Improve Information in Stress Concentration Zones 131
4.11 Fringe Tracing Algorithm 133
4.12 Ordering of Fringes 135
4.13 Closure 136
Exercises 137
References 138

5 Phase Shifting, Polarization Stepping and Fourier Transform Methods 141

5.1 Introduction 141
5.2 Early Attempts for Automated Polariscopes 142
5.3 Phase Shifting in Photoelasticity 144
5.4 Intensity of Light Transmitted for a Generic Arrangement of a Plane Polariscope 146
5.5 Intensity of Light Transmitted for a Generic Arrangement of a Circular Polariscope 149
5.6 Evaluation of Fractional Fringe Order along an Isoclinic Contour 152
5.7 Whole Field Evaluation of Photoelastic Data by Using a Plane Polariscope 153
5.8 Whole Field Evaluation of Photoelastic Data by Using a Circular Polariscope 155

5.8.1 The Generic Procedure 155
5.8.2 Calculation and Representation of Whole Field Data . 159
5.8.3 Parameters Affecting the Generation of Phase Map and its Solution 162
Influence of local oscillations of isoclinic parameter on fractional retardation calculation 162
Importance of isoclinic parameter representing either σ_1 or σ_2 direction over the domain 163
Ambiguity in experimentally evaluating the isoclinic parameter 167
Interactive approach to obtain a good phase map . . . 169
5.9 Error Sources and Methods to Minimise Their Influence . . . 172
5.9.1 Influence of Error in Measuring Intensities 174
5.9.2 Errors Due to Mismatch of Quarter-Wave Plates . . . 177
5.10 Evaluation of Isoclinic Value by Phase Shifting Technique . 181
5.10.1 Use of Two Loads to Get Continuous Isoclinic Contours 181
5.10.2 Use of Multiple Wavelengths to Get Continuous Isoclinic Contours 183
5.11 Polarization Stepping for Isoclinic Determination 185
5.12 Fourier Transform Methods for Photoelastic Data Acquisition 188
5.12.1 Use of Carrier Fringes 188
5.12.2 Use of Multiple Polarization Stepped Images 189
5.12.3 Use of Load Stepping 191
5.13 Comparative Evaluation of Phase Shifting, Polarization Stepping and Fourier Transform Techniques 192
5.14 Closure 193
Exercises 193
References 194

6 Phase Unwrapping and Optically Enhanced Tiling in Digital Photoelasticity 199

6.1 Introduction 199
6.2 Boundary Detection 200
6.3 Noise Removal in Phase Maps 201
6.4 Algorithm for Phase Unwrapping 202
6.5 Representation of the Unwrapped Phase. 205
6.5.1 Three-Dimensional Plots 205
6.5.2 Total Fringe Order Viewing on the Image 206
6.6 Parameters Affecting Phase Unwrapping 207
6.6.1 Influence of the Selection of the Phase Unwrapping Threshold 207
6.6.2 Influence of the Location of the Primary Seed Point. . 209
6.7 Use of Tiling Procedure for Phase Unwrapping. 210

6.8 Digital Magnification of High Fringe Density Zones 211
6.8.1 Replication 211
6.8.2 Linear Interpolation 212
6.8.3 Higher Order Interpolation 212
6.9 Optically Enhanced Tiling (OET) 212
6.10 Cementing of a Tile 213
6.11 OET Applied to a Circular Disc Under Diametral Compression 215
6.12 OET Applied to a Ring Under Diametral Compression 217
6.13 Closure 219
Exercises 220
References 220

7 Colour Image Processing Techniques 221

7.1 Introduction 221
7.2 Colour Models 223
7.2.1 RGB Model 224
7.2.2 HSI Model 225
7.3 Colour Image Processing Systems 226
7.3.1 Hardware 226
Transmission Photoelasticity 227
Reflection Photoelasticity 227
7.3.2 Software 228
7.4 Typical Spectral Response of a Colour Camera 228
7.5 Intensity of Light Transmitted in White Light for Various Polariscope Arrangements 230
7.6 Three Fringe Photoelasticity (TFP) 234
7.6.1 Calibration 235
7.6.2 Methodology 240
7.6.3 Application to the Problem of a Circular Disc Under Diametral Compression 240
7.7 Green Image Plane as a Green Filter 243
7.8 Phase Shifting in Colour Domain 244
7.8.1 Transmission Photoelasticity 244
7.8.2 Reflection Photoelasticity 246
7.9 Spectral Content Analysis (SCA) 249
7.10 Digital Spectral Content Analysis (DSCA) 252
7.11 Hybrid Techniques 252
7.11.1 Polarization Stepping in Colour Domain 253
7.12 Tricolour Photoelastic Method 254
7.13 Closure 258
Exercises 261
References 261

8 Evaluation of Contact Stress Parameters and Fracture Parameters 265

8.1 Introduction 265
8.2 Basic Data Required and its Digital Acquisition 266
8.2.1 Conversion of Pixel Co-ordinates to Model Co-ordinates 266
8.2.2 Rotational Transformation 267
8.3 Stresses in Terms of Contact Length and Geometrical and Elastic Properties of the Bodies in Contact 268
8.4 Evaluation of Contact Stress Parameters by Least Squares Analysis 270
8.4.1 Validation for Hertzian and Non-Hertzian Contact . . 272
8.5 Developments in the Description of the Stress Field Equations in the Neighbourhood of a Crack-tip 273
8.5.1 Mode-I Stress Field Equations 273
8.5.2 Mixed-Mode (Combination of Mode-I and Mode-II) Stress Field Equations 276
8.5.3 Equivalence Between the Multi-Parameter Stress Field Equations 278
8.6 Developments in SIF Evaluation Methodology 278
8.7 Evaluation of Mixed-Mode Stress Field Parameters Using Least Squares Technique 281
8.8 Experimental Validation of the Methodology 284
8.8.1 Mode-I Loading 284
8.8.2 Mixed-Mode Loading 288
8.9 Contact Stress and Fracture Analysis of a Spur Gear 290
8.9.1 Loading Frame Design 290
8.9.2 Evaluation of Contact Parameters 292
Measurement of radius of curvature at the point of contact 292
Experimental results 293
8.9.3 Evaluation of Fracture Parameters 294
8.10 Closure 298
Exercises 299
References 299

9 Stress Separation Techniques 303

9.1 Introduction 303
9.2 Oblique Incidence Method 304
9.2.1 Secondary Principal Stresses 304
9.2.2 The Methodology 305
9.3 Shear Difference Technique 307
9.3.1 Conventional Method 307
9.3.2 Improvement by Tesar 308

9.4 Survey of Numerical Methods 309
9.4.1 Integration of Compatibility Condition 309
Finite difference approach 309
9.4.2 Integration of Stress Difference Equations 312
9.4.3 Least Squares Method 314
9.4.4 Hybrid Techniques 316
9.4.5 Methods Using Only Isochromatic Data 317
9.5 Stress Separation by Combined Phase Shifting and FEM . . 318
9.5.1 Finite Element Formulation 319
9.5.2 Meaningful Discretization of the Domain 323
9.5.3 Plotting of Fringe Contours from FE Results 324
9.5.4 Influence of Error in Fringe Data 327
9.5.5 Application of the Technique to the Problem of Plate with a Hole 330
9.6 Use of Integrated Photoelasticity Concepts for Stress Separation 332
9.6.1 Least Squares Algorithm 333
9.6.2 Design of the Loading Frame 333
9.6.3 Application to the Problem of Disc under Diametral Compression 334
9.7 Stress Separation in Three-Dimensional Photoelasticity . . . 337
9.8 Stress Separation in Reflection Photoelasticity 342
9.8 Closure 344
Exercises 344
References 345

10 Fusion of Digital Photoelasticity, Rapid Prototyping and Rapid Tooling Technologies 347

10.1 Introduction 347
10.2 Difficulties in Conventional Three-Dimensional Photoelasticity 348
10.3 Rapid Prototyping in Model Making 348
10.3.1 Software Issues in RP 349
10.3.2 Stereolithography Process 351
10.3.3 Solid Ground Curing 354
10.3.4 Fused Deposition Modelling 355
10.4 Direct Analysis of RP Models by Photoelastic Coatings . . . 356
10.4.1 Experimental Results 357
10.4.2 Analysis of the Results. 358
Evaluation of Young's modulus by tensile test 358
Study on the seepage of the adhesive 359
Numerical simulation of fringe patterns 359
10.4.3 Recommendations. 361
10.5 Direct Use of RP Models for Transmission Photoelastic Analysis. 361

10.6 Rapid Tooling for Model Making 363
10.6.1 Basic Steps in Rapid Tooling 363
10.6.2 Digital Photoelastic Characterisation of the Process . . 364
10.7 Closure 365
Exercises 366
References 366

11 Recent Developments and Future Trends 369

11.1 Introduction 369
11.2 Evaluation of Characteristic Parameters 369
11.2.1 Srinath and Keshavan's Method 370
11.2.2 Whole Field Determination of Characteristic Parameters by Phase Shifting 371
Development of relevant equations 371
Experimental evaluation of characteristic parameters . 375
Whole field theoretical evaluation of characteristic parameters 375
11.3 Tensorial Tomography 378
11.4 Developments in DIP Hardware 383
11.5 Developments in DIP Software 384
11.5.1 Development of a Device Independent Software . . . 385
Selection of software features 385
FRN_DAT software 385
An application 386
11.5.2 Future Possibility 387
11.6 Digital Dynamic Photoelasticity 388
11.6.1 Classification of High, Very-high and Ultra-high-speed Photography 388
11.6.2 Classical Methods for High-speed Photography . . . 388
11.6.3 Digital Dynamic Recording 391
11.7 Application to Composites 395
11.7.1 Photo-Orthotropic Elasticity Theories 396
Stress-Optic law 396
Strain-Optic law 397
11.7.2 Calibration of Photo-Orthotropic Composites 398
11.7.3 Influence of Residual Birefringence 399
11.7.4 Separation of Stresses in Photo-Orthotropic Elasticity 400
11.7.5 Application of Digital Photoelasticity to Composites 400
11.8 Closure 401
Exercises 401
References 401

Index 405

1 Transmission Photoelasticity

1.1 Introduction

Photoelasticity is an experimental method for analysing stress or strain fields in mechanics. The technique of photoelasticity is very well developed and many standard textbooks [1–7] have been written. An early description of the method of photoelasticity was provided by Coker and Filon [1] in 1931. Then, in 1937, Oppel [8] introduced the concept of frozen stress photoelasticity, which has facilitated the analysis of three-dimensional problems with the use of two-dimensional concepts. This is achieved by initially stress freezing the model and then mechanically slicing it. The mechanical slicing was replaced by optical slicing with the use of scattered light by Weller [9] in 1939. A new branch of photoelasticity viz., the *scattered light photoelasticity* was then developed [10]. Analysis of a three-dimensional model as a whole was proposed by Aben [11] and the technique is known as *integrated photoelasticity*.

In this chapter, the basic principles of transmission light photoelasticity are discussed. The various arrangements of the polariscope, use of monochromatic and white lights are also discussed. Use of Jones calculus for analysing various optical arrangements of a polariscope is presented. The role of digital computers on the analysis of photoelastic data and the use of computer graphics for fringe reconstruction are highlighted. A comprehensive discussion is also presented on the ordering of fringes.

1.2 Physical Principle Used in Photoelasticity

Certain non-crystalline transparent materials, notably some polymeric plastics, are optically isotropic under normal conditions but become doubly refractive or birefringent when stressed. This effect normally persists while the loads are maintained but vanishes almost instantaneously or after some interval of time depending on the material and conditions of loading when the loads are removed.

This phenomenon of temporary or artificial birefringence was first observed by Brewster [12] in 1816. This is the physical characteristic on which photoelasticity is based.

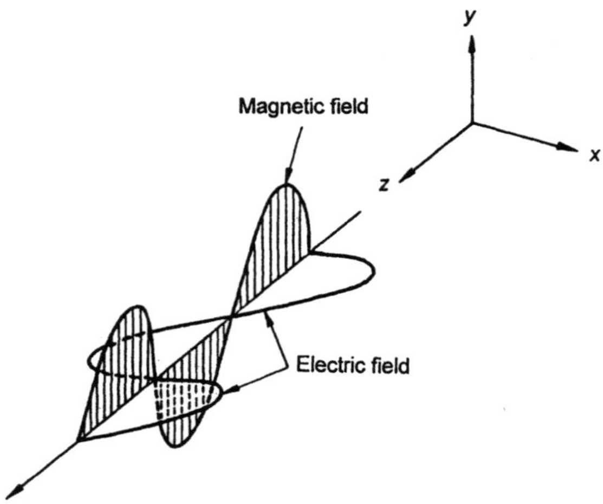

Fig. 1.1. The electric and magnetic vectors associated with a plane electromagnetic wave

1.3 Nature of Light

Maxwell's electromagnetic theory predicts the presence of two vector fields in light waves, viz., an electric field (***E***) and a magnetic field (***H***) (Fig. 1.1). This theory was a significant step in the understanding of light, as these fields can propagate through space unsupported by any known matter. The need of the hypothetical ether, which was used in earlier studies, was eliminated. The magnetic and electric vectors are in phase, perpendicular to each other and at right angles to the direction of propagation. Either of these vectors can be taken as the fundamental light vector. In this book the electric vector is taken as the fundamental light vector. The instantaneous magnitude E of the electric vector as observed at a fixed point along the direction of propagation can be represented as

$$E = a\cos\frac{2\pi}{\lambda}ct = a\cos 2\pi ft = a\cos\omega t \tag{1.1}$$

where, λ is wavelength of light, c is velocity of propagation, t is time, a is amplitude, f is the frequency of light and ω is the circular frequency of the light. Instead of a cosine function, a sine function can also be used to represent a plane harmonic wave i.e.,

$$E = a\sin\omega t \tag{1.2}$$

The electromagnetic theory also unites light with all the other invisible entities of the electromagnetic spectrum such as gamma rays, ultraviolet rays, infra red rays etc.

The frequency f of a light wave determines that quality which the eye recognises as colour. Visible range of the spectrum is a small band centred about a wavelength of approximately 550 nm. The lowest frequency which the human eye can recognise as light is about 3.9×10^{14} Hz and this corresponds to deep red colour. The highest frequency is about 7.7×10^{14} Hz corresponding to deep violet. Light from a source that emits a continuous spectrum with nearly equal energy for every wavelength is interpreted as *white light*. Light of a single wavelength is known as a *monochromatic light*.

1.4 Polarization

Most light sources consist of a large number of randomly oriented atomic or molecular emitters. The light rays emitted from such a source will have no preferred orientation and the tip of the light-vector describes a random vibratory motion in a plane transverse to the direction of propagation (Fig. 1.2). If the tip of the light vector is forced to follow a definite law, the light is said to be polarised.

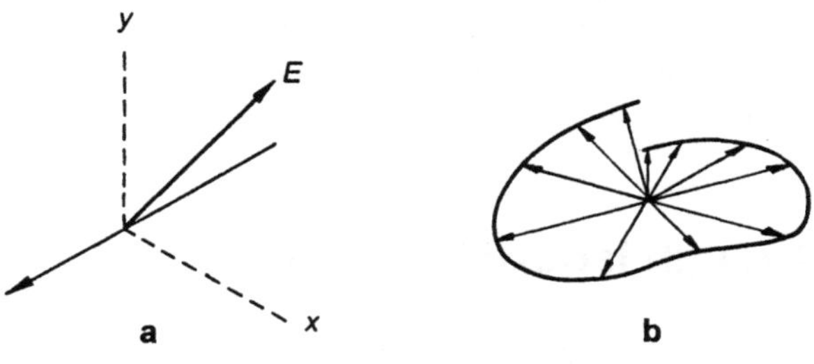

Fig. 1.2. **a** Representation of light as a vector. **b** The light vector magnitude and its direction change randomly in a normal light source

For example, if the tip is constrained to lie on the circumference of a circle, it is said to be circularly polarised. If the tip describes an ellipse, it is said to be elliptically polarised, Fig. 1.3. If the light vector is parallel to a given direction in the wave front, it is said to be linearly or plane polarised. Plane polarised light can be produced from a natural source by the following ways. (1) Reflection (2) Scattering (3) Use of Polaroid sheets or (4) Nicol's prism. Using the properties of birefringence, a linearly polarised light can be changed to a circular or elliptically polarised light (Sect. 1.8). If the tip of the light vector describes a counter-clockwise motion, then it is said to be right-handedly polarised (Fig. 1.3a). On the other hand, if the path is traversed in a clockwise direction, then it is said to be left-handedly polarised (Fig. 1.3b).

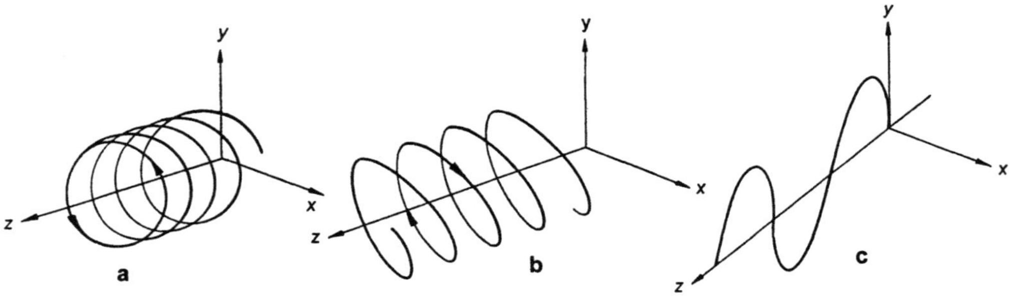

Fig. 1.3. Forms of polarised light: **a** circularly polarised **b** elliptically polarised **c** linearly polarised. Note that **a** is right-handedly polarised and **b** is left-handedly polarised

Large fields of commercial diffused light polariscopes are made of Polaroid sheets. These sheets are made from a sheet of polyvinyl alcohol in which the long chain molecules are oriented by stretching, thereby achieving birefringence. These long chains absorb pigment molecules of iodine by selective attachments. The selective absorption of one of the refracted rays occurs because of the orientation of the chemical bonds at the attachment sites. At the extreme blue end of the spectrum, the polarising action is not very effective and when a fairly powerful white light is viewed through two crossed Polaroid's, the field appears blue.

1.5 Passage of Light Through Isotropic Media

The laws governing the passage of light through isotropic media are well known as Snell's laws of refraction and reflection. These laws are:

1. The normal to the incident wave, the normal to the interface, and the normals to the reflected and refracted waves, all lie in one plane.
2. The angle of incidence is equal to the angle of reflection, Fig. 1.4.
3. The ratio of the sine of the angle of incidence to the sine of the angle of refraction is a constant for two given isotropic media. This is known as the relative refractive index and if the medium one is vacuum then it is absolute refractive index. Thus, if c is the velocity of light in vacuum, then

$$\frac{\sin i}{\sin r} = \frac{v_1}{v_2} = n_{12},\ n_1 = \frac{c}{v_1},\ n_2 = \frac{c}{v_2} \tag{1.3}$$

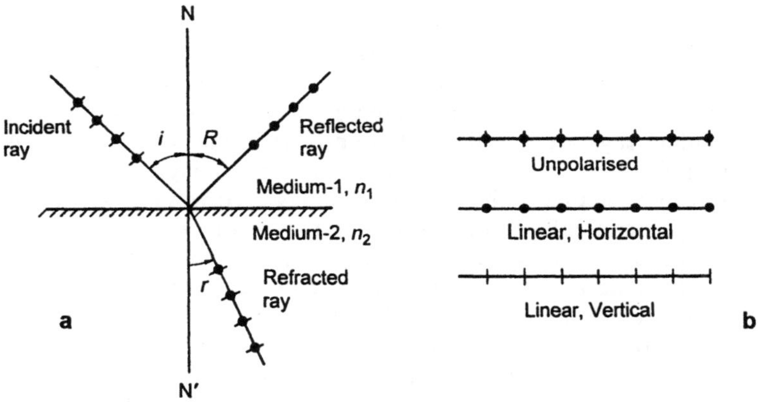

Fig. 1.4. a Passage of light through an isotropic medium. **b** Schematic representation of unpolarised and linearly polarised light

1.6 Passage of Light Through a Crystalline Medium

The crystalline media are optically anisotropic and are characterised by the fact that a single incident ray will give rise to two refracted rays, ordinary 'o' and extraordinary 'e' (Fig. 1.5), thus exhibiting what is generally known as double refraction. One of the rays is extraordinary in the sense that it manages to violate Snell's Law under suitable circumstances. This ray need not be confined to the plane of incidence. Further, its velocity changes in a continuous way with angle of incidence. The two indices of refraction for

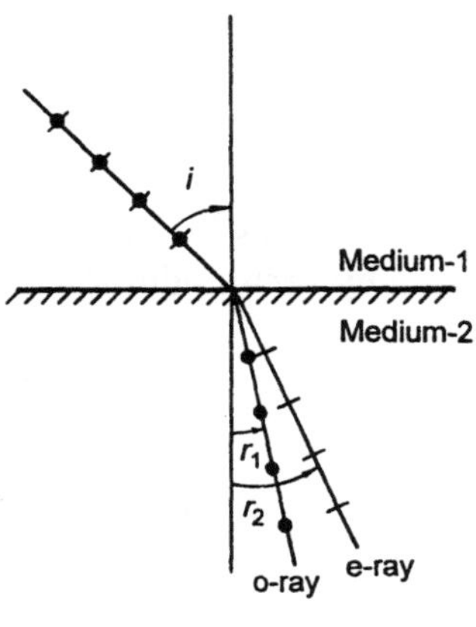

Fig. 1.5. Passage of light through a crystalline medium

rays 'e' and 'o' are equal only in the direction of an optic axis. An isotropic medium can transmit common light, whereas the light travelling though a crystal is always polarised. The ordinary and extra-ordinary rays are plane polarised and their planes of polarization are perpendicular to each other.

When the incident rays are parallel to the optic axis, the 'o' and 'e' rays have the same refractive index and they behave as though they travel through an isotropic medium obeying Snell's law. However, when the incident rays are perpendicular to the optic axis, the 'e' ray will travel faster than the 'o' ray because of its lower refractive index but it will travel in the same direction. The general case is when the incident rays are at some angle between 0° and 90° from the optic axis, the 'o' ray will travel undeviated and be refracted according to Snell's law. The 'e' ray will deviate from the 'o' ray because of the different indices with direction. The direction of deviation will be away from the optic axis and generally out of the plane of incidence.

The photoelastic phenomena observed is due to the situation in which the incident rays are perpendicular to the optic axis. The principal stress directions act as polarising axes at the point of interest. The ordinary and extra-ordinary rays travel in the same direction but with different velocities. The two rays are polarised and the polarising axes are mutually perpendicular. When the rays emerge out, there exists a relative retardation between the rays, which contribute to the formation of fringes.

1.7 Light Ellipse

When the incident light is perpendicular to the optic axis, the light emerging out of a crystal plate has two mutually perpendicular plane polarised lights of different phases. Let these be represented as

$$\begin{aligned} E_x &= a_x \cos(\omega t + \alpha_1) \\ E_y &= a_y \cos(\omega t + \alpha_2) \end{aligned} \tag{1.4}$$

Let the relative phase difference between these two vibrations be $\alpha_1 - \alpha_2 = \delta$. The magnitude of the resulting light vector is given by vector addition as

$$E = \sqrt{E_x^2 + E_y^2} \tag{1.5}$$

The trace of the tip of the resulting electric vector on a plane perpendicular to the axis of propagation at a point can be obtained by eliminating time from Eq. (1.4).

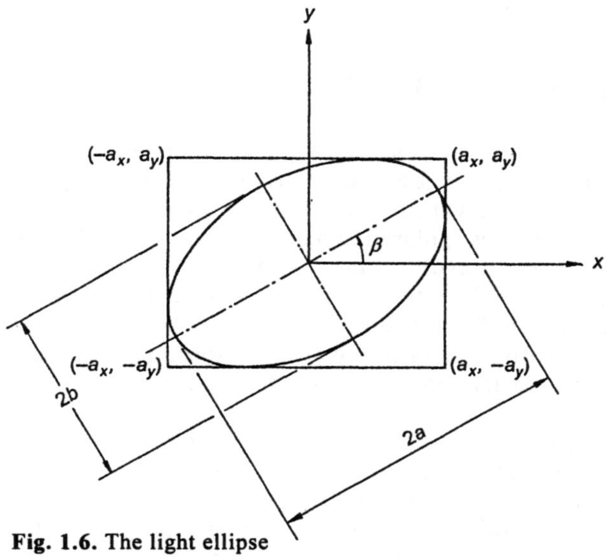

Fig. 1.6. The light ellipse

$$\frac{E_x^2}{a_x^2}+\frac{E_y^2}{a_y^2}-\frac{2E_xE_y}{a_xa_y}\cos\delta=\sin^2\delta \tag{1.6}$$

This is the general equation of an ellipse with E_x and E_y as the rectangular axis. The azimuth β (Fig. 1.6) of the ellipse with the horizontal is obtained as

$$\tan 2\beta=\frac{2a_xa_y}{a_x^2-a_y^2}\cos\delta \tag{1.7}$$

If a and b are the semi-major and minor axes of the ellipse, these values are given as

$$\begin{aligned} a^2&=a_x^2\cos^2\beta+a_y^2\sin^2\beta+a_xa_y\sin 2\beta\cos\delta \\ b^2&=a_x^2\sin^2\beta+a_y^2\cos^2\beta-a_xa_y\sin 2\beta\cos\delta \end{aligned} \tag{1.8}$$

The elliptically polarised light becomes a circularly polarised light when $\delta=\pi/2,\ 3\pi/2,\ \ldots$ and the amplitudes a_x and a_y are equal. If the amplitudes are not equal, then one gets an elliptically polarised light with the azimuth β as zero. This result indicates that the two vibrating components along the major and minor axes of the ellipse are at a phase difference of $\pi/2$. One gets a linearly polarised light if the phase difference between the waves are $\delta=0,\ \pi,\ 2\pi,\ \ldots$.

1.8
Retardation Plates and Wave Plates

Consider a crystalline plate of thickness h. Let a plane-polarised light be incident normally as shown in Fig. 1.7. Let the polarising axes of the plate be oriented at angles θ and $\theta + \pi/2$ with the horizontal. Since the incident ray is perpendicular to the optic axes (polarising axes); two rays namely, the ordinary and the extra-ordinary travel in the same direction but with different velocities of v_1 and v_2. These two rays will have a net phase difference of δ upon emergence from the crystal plate. The two waves shown in Fig. 1.7b are representing the ordinary and extraordinary rays and not to be confused with the electric and magnetic field of a light vector shown in Fig. 1.1.

The purpose of this section is to show the relative retardation (δ) in terms of the thickness and optical properties of the crystal plate. Since the velocities of propagation within the crystal is different for these two rays, they will take respectively h/v_1 and h/v_2 seconds to traverse the plate. This time difference contributes to the phase difference. Let the frequency of the light be f, then

$$\begin{aligned}\delta &= 2\pi f\left(\frac{h}{v_1}-\frac{h}{v_2}\right) = 2\pi h\frac{c}{\lambda}\left(\frac{1}{v_1}-\frac{1}{v_2}\right)\\ &= \frac{2\pi h}{\lambda}(n_1 - n_2)\end{aligned} \tag{1.9}$$

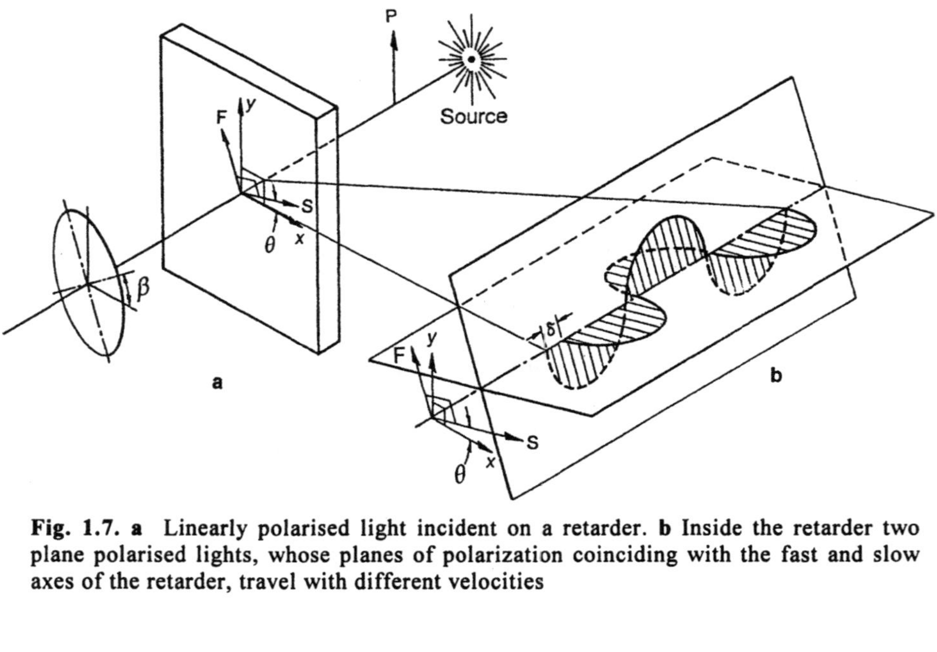

Fig. 1.7. a Linearly polarised light incident on a retarder. **b** Inside the retarder two plane polarised lights, whose planes of polarization coinciding with the fast and slow axes of the retarder, travel with different velocities

The emerging light is in general, elliptically polarised. If the thickness is such as to produce a phase difference of $\pi/2$ radian, then, it is a quarter-wave plate ($\lambda/4$) and if it is π radians, it is a half-wave plate. If the retardation is 2π, then, one gets a full-wave plate and the incident light is unaltered. Usually, these are made of mica or quartz and $\lambda/4$ plates are of the thickness of a millimetre. It is to be noted that a crystal plate can be called a quarter-wave plate, half-wave plate or a full-wave plate only for a particular wavelength. Wave-plates or retarders have two polarising axes and one of these axes is labelled as a fast axis (F) and the other as the slow axis (S).

1.9 Stress-Optic Law

Consider a transparent model made of high polymer subjected to a plane state of stress. Let the state of stress at a point be characterised by the principal stresses σ_1, σ_2 and their orientation with reference to a set of axes. When the model is stressed, experiments show that the model becomes doubly refractive and the directions of the polarising axes in the plane of the model at any point coincide with the directions of the principal stress axes at that point. Let n_1 and n_2 be the refractive indices for vibrations corresponding to these two directions. Let n be the refractive index in the unstressed state. Maxwell formulated in 1852 the relations between stresses and the indices of refraction as

$$\begin{aligned} n_1 - n &= c_1\sigma_1 - c_2\sigma_2 \\ n_2 - n &= c_1\sigma_2 - c_2\sigma_1 \end{aligned} \tag{1.10}$$

where, c_1 is called the direct stress-optic coefficient and c_2 is the transverse stress optic coefficient.

If a plane polarised light is incident on the model, then as discussed in Sect. 1.8, the relative retardation can be obtained as a function of the refractive indices as

$$\delta = \frac{2\pi h}{\lambda}(n_1 - n_2) \tag{1.11}$$

Substituting Eq. (1.10) in Eq. (1.11),

$$\delta = \frac{2\pi h}{\lambda}(c_1 + c_2)(\sigma_1 - \sigma_2) \tag{1.12}$$

This equation indicates that in a photoelastic model, the relative retardation changes from point to point depending on the stress level. If $c_1 + c_2$ is replaced by C, the relative retardation is given as

$$\delta = \frac{2\pi h}{\lambda}C(\sigma_1 - \sigma_2) \tag{1.13}$$

which can be rewritten in terms of fringe order N as

$$N = \frac{\delta}{2\pi} = h\frac{C}{\lambda}(\sigma_1 - \sigma_2) \tag{1.14}$$

The relative stress-optic coefficient C is usually assumed to be a constant for a material. However, various studies have shown that this coefficient depends on wavelength and should be used with care.

The Eq. (1.14) can be recast as

$$\sigma_1 - \sigma_2 = \frac{NF_\sigma}{h} \tag{1.15}$$

where

$$F_\sigma = \frac{\lambda}{C} \tag{1.16}$$

which is known as the material stress fringe value with the units N/mm/fringe. Equation (1.15) is famously known as stress-optic law as it relates the stress information to optical measurement. The principal stresses are labelled such that σ_1 is always algebraically greater than σ_2. In view of it, the L.H.S of Eq. (1.15) is always positive. Thus, in photoelasticity fringe order N is always positive. If the fringe order and the material stress fringe values are known, then, one can obtain the principal stress difference.

Determination of material stress fringe value is known as calibration of a photoelastic material. Material stress fringe value has to be evaluated with at least two to three decimal places accuracy, as it is the only parameter which links the optical information to stresses. Equation (1.16) shows the dependence of F_σ on the wavelength. Equation (1.15) implicitly gives the indication that F_σ and $(\sigma_1 - \sigma_2)$ are linearly related. However, at higher stress levels, the relationship is non-linear and Eq. (1.15) cannot be used. Therefore, Eq. (1.15) should be used with care.

1.10 Plane Polariscope

In photoelasticity, stress field is evaluated through the use of light. Typically for a known incident light, the emergent light is analysed for stress information. For example, while formulating the stress-optic law, a plane-polarised light was considered to be incident on the model. Since a polarised light is used, the basic equipment used in photoelasticity is called a polariscope.

One of the simplest optical arrangements possible is a plane polariscope. Figure 1.8 shows the typical arrangement. Let the light source be a monochromatic one. A circular disc under diametral compression is kept in the field of view. The incident light on the model is plane polarised. As it

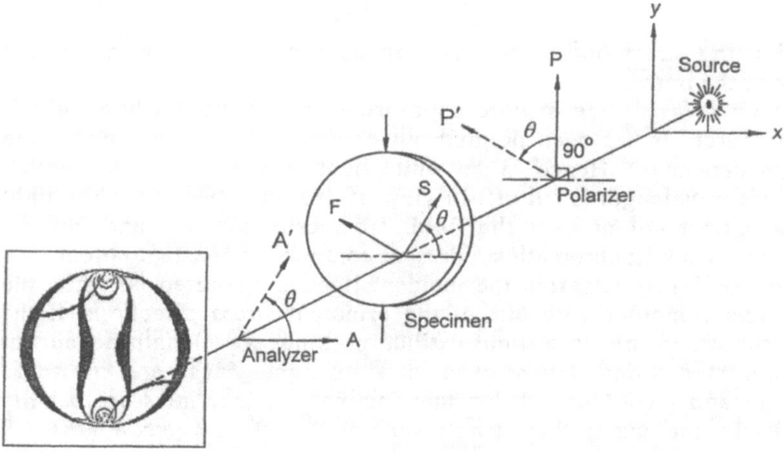

Fig. 1.8. Arrangement of optical elements in a plane polariscope set to dark-field arrangement. The fringe patterns correspond to the position P′ and A′ of the polarizer and analyzer respectively. One observes both isochromatics and θ-degree isoclinic ($\theta > 45$-deg) on the screen. The background field is shown as white for easy visualisation of the model boundary

passes through the model, the state of polarization changes from point to point depending on the principal stress direction and the magnitude of principal stress difference. The information about the stress field can be obtained if the state of polarization of the emergent light is studied. This is easily achieved by introducing a polarizer at 0°. Since this optical element helps to analyse the emergent light, it is known as an analyzer. With the introduction of the analyzer, fringe contours that are black appear on the screen. The fringe contours correspond to those points where intensity of light transmitted is zero. Since the analyzer is kept at 0°, this is possible only if the emergent light from the model has its plane of polarization along the vertical. Thus, to give physical meaning to the fringe contours, we have to identify the conditions under which the incident plane polarised light is unaltered as it passes through the model. It is to be noted that, in this optical arrangement, the polarizer and analyzer are crossed and the background field is dark.

The incident light is unaltered on all those points where the model behaves like a full-wave plate. This happens when the principal stress difference $(\sigma_1 - \sigma_2)$ is such as to cause a relative phase difference of $2m\pi$ ($m = 0, 1, 2, \ldots$), where m is an integer. Since stress is continuous, one observes a collection of points forming contours that satisfy the above condition and the respective fringe field is known as *isochromatics*. Though in a mathematical sense, the fringe field should appear as line contours, due to deficiencies of the recording medium, one generally observes a band representing the fringe. Thus, by knowing isochromatic fringe order at a point and using stress-optic law, one can determine the principal stress difference at the point.

The term *isochromatics* is more appropriate to use when white light is used as a source. It has been pointed out earlier that the wave plates are wavelength dependent. Hence, when white light is incident on the model, only a single wavelength is cut off. In view of this one observes white light minus the extinct colour over the field. *Iso* means constant and *chroma* means colour. Thus isochromatics represent contours of constant colour.

Another possibility, wherein the incident light is unaltered is, when the polarizer axis coincides with one of the principal stress directions at the point of interest. In this case, light extinction is not wavelength dependent and one observes a dark fringe even in white light. These are known as *isoclinics* meaning contours of constant inclination. Isoclinics are usually numbered with the angles they denote such as 0°, 10°, 15° etc. Figure 1.9 shows the isoclinics and isochromatics observed for a circular disc under diametral compression in a plane polariscope. Figure 1.9a corresponds to 22.5° isoclinic and Fig. 1.9b gives the –22.5° isoclinic. The principal stress direction on all points lying on an isoclinic is a constant and is indicated in Fig. 1.10.

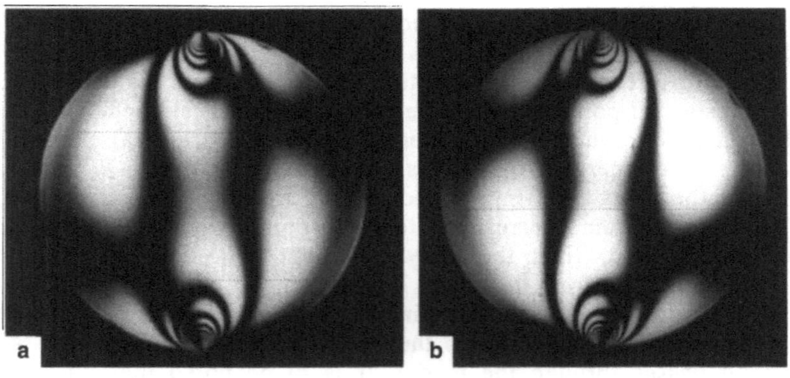

Fig. 1.9. Dark-field fringe patterns of a circular disc under diametral compression observed in a plane-polariscope: **a** 22.5° isoclinic **b** –22.5° or 67.5° isoclinic

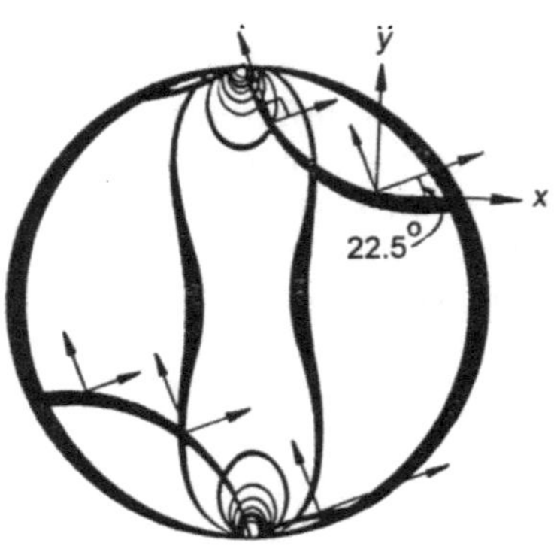

Fig. 1.10. All points lying on an isoclinic have the same orientation of the principal stress direction

Thus, in a plane polariscope, one has two sets of contours, namely, *isochromatics* and *isoclinics* superposed over each other. Leaving the zeroth fringe order isochromatics, these two contours can be distinguished by using a white light. To distinguish an isoclinic from a zeroth fringe order, rotate the polarizer and analyzer in crossed combination. The zeroth fringe order will not move whereas the isoclinics will move.

1.10.1 Analysis by Trigonometric Resolution

Let the linearly polarised light coming out of the polarizer be $ke^{i\omega t}$. For simplicity, let us consider only the real part:

$$E_1 = k\cos\omega t \tag{1.17}$$

Here E_1 is the component of light vector along the polarizer axis. Similarly, let the components of light vector be indicated as shown in Fig. 1.11. Upon entering the model, the light vector gets resolved along the principal stress axes.

$$\begin{aligned} E_2 &= k\sin\theta\cos\omega t \\ E_3 &= k\cos\theta\cos\omega t \end{aligned} \tag{1.18}$$

The model introduces a phase difference of δ and let E_2 be the slow and E_3 be the fast axes of the model. Splitting the relative retardation uniformly between the fast and slow axes, one gets

$$\begin{aligned} E_4 &= k\sin\theta\cos\left(\omega t - \frac{\delta}{2}\right) \\ E_5 &= k\cos\theta\cos\left(\omega t + \frac{\delta}{2}\right) \end{aligned} \tag{1.19}$$

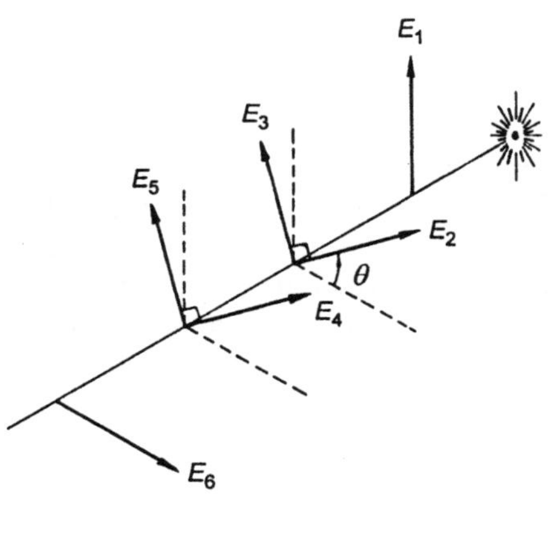

Fig. 1.11. Light vector nomenclature used for trigonometric resolution

Analyzer will transmit only that component which is along its axis. From vector algebra,

$$\begin{aligned} E_6 &= E_4 \cos\theta - E_5 \sin\theta \\ &= k \sin\theta \cos\theta \cos\left(\omega t - \frac{\delta}{2}\right) - k \cos\theta \sin\theta \cos\left(\omega t + \frac{\delta}{2}\right) \\ &= \frac{k}{2} \sin 2\theta \left[\cos\left(\omega t - \frac{\delta}{2}\right) - \cos\left(\omega t + \frac{\delta}{2}\right)\right] \\ &= k \sin 2\theta \sin\frac{\delta}{2} \sin\omega t \end{aligned} \tag{1.20}$$

The human eye and all other light sensing instruments respond to the intensity of light, which is proportional to the square of the amplitude. The rapid time variations cannot be detected and for a sodium light, the frequency f is 5.1×10^{14} Hz. Hence, the time dependent component is usually not considered. The intensity of light is obtained as

$$I_p = I_a \sin^2 2\theta \sin^2\frac{\delta}{2} \tag{1.21}$$

where, I_a accounts for the proportionality constant and the amplitude of incident light. The intensity of light coming out of the analyzer is zero when (a) $\theta = 0$ or $\pi/2$ or (b) $\delta = 2m\pi$ (m = 0, 1, 2, ...). Thus the condition 'a' corresponds to isoclinics and 'b' corresponds to isochromatics.

1.11 Jones Calculus

In general, an optical element in a polariscope introduces a rotation and retardation. In Jones calculus, these basic operations are represented as matrices.

1.11.1 Rotation Matrix

This matrix is useful to find the components of a vector if the reference axes are rotated by an arbitrary angle θ. For example, let x, y represent the initial reference axes and the final reference axes be x', y' (Fig. 1.12). Let the components of the light vector along x and y directions be u, v; then the components u' and v' along the directions x' and y' are obtained as follows:

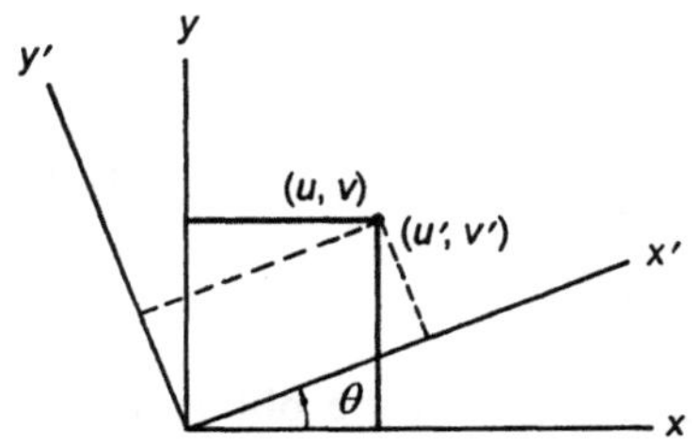

Fig. 1.12. Transformation of co-ordinate systems

$$\begin{aligned} u' &= u\cos\theta + v\sin\theta \\ v' &= -u\sin\theta + v\cos\theta \end{aligned} \tag{1.22}$$

This can be written in a matrix form as

$$\begin{Bmatrix} u' \\ v' \end{Bmatrix} = \begin{bmatrix} \cos\theta & \sin\theta \\ -\sin\theta & \cos\theta \end{bmatrix} \begin{Bmatrix} u \\ v \end{Bmatrix} \tag{1.23}$$

The following matrix is referred to as the rotation matrix:

$$\begin{bmatrix} \cos\theta & \sin\theta \\ -\sin\theta & \cos\theta \end{bmatrix} \tag{1.24}$$

The usual convention of measuring angles is valid for constructing a rotation matrix and the angle is positive if it is measured counter-clockwise.

1.11.2 Retardation Matrix

A doubly refracting medium introduces a relative retardation δ between the vibrating components along its axes. Let u and v be the vibrating components entering the medium with certain retardations α_1 and α_2 respectively.

$$\begin{aligned} u &= a_1\cos(\omega t + \alpha_1) \\ v &= a_2\cos(\omega t + \alpha_2) \end{aligned} \tag{1.25}$$

If u is the slow axis and v is the fast axis, then, the light coming out of the medium can be represented as

$$u' = a_1 \cos\left(\omega t + \alpha_1 - \frac{\delta}{2}\right)$$
$$v' = a_2 \cos\left(\omega t + \alpha_2 + \frac{\delta}{2}\right) \tag{1.26}$$

Using complex number notations, the emerging light can be obtained as a matrix operation of the incident light as

$$\begin{Bmatrix} u' \\ v' \end{Bmatrix} = \Re \begin{bmatrix} e^{-i\delta/2} & 0 \\ 0 & e^{i\delta/2} \end{bmatrix} \begin{Bmatrix} a_1 e^{i\alpha_1} \\ a_2 e^{i\alpha_2} \end{Bmatrix} e^{i\omega t} \tag{1.27}$$

The first matrix on the right side of Eq. (1.27) is called the retardation matrix. On the explicit understanding that we deal with real parts only, the notation $\Re$ is usually left out. It is to be noted that in the above representation, the u axis is considered as the slow axis.

1.11.3 Representation of a Retarder

As mentioned before, a retarder introduces a rotation and retardation. Let the slow axis of the retarder be oriented at an angle θ and let the total relative retardation introduced by the retarder be δ. The incident light has to be rotated by an angle θ and then the relative retardation of δ is to be introduced. The matrix representation of these operations is as follows:

$$\begin{bmatrix} e^{-i\delta/2} & 0 \\ 0 & e^{i\delta/2} \end{bmatrix} \begin{bmatrix} \cos\theta & \sin\theta \\ -\sin\theta & \cos\theta \end{bmatrix} \tag{1.28}$$

If these matrices are operated on the incident light one will get the emerging components along the retarder axes only. However, it becomes convenient if the emerging light is also represented with respect to the original reference axes. This can be achieved by post-multiplying Eq. (1.28) by another rotation matrix as follows:

$$\begin{bmatrix} \cos\theta & -\sin\theta \\ \sin\theta & \cos\theta \end{bmatrix} \begin{bmatrix} e^{-i\delta/2} & 0 \\ 0 & e^{i\delta/2} \end{bmatrix} \begin{bmatrix} \cos\theta & \sin\theta \\ -\sin\theta & \cos\theta \end{bmatrix} \tag{1.29}$$

As several optical elements of a polariscope can be represented in general as a retarder, it is convenient to obtain the product of all these matrices in Eq. (1.29) for later use. The representation of a retarder matrix is

$$\begin{bmatrix} \cos\frac{\delta}{2} - \mathrm{i}\sin\frac{\delta}{2}\cos 2\theta & -\mathrm{i}\sin\frac{\delta}{2}\sin 2\theta \\ -\mathrm{i}\sin\frac{\delta}{2}\sin 2\theta & \cos\frac{\delta}{2} + \mathrm{i}\sin\frac{\delta}{2}\cos 2\theta \end{bmatrix} \tag{1.30}$$

1.11.4 Polarizer

A polarizer allows only the component of light along its axis. For a polarizer kept at 90° with respect to the horizontal, the matrix representation is

$$\begin{Bmatrix} 0 \\ 1 \end{Bmatrix} \tag{1.31}$$

1.11.5 Quarter-Wave Plate

Quarter-wave plates used in a circular polariscope are retarders giving a retardation of $\pi/2$ radians. For a quarter-wave plate whose slow axis is at 135°, the matrix representation is obtained by substituting $\theta = 135°$ and $\delta = \pi/2$ in Eq. (1.30), one gets

$$\frac{1}{\sqrt{2}}\begin{bmatrix} 1 & \mathrm{i} \\ \mathrm{i} & 1 \end{bmatrix} \tag{1.32}$$

1.12 Analysis of Plane Polariscope by Jones Calculus

Using Jones calculus, the components of light vector along the analyzer axis and perpendicular to the analyzer axis for the polariscope arrangement are obtained as

$$\begin{Bmatrix} E_x \\ E_y \end{Bmatrix} = \begin{bmatrix} \cos\frac{\delta}{2} - \mathrm{i}\sin\frac{\delta}{2}\cos 2\theta & -\mathrm{i}\sin\frac{\delta}{2}\sin 2\theta \\ -\mathrm{i}\sin\frac{\delta}{2}\sin 2\theta & \cos\frac{\delta}{2} + \mathrm{i}\sin\frac{\delta}{2}\cos 2\theta \end{bmatrix} \begin{Bmatrix} 0 \\ 1 \end{Bmatrix} k e^{\mathrm{i}\omega t} \tag{1.33}$$

where $ke^{\mathrm{i}\omega t}$ is the incident light vector, E_x and E_y are the components of light vector along the analyzer axis and perpendicular to the analyzer axis respectively. To find the intensity of light transmitted one can first find the real part of E_x and then use the procedure mentioned in Sect. 1.10.1. Alternately, one can also find the intensity by considering the product $E_x E_x^*$ where

E_x^* denotes the complex conjugate of E_x. By both these approaches, the intensity of light transmitted turns out to be,

$$I_p = I_a \sin^2\frac{\delta}{2}\sin^2 2\theta \tag{1.34}$$

which is same as Eq. (1.21)

From the foregoing analysis it is quite clear how the calculations become very simple if Jones calculus is used. The advantage will be more clearly seen if several optical elements need to be analysed, such as, in a circular polariscope. Henceforth, only Jones calculus will be used in this book.

1.13 Circular Polariscope

Figure 1.13 shows a general arrangement for a circular polariscope. In this a circularly polarised light is used to reveal the stress information in the model. This is achieved by introducing a quarter-wave plate with one of its axis at 45° after the polarizer. The emergent light is analysed using a combination of a quarter-wave plate and an analyzer. The basic arrangement shown in Fig. 1.13 cuts the light beyond the model boundary and the background appears dark and hence the arrangement is known as a dark-field arrangement. Inside the model boundary, the intensity of light transmitted is governed by the stress field and this can be obtained by Jones calculus.

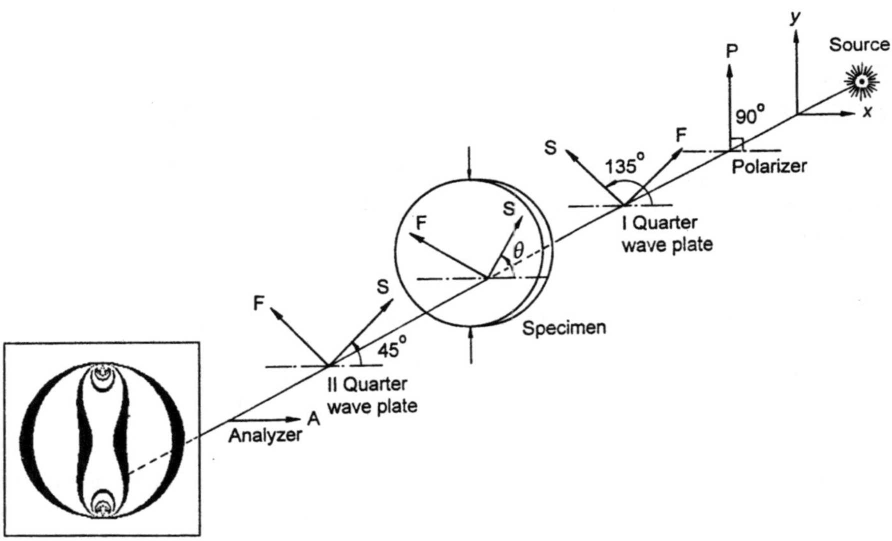

Fig. 1.13. Arrangement of optical elements in a circular polariscope. The background field is shown as white for easy visualisation of the model boundary

The matrix representations for the polarizer, quarter-wave plate whose slow axis at 135°, and a model whose slow axis at angle θ have already been discussed. On similar lines the matrix representation of the quarter-wave plate after the model (Fig. 1.13) can be obtained by taking angle θ as 45° and δ as $\pi/2$ in Eq. (1.30) and one obtains

$$\frac{1}{\sqrt{2}}\begin{bmatrix} 1 & -\mathrm{i} \\ -\mathrm{i} & 1 \end{bmatrix} \tag{1.35}$$

To get the components of light along the analyzer axis and perpendicular to it, the individual matrices of the various optical elements have to be multiplied in the order they are placed in the polariscope as

$$\begin{Bmatrix} E_x \\ E_y \end{Bmatrix} = \frac{1}{2}\begin{bmatrix} 1 & -\mathrm{i} \\ -\mathrm{i} & 1 \end{bmatrix}\begin{bmatrix} \cos\frac{\delta}{2} - \mathrm{i}\sin\frac{\delta}{2}\cos 2\theta & -\mathrm{i}\sin\frac{\delta}{2}\sin 2\theta \\ -\mathrm{i}\sin\frac{\delta}{2}\sin 2\theta & \cos\frac{\delta}{2} + \mathrm{i}\sin\frac{\delta}{2}\cos 2\theta \end{bmatrix} \times \begin{bmatrix} 1 & \mathrm{i} \\ \mathrm{i} & 1 \end{bmatrix}\begin{Bmatrix} 0 \\ 1 \end{Bmatrix} k e^{\mathrm{i}\omega t} \tag{1.36}$$

Upon simplification one gets

$$\begin{Bmatrix} E_x \\ E_y \end{Bmatrix} = \begin{Bmatrix} \sin\frac{\delta}{2} e^{-\mathrm{i}2\theta} \\ \cos\frac{\delta}{2} \end{Bmatrix} k e^{\mathrm{i}\omega t} \tag{1.37}$$

Let the intensity of light transmitted in dark-field be denoted as I_d and is obtained as the product of $E_x E_x^*$ which is simplified as

$$I_\mathrm{d} = I_\mathrm{a}\sin^2\frac{\delta}{2} \tag{1.38}$$

where I_a accounts for the amplitude of the incident light vector. By rotating the analyzer and keeping it along the vertical, one gets a background that is bright. The fringes in the model also shift when the analyzer is rotated. In this case, the analyzer will transmit the E_y component of Eq. (1.37) and denoting the intensity of light as I_ℓ, one gets

$$I_\ell = I_\mathrm{a}\cos^2\frac{\delta}{2} \tag{1.39}$$

It is to be noted that for both dark and bright-fields, the intensity equations

Table 1.1. Polariscope arrangements

Set up	Polarizer and Analyzer	Quarter-wave Plates	Background Field
Plane Polariscope	Crossed	None	dark
Circular Polariscope	Crossed	Crossed	dark†
	Parallel	Parallel	dark
	Crossed	Parallel	bright
	Parallel	Crossed	bright†

† Preferred optical arrangements to minimise the influence of quarter-wave plate error.

are independent of θ and hence the extinction condition is only a function of δ and thus only isochromatics will be seen. This is a significant achievement in separating isoclinics and isochromatics. Although several combinations of the optical elements can give either dark or bright-fields, the popularly known combinations are given in Table 1.1.

In dark-field arrangement, intensity is zero when $\delta = 2m\pi$ ($m = 0, 1, 2, \ldots$) and fringes correspond to 0, 1, 2, ... etc. In bright-field arrangement, intensity is zero when $\delta = (2m + 1)\pi$, i.e., when the retardation is an odd multiple of half-wave lengths. The fringes correspond to 0.5, 1.5, 2.5 etc. Figure 1.14 shows that isochromatics alone are observed in a circular polariscope for dark and bright-field arrangements. One can also observe the fringe shift between the dark and bright-field images.

1.14 Use of White Light

Most polariscopes come with a monochromatic light source as well as a white source. Using white light in a plane polariscope, it is easy to identify

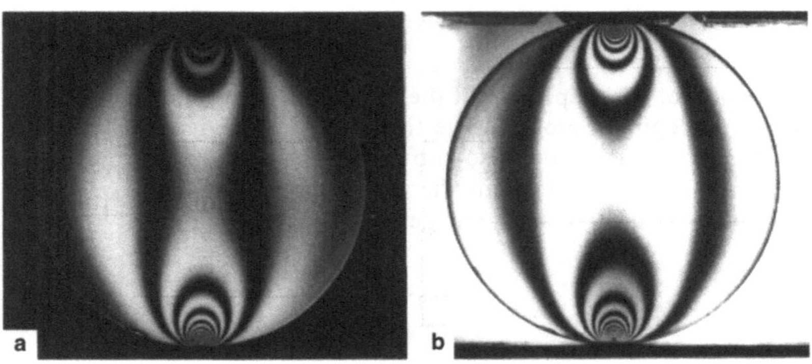

Fig. 1.14. **a** Dark-field isochromatics. **b** Bright-field isochromatics

an isoclinic from an isochromatic, since an isochromatic appears coloured. The colour sequence is an indication of the retardation level. This colour sequence can be used to identify whether fringe orders are increasing or decreasing in any particular direction. Table 1.2 summarises the colour sequence observed in a dark-field polariscope along with the approximate fringe orders. In conventional photoelasticity, quantitative measurements are usually made using a monochromatic light.

1.15 Determination of Isoclinic and Isochromatic Fringe Order at a Point

1.15.1 Ordering of Isoclinics

Isoclinics represent the contours along which the orientation of the principal stress direction is a constant. In a plane polariscope (dark-field arrangement), one observes isochromatics superimposed with a particular degree isoclinic. The principal stress directions coincide with the polarizer and analyzer directions and in most polariscopes, the orientation of the polarizer axis with the vertical is referred to as the isoclinic angle. Thus, the ordering of an isoclinic is very simple and straightforward.

Table 1.2. Sequence of colours produced in dark-field polariscope with white light

Colour Observed	Retardation (nm)	Approximate Fringe Order
Black	0	0.00
Grey	160	0.28
White	260	0.45
Yellow	350	0.60
Orange	460	0.79
Dull Red	520	0.90
Tint of Passage	577	1.00
Blue	620	1.06
Blue-green	700	1.20
Green-Yellow	800	1.39
Orange	940	1.63
Rose Red	1050	1.82
Tint of Passage	1150	2.00
Green	1350	2.35
Green-yellow	1450	2.50
Orange-red	1550	2.65
Tint of passage	1730	3.00

In order to determine the direction of the principal stresses at a desired point in a stressed model, the model is viewed through a plane polariscope and the polarizer and analyzer are rotated in unison until the dark band representing the isoclinic passes through the point of interest. The orientations of the polarizer and analyzer coincide with the principal stress directions at the point. It is to be noted that an ambiguity exists in identifying whether the isoclinic represents σ_1 or σ_2 directions. To remove the ambiguity, the polariscope needs to be calibrated (Sect. 1.20).

1.15.2 Ordering of Isochromatics

Ordering of isochromatics is one of the crucial steps in the process of determining quantitative information from the fringe field. Experience and intuition have guided the numbering of isochromatic fringes and it is not an easy task to order the fringes from a given dark/bright-field black & white photograph. The experimentalists in the past have attempted to use several auxiliary methods to assist fringe ordering. One of the simplest approaches is to use a white light source, as the colour code can be used (Table 1.2) to identify the fringe orders and also the fringe gradient direction. The method is advantageous up to the third order fringe in the field, as beyond this the colours merge. Further guidelines on fringe ordering are given in Sect. 1.18.

Let us now look at a scenario wherein at the point of interest no bright or dark-field fringe passes through but the fringe gradient information is known. The question is how to find the isochromatic fringe order? One of the simplest approaches is as follows. Using a dark-field set-up and a bright-field set-up, two sets of isochromatic fringe patterns over the field of view be recorded. The dark-field set-up gives fringes representing an integral number of fringe orders. The bright-field set-up gives fringes representing odd multiples of half integer. With these, one can determine the retardation existing at the point graphically. However, situations often exist where the model under the given loading may not reveal a sufficient number of fringes for a graphical analysis. In such situations, one adopts what is gênerally known as the compensation technique to determine the fractional fringe order existing at a point.

1.16 Tardy's Method of Compensation

Compensation techniques are basically point by point techniques. The basic principle is that, by external means, the retardation provided by the model is compensated such that a fringe passes through the point of interest. The additional retardation added or subtracted is known as fractional retardation. One can use a compensator such as a Babinet-Soleil compensator for this purpose or use the analyzer itself as the compensator. The use of the analyzer as a compensator is known as Tardy's method of compensation.

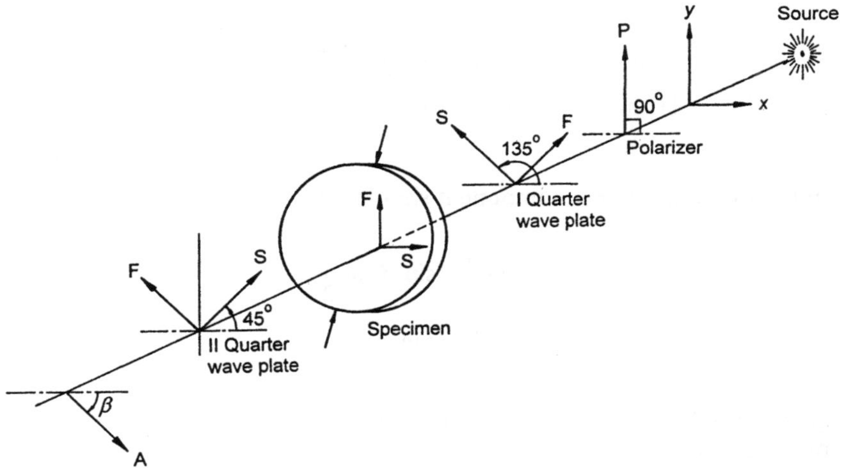

Fig. 1.15. Optical arrangement to employ Tardy's method of compensation

One of the basic requirements in all the compensation techniques is that the principal stress directions at the point of interest be known to add or subtract a retardation. Thus, in Tardy's method of compensation, initially, the principal stress directions at the point of interest is determined using a plane polariscope. A circular polariscope is then formed such that the polarizer is kept at the isoclinic angle and all the other optical arrangements are appropriately arranged. At this stage, if the optical elements are correctly aligned, there should be no difference in the isochromatic field compared to the conventional arrangement (Table 1.1). The analyzer alone is then rotated until a fringe passes through the point of interest. The rotation given to the analyzer can be related to the fractional retardation by using Jones calculus as follows.

Instead of rotating the entire optical set-up to the isoclinic angle, in Fig. 1.15, the model is rotated such that one of its axes at the point of interest coincides with the polarizer. Let the analyzer be rotated by an angle β to get light extinction. Then by Jones calculus, the intensity of light along the analyzer axis and perpendicular to it can be obtained as

$$\begin{Bmatrix} E_{-\beta} \\ E_{-\beta+\pi/2} \end{Bmatrix} = \frac{1}{2}\begin{bmatrix} \cos\beta & -\sin\beta \\ \sin\beta & \cos\beta \end{bmatrix}\begin{bmatrix} 1 & -\mathrm{i} \\ -\mathrm{i} & 1 \end{bmatrix}\begin{bmatrix} \cos\frac{\delta}{2} - \mathrm{i}\sin\frac{\delta}{2} & 0 \\ 0 & \cos\frac{\delta}{2} + \mathrm{i}\sin\frac{\delta}{2} \end{bmatrix} \tag{1.40}$$

$$\times \begin{bmatrix} 1 & \mathrm{i} \\ \mathrm{i} & 1 \end{bmatrix}\begin{Bmatrix} 0 \\ 1 \end{Bmatrix} k e^{\mathrm{i}\omega t}$$

Upon simplification one gets

$$\begin{Bmatrix} E_{-\beta} \\ E_{-\beta+\pi/2} \end{Bmatrix} = \begin{bmatrix} \cos\beta & -\sin\beta \\ \sin\beta & \cos\beta \end{bmatrix} \begin{Bmatrix} \sin\dfrac{\delta}{2} \\ \cos\dfrac{\delta}{2} \end{Bmatrix} k e^{\mathrm{i}\omega t} \tag{1.41}$$

For light extinction, the $E_{-\beta}$ component should be zero. This occurs when,

$$\begin{aligned} \cos\beta \sin\frac{\delta}{2} &= \sin\beta \cos\frac{\delta}{2} \\ \tan\frac{\delta}{2} &= \tan\beta \end{aligned} \tag{1.42}$$

This gives, $\delta = 2\beta$. If β is measured in degrees, the fractional fringe order δ_{N} is obtained as

$$\delta_{\mathrm{N}} = \pm\frac{|\beta|}{180^\circ} \tag{1.43}$$

Unlike the full fringe order N, the fractional fringe order has a sign. The sign convention is not decided by the sign of β but by the physics of the problem. The fractional fringe order δ_{N} is taken as positive, if a lower fringe order moves to the point of interest and vice versa.

Tardy's method of compensation is quite a useful technique and the fringe order can be measured with a high degree of accuracy.

1.17 Calibration of Photoelastic Model Materials

The stress-fringe values of model materials vary with time and also from batch to batch. Hence, it is necessary to calibrate each sheet or casting at the time of the test. Calibration is performed on simple specimens for which closed form stress field solution is known. Although the stress fields for simple tension or beam under pure bending are known, the use of a circular disc under diametral compression is preferred for calibration. Circular disc is preferred, because the specimen is compact, easy to machine and it can also be easily loaded. The stress field in a circular disc cannot be obtained using principles of *Strength of Materials* but one has to use the principles of *Theory of Elasticity*.

1.17.1 Stress Field in a Circular Disc Under Diametral Compression

The stress field in a circular disc under diametral compression can be obtained [2] starting from Bousinesque's solution for a compressive load on a half plane and then applying the principle of superposition appropriately. With the centre of the disc as the origin, the stress field is obtained as

$$\begin{Bmatrix} \sigma_x \\ \sigma_y \\ \tau_{xy} \end{Bmatrix} = -\frac{2P}{\pi t} \begin{Bmatrix} \dfrac{(R-y)x^2}{r_1^4} + \dfrac{(R+y)x^2}{r_2^4} - \dfrac{1}{D} \\ \dfrac{(R-y)^3}{r_1^4} + \dfrac{(R+y)^3}{r_2^4} - \dfrac{1}{D} \\ \dfrac{(R+y)^2 x}{r_2^4} - \dfrac{(R-y)^2 x}{r_1^4} \end{Bmatrix} \tag{1.44}$$

where $r_1^2 = x^2 + (R - y)^2$ and $r_2^2 = x^2 + (R + y)^2$, R denotes the radius of the disc, D represents its diameter, t is the thickness of the disc and P is the compressive load applied. The principle stress difference $(\sigma_1 - \sigma_2)$ at any point in the disc can be expressed as

$$(\sigma_1 - \sigma_2) = \frac{4PR}{\pi t} \frac{R^2 - (x^2 + y^2)}{(x^2 + y^2 + R^2)^2 - 4y^2 R^2} \tag{1.45}$$

At the centre of the disc due to symmetry, the shear stress is zero and the principal stress difference is obtained as

$$(\sigma_1 - \sigma_2) = \frac{8P}{\pi D t} \tag{1.46}$$

1.17.2 Conventional Method

Combining Eq. (1.15) and Eq. (1.46) one gets

$$F_\sigma = \frac{8P}{\pi D N} \tag{1.47}$$

Equation (1.47) forms the basis of the conventional approach. A circular disc is to be cut from the sheet to be calibrated. The ratio of diameter to thickness of the disc should be chosen so that the disc does not buckle under the load. Fringe order at the centre of the disc is obtained using Tardy's method of compensation. The load is first increased and then decreased incrementally. A graph is drawn between the load P and fringe order N. A best-fit straight line is then drawn through the points (graphical approach to least squares) and the slope of the line P/N is used in Eq. (1.47) to evaluate the material fringe value.

Since the principal stress directions at the centre of the disc are along the basic configuration of the polariscope (the isoclinic is of zero degree) while applying Tardy's method of compensation, it is enough that one just rotates the analyzer appropriately to get light extinction (at this stage, the fringes resemble the number eight) and note the angle β. This simplified procedure coupled with a high accuracy possible with Tardy method of compensation has attracted the users to adopt the circular disc as a calibration model.

1.17.3 Sampled Linear Least Squares Method

1.17.3.1 Need for a Better Methodology

In a photoelastic experiment, material fringe value is a very crucial parameter since this is the only parameter, which relates the optical phenomena to the stresses. Up to two or three decimal places accuracy may be required. Though photoelasticity is a whole field technique, data from only one point is used in the conventional method. Due to the spread of the applied loads, Coker and Filon have shown that the agreement between the theoretical and experimental value at the centre of the disc is off by about 4 percent. Durelli and Mulzat [13] have shown that when the deflections become large, which is common in stress freezing experiments, the lack of agreement between theory and experiment becomes even larger than 4%. However, in the zone r/R = 0.3 to 0.5, the theoretical and experimental results are in good agreement. Hence, the use of field data will be more suitable to evaluate the material stress fringe value accurately.

1.17.3.2 Use of Whole-Field Data to Evaluate Material Fringe Value

The basic idea is to use as many data points as possible from the field to evaluate the material stress fringe value. It is worthwhile to account for residual birefringence also in the analysis. Let the residual birefringence expressed in fringe orders be assumed as a linear function in x and y as [14]

$$N_r(x,y) = Ax + By + C \tag{1.48}$$

The fringe order at any point is the sum of the fringe order calculated by theory and the residual fringe order N_r. Using the stress-optic relation and Eq. (1.48)

$$N(x,y) = \frac{(\sigma_1 - \sigma_2)t}{F_\sigma} + N_r(x,y) \tag{1.49}$$

For computer implementation, it is desirable to express the product $t(\sigma_1 - \sigma_2)$ in terms of x, y as

$$S(x,y)=\frac{4PR}{\pi}\frac{R^2-(x^2+y^2)}{(x^2+y^2+R^2)^2-4y^2R^2} \tag{1.50}$$

For any point (x_m, y_m) in the field, Eq. (1.50) can be written as

$$N_m(x_m,y_m)=\frac{1}{F_\sigma}S_m(x_m,y_m)+Ax_m+By_m+C \tag{1.51}$$

From photoelastic data, x_m, y_m and N_m can be determined. In Eq. (1.51), N_m is expressed, as a function of $1/F_\sigma$, A, B and C and these are to be determined. Since there are only four unknowns, four equations are sufficient to solve for them. Thus, in principle, any four points in the field could be used to construct these equations. However, the solution thus obtained may have a strong bearing on the choice of the data points. On the other hand, if one takes several points in the field, one gets over determined set of equations i.e.; the number of equations is more than the number of unknowns. The usual method to solve such a system of equations is to obtain a new set of equations, using the least squares criteria, which are equal to the number of unknowns. If there are totally M data points ($M > 4$), the cumulative error term is obtained as

$$e=\sum_{m=1}^{M}\left[\frac{1}{F_\sigma}S_m(x_m,y_m)+Ax_m+By_m+C-N_m\right]^2 \tag{1.52}$$

The least squares criteria requires that

$$\frac{\partial e}{\partial(1/F_\sigma)}=\frac{\partial e}{\partial A}=\frac{\partial e}{\partial B}=\frac{\partial e}{\partial C}=0 \tag{1.53}$$

The application of the above conditions gives four equations and by solving it, one can get a unique solution.

$$\begin{aligned}
&\sum_{m=1}^{M}2\left[\frac{1}{F_\sigma}S_m(x_m,y_m)+Ax_m+By_m+C-N_m\right]S_m(x_m,y_m)=0\\
&\sum_{m=1}^{M}2\left[\frac{1}{F_\sigma}S_m(x_m,y_m)+Ax_m+By_m+C-N_m\right]x_m=0\\
&\sum_{m=1}^{M}2\left[\frac{1}{F_\sigma}S_m(x_m,y_m)+Ax_m+By_m+C-N_m\right]y_m=0\\
&\sum_{m=1}^{M}2\left[\frac{1}{F_\sigma}S_m(x_m,y_m)+Ax_m+By_m+C-N_m\right]=0
\end{aligned} \tag{1.54}$$

In matrix notation, the above set of equations can be written as

$$2\left\{[b]^T[b]\{u\}-[b]^T\{N\}\right\}=0 \tag{1.55}$$

which reduces to

$$[b]^T[b]\{u\}=[b]^T\{N\} \tag{1.56}$$

where,

$$[b]=\begin{bmatrix} S_1 & x_1 & y_1 & 1 \\ S_2 & x_2 & y_2 & 1 \\ \vdots & \vdots & \vdots & \vdots \\ S_M & x_M & y_M & 1 \end{bmatrix},\ \{u\}=\begin{Bmatrix} 1/F_\sigma \\ A \\ B \\ C \end{Bmatrix},\text{ and }\{N\}=\begin{Bmatrix} N_1 \\ N_2 \\ \vdots \\ N_M \end{Bmatrix} \tag{1.57}$$

The unknown coefficient vector $\{u\}$ can be easily evaluated using the standard Gaussian elimination procedure. The construction of Eq. (1.56) is very simple from experimental data and hence the method has found wide acceptance.

It is desirable that the final results are nearly independent of the choice of data points from the field. In order to achieve this, the least squares technique has to be combined with a random sampling process. This can be done by initially collecting a large number of data points. Out of them select a small subset of data points in random order and least squares analysis be carried out for each of these subsets. The converged values of the parameters are then averaged for theoretical reconstruction. Collection of 40 data points from the field with 20 data points for each subset repeated six times is adequate for parameter estimation [14]. A program code implementing the sampled least squares technique is enclosed in the CD under the name *fsigma.c.*

Least squares technique by itself does not guarantee that the result is physically admissible. Theoretical reconstruction of fringe patterns after evaluating the parameters in a least squares sense is very important and the evaluation of the parameter values is complete only if the reconstructed fringe patterns agree well with the experimentally obtained ones.

1.17.4 Theoretical Reconstruction of Fringe Patterns

The isochromatic patterns correspond to the loci of maximum shear stress, which is expressed in terms of the Cartesian stress components as

$$(2\tau_m)^2=(\sigma_x-\sigma_y)^2+(2\tau_{xy})^2 \tag{1.58}$$

Using the stress-optic law one obtains

$$\left(\frac{NF_\sigma}{t}\right)^2 = (\sigma_x - \sigma_y)^2 + (2\tau_{xy})^2 \tag{1.59}$$

One approach to plot fringe skeletons is to find the co-ordinates of fringe points in the entire field corresponding to a particular fringe order and join them appropriately. However, this approach is quite difficult, as in general, the Eq. (1.59) is non-linear in terms of x and y. A scanning approach will be quite effective. In this, the fringe order at every point forming the grid is to be evaluated. If the fringe order evaluated is close to a dark-field or a bright-field fringe, the corresponding grid point is selected as fringe point. In digital computations, it is usual to use a tolerance value for checking whether it is close to a dark or bright-field fringe. A tolerance value of the order of ± 0.1 fringe orders is sufficient for most studies. Usage of the tolerance helps to bring out the features of the experimentally observed fringes, namely, the thickness variation of the fringes. The fringe quality will be good if a very fine grid of points is chosen for fringe order evaluation. It is recommended to use pixel increments of the video monitor for selecting the scanning grid. A program code implementing the scheme is enclosed in the CD under the name *reconstr.c*.

1.18 Further Comments on Fringe Ordering

In a first course in Mechanics of solids, the student is trained in finding out stress components, principal stresses, displacements etc. at a *point* only, for various types of simple problem situations. The concept of *whole field* information itself is new to the student when he comes for learning Photoelasticity.

Insight into fringe ordering could be gained by knowing the nature of fringes for problems for which stress field is known. Since isochromatics represent contours of $(\sigma_1 - \sigma_2)$, for several known problem situations, one can identify the nature of fringe contours. Let us consider the problem of a uniform plate under tension. Here, only σ_y exists and this itself is σ_1 and σ_2 is zero. Thus, isochromatics should correspond to contours of σ_y. As the stress field is uniform, every point will behave in a similar manner and light extinction will occur uniformly. In a white light source, one will see uniform colour and by knowing the colour sequence, the fringe order can be identified. However, in a monochromatic source one cannot distinguish between the order of fringes unless the model is gradually loaded and the fringe formation is closely monitored.

In the case of a beam under pure bending, the contours of isochromatics correspond to contours of σ_x and one will observe discrete horizontal lines parallel to the x-axis. The zeroth fringe order coincides with the neutral axis and for any increase in the bending moment, the zeroth fringe order will not move. This is the property of zeroth fringe order and can be used to distinguish it from isoclinics while using a monochromatic light source. Since the fringes do not have a sign, one cannot say which portion of the beam is under compression and which is in tension. One needs auxiliary methods to achieve this, which is discussed in Sect. 1.19.

On similar lines, one can plot fringe fields for several problems that have closed form solution from theory of elasticity such as a concentrated load on a semi-infinite plane, problem of thick cylinder, ring under diametral compression etc. This is left as an exercise to the reader. General guidelines for fringe ordering could be formulated on the basis of properties of isochromatic fringe field, isoclinic fringe field and also certain principles from mechanics of solids.

Though several textbooks have been written on photoelasticity, very little attention has been paid in these to present a comprehensive approach on fringe ordering. The problem of a circular disc under diametral compression has received wide attention in photoelasticity literature. It has served as a benchmark problem to verify the data acquisition and stress tensor evaluation procedures in photoelasticity. This problem received wide attention mainly because the model is easy to make and the loading conditions as demanded in theoretical assumptions can be achieved easily. However, the problem is not ideally suited for learning fringe ordering as the fringe field is very simple. Properties of fringe field could be best illustrated with the example of a ring under diametral compression.

A ring under diametral compression represents various important aspects of a general fringe field such as *source, sink, saddle point, singular point* and *isotropic point*. The first three correspond to isochromatic fringe field and the rest correspond to isoclinic fringe field. Figure 1.16 shows the *source, sink* and *saddle points* for the problem of a ring under diametral compression. It is to be noted that there is no standard procedure to order fringes. Only certain broad guidelines are highlighted. The nuances of fringe ordering and fringe field features can be appreciated better by using PHOTOSOFT_H [15].

1.18.1 Properties of Isochromatic Fringe Field

The load application points, stress concentration zones and the tips of cracks are stress raisers and they act like a *source* in the fringe field. As the load is increased, fringe orders appear at source and propagate to other areas. A *sink* represents the point at which fringes vanish. It is at a lower fringe order compared to the neighbourhood but not necessarily zero. One of the techniques (at least in static experiments) usually adopted is to increase the load gradually and to identify the points at which the fringes originate and vanish. A *saddle point* is bounded by two families of isochromatic fringes, one representing a fringe order slightly greater and the

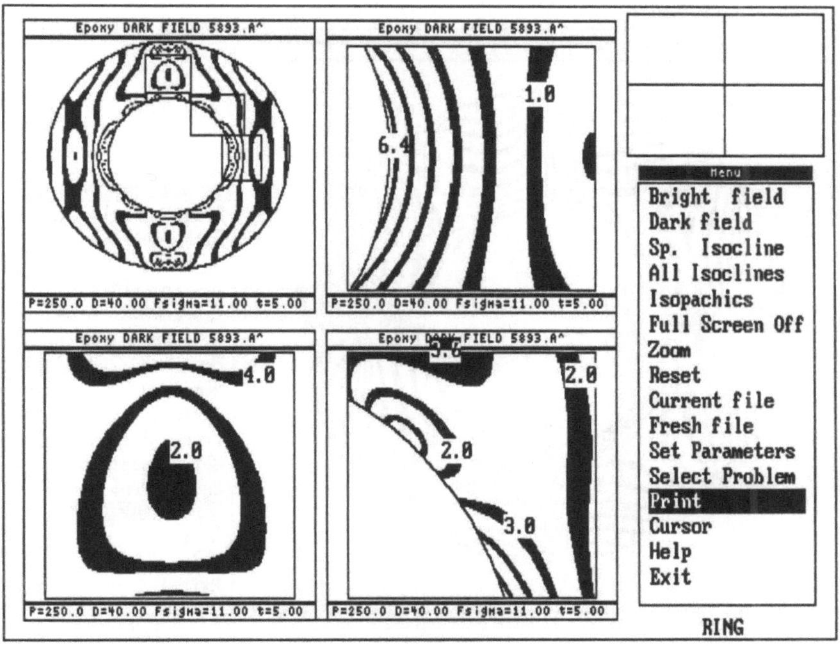

Fig. 1.16. Various fringe features of the isochromatic fringe field. **a** Overall isochromatic field of a ring under diametral compression **b** Source **c** Sink and **c** Saddle point (Courtesy: PHOTOSOFT_H)

other slightly smaller than the one corresponding to the saddle point. From the foregoing definition, it is seen that the centre of the disc under diametral compression is a *saddle point.*

1.18.2 Properties of Isoclinic Fringe Field

Certain features of the isoclinic fringe field come in handy for ordering isochromatics. One of the important features of the isoclinic fringe field is the identification of an *isotropic point.* An *isotropic point* is one, through which all isoclinics pass through. The point is termed as isotropic because at these points, the loaded polymer does not behave like a crystal but behaves like an optically isotropic material. An *isotropic point* lying on the free surface is termed as a *singular point.* The sign of the boundary stress changes when it crosses a singular point. All the isoclinics also pass through the load application points but these are not isotropic points.

Two different forms of isotropic points can be distinguished. The isoclinics rotate about an isotropic point when the crossed Polaroid's are

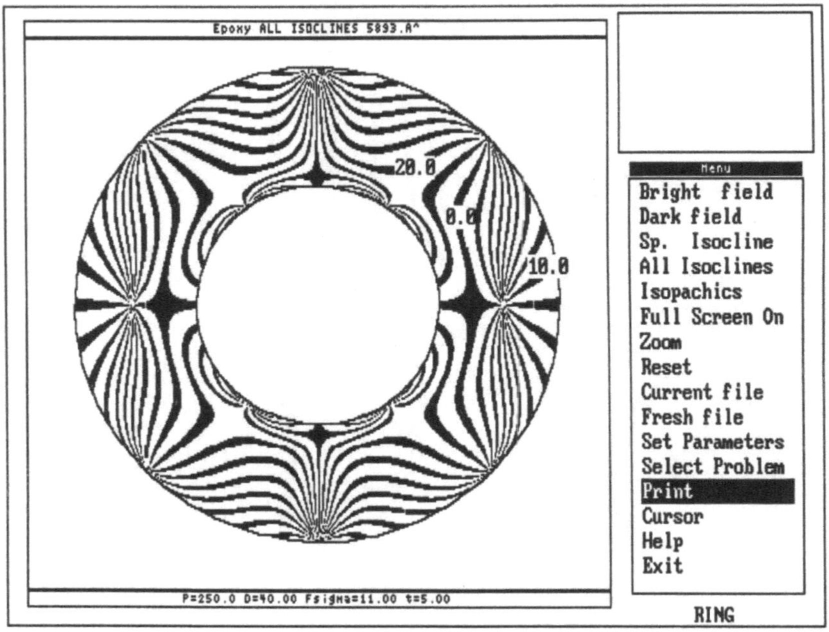

Fig. 1.17. Various fringe features of the isoclinic fringe field. Isoclinics in steps of 10 degrees interval are plotted for a ring under diametral compression (Courtesy: PHOTOSOFT_H)

rotated. When the isoclinic lines rotate in the same sense, as the Polaroid's, the corresponding isotropic point is termed as a positive isotropic point. Otherwise it is termed as a negative isotropic point. A singular point at a load-free edge can only be a negative isotropic point. When the isoclinics pass through several isotropic points in a sequence, the points must be alternatively positive and negative. Figure 1.17 shows the isoclinics in steps of 10 degrees for the problem of a ring under diametral compression. One can observe four singular points each on the inner and outer boundaries of the ring and two isotropic points along the horizontal diameter.

1.18.3
Use of Fringe Field Properties to Identify Fringe Ordering

Fringe ordering can be made simpler if one knows the fringe gradient direction and the order of at least one fringe forming the fringe packet. The fringe gradient directions can be established if one understands the various features of the isochromatic fringe field such as *source, sink,* or *saddle point*. Identification of zeroth order fringe is possible with the knowledge of isoclinic fringe field namely *isotropic* and *singular points.*

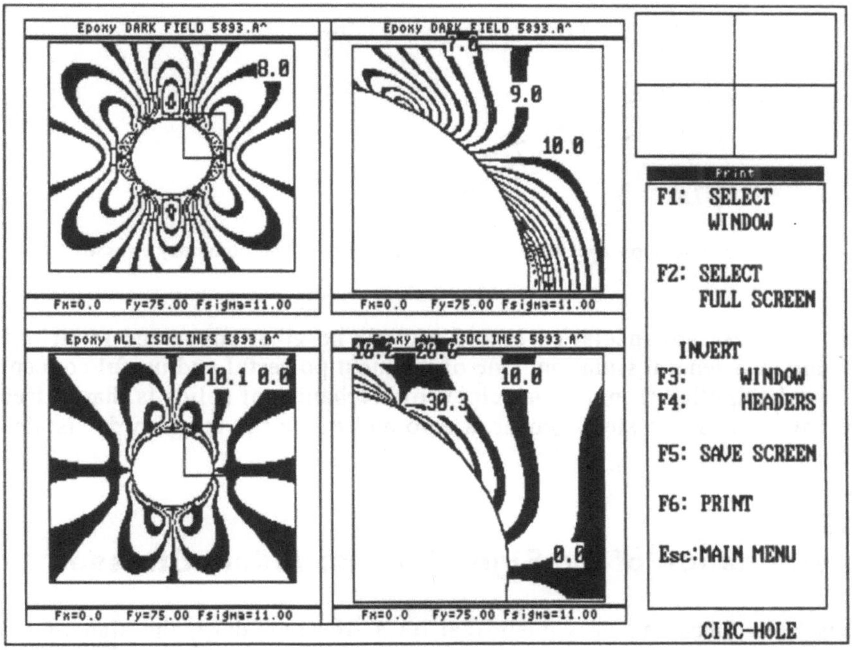

Fig. 1.18. Singular point in a plate with a hole. **a** Overall isochromatic fringe field **b** Close up of isochromatics at the singular point **c** Overall isoclinic field **d** Close up of isoclinics at the singular point (Courtesy: PHOTOSOFT_H)

A hole in a tensile strip acts like a stress raiser and on the boundary of the hole, the tangential stress (also the fringe order) changes its sign four times. Whenever there is a reversal of sign, it has to pass through zero and at such points the fringe order is zero. Figure 1.18 shows the *singular point* for the problem of a hole in a tension strip and also shows the corresponding isochromatic fringe field. Four such points can be found on the boundary of the circular hole.

In certain problems wherein the loading and geometry could be such that no zeroth fringe order exists. These are special class of problems and a case in point is the internally pressurised thick ring.

1.18.4 Role of Principles of Solid Mechanics in Fringe Ordering

In most problems, the identification of zeroth order fringe is possible based on the mechanics of the problem. In certain special situations, one has a zeroth line. For example, in the problem of a beam under pure bending, the neutral axis is a zeroth line. In the case of the circular disc, the outer boundary, except the load application points, has the fringe order value as

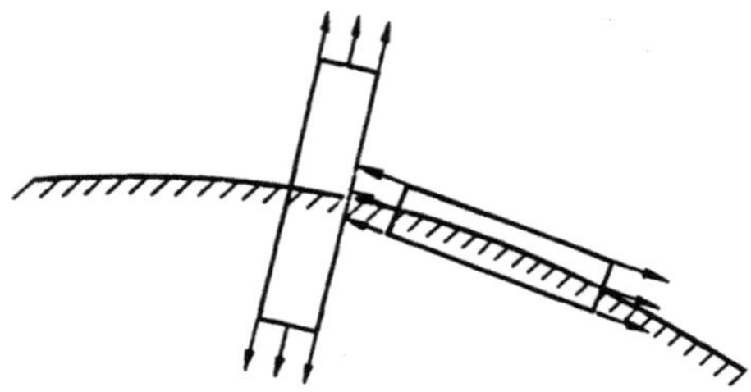

Fig. 1.19. Use of a tension specimen to identify the sign of the boundary stresses

zero. But these are special cases and provide no great advantage for fringe ordering in a general situation. One of the most powerful and useful concept that can be applied from the principles of mechanics of solids is that at free external corners, the stress tensor is zero and hence the fringe order is also zero.

1.19 Determination of the Sign of the Boundary Stresses

It has been discussed previously that on a free boundary, the sign of the boundary stresses change at the singular points. How to find out the sign of these stresses? One simple approach is to apply an external compression perpendicular to the boundary by a fingernail and observe the change in the boundary fringe order. This is also referred to as a nail test. One can also use a sharp edge to do this. It is well known that on a free boundary, the stresses can only be tangential. Let us for example, assume that initially the boundary stress (σ_t) is positive. The fringe order is proportional to σ_t. If a compressive load ($-\sigma_c$) is applied, then the fringe order is proportional to ($\sigma_t + \sigma_c$) and thus the fringe order should increase. Usually, the change in fringe order may be small and one has to use colour code to observe the change.

One can also use another approach wherein a simple tension specimen is kept tangential or perpendicular to the boundary (Fig. 1.19). If it is kept tangential and if one observes an increase in fringe orders, then the boundary stress is positive. The test could be used to identify the tension or compressive sides of a beam under bending.

1.20 Resolving the Ambiguity on the Principal Stress Direction

It has been pointed out earlier, that the isoclinic fringe parameter gives one of the principal stress directions and does not by itself says whether it is corresponding to either σ_1 or σ_2. It is noteworthy to point out that such an ambiguity exists even in the theoretical calculation of the principal stress directions by the famous formula

$$\tan 2\theta = \frac{2\tau_{xy}}{(\sigma_x - \sigma_y)} \tag{1.60}$$

Determination of θ by Eq. (1.60) involves inverse trigonometric operations and the solution is multi-valued. Unless an auxiliary method, such as a Mohr's circle is used, one cannot find σ_i and its associated direction θ_i. However, if the problem of principal stresses and their orientations is solved as an eigen value problem, the ambiguity can be resolved mathematically.

In photoelasticity, the ambiguity of principal stress direction is resolved as follows. This is achieved by calibrating the polariscope by using a tension specimen. A tension specimen is viewed in a circular polariscope with the axis of the tensile loading along the polarizer axis. The analyzer is then rotated anti-clockwise. Let us say that the fringe order increases by doing so. If a similar variation of fringe order is also observed in a practical problem, then the polarizer axis is the major principal axis.

1.21 Introduction to Three-Dimensional Photoelasticity and Integrated Photoelasticity

The interpretation of fringe patterns such as isochromatics and isoclinics to difference in principal stresses and their orientation became simple and straightforward in two-dimensional photoelasticity, mainly because, the model behaved like a simple retarder at the point of interest. In a general three-dimensional problem, the principal stresses vary continuously and also their directions need not remain in one plane. However, it is well understood that only the components of stresses that lie in a plane perpendicular to the light path contribute to photoelastic effect. From these stress components, one can find pseudo principal stress difference and their orientations and these are referred as secondary principal stresses (Chap. 9) and their directions in photoelastic literature. In a three-dimensional model, these secondary principal stress difference and their orientation continuously vary along the light path (Fig. 1.20) and one will observe the integrated effect of these variations as fringe patterns. Thus, one to one correspondence between the fringes observed and the state of stress of a particular point along the light path cannot be established.

Two approaches are used to analyse such problem situations. One of these uses concepts of two-dimensional photoelasticity for solving the three dimensional problem by appropriately slicing the model and is experimen-

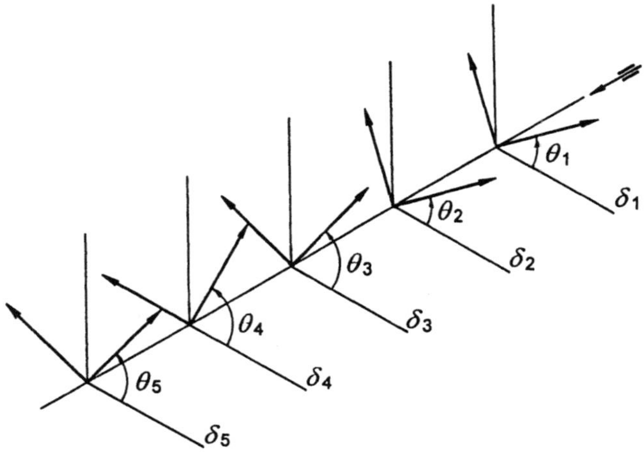

Fig. 1.20. Figure showing the variation of principal stress direction along the light path in a three-dimensional model

tally intensive. The other approach attempts to relate the integrated retardation pattern to the stress distribution along the light path and is mathematically intensive.

1.21.1 Conventional Three-Dimensional Photoelasticity

Among the various experimental techniques, only photoelasticity offers techniques for measuring the stress field interior to the model. Success of three-dimensional photoelasticity became possible with the developments in material research, which made it possible for the use of a unique process named as *stress freezing*. In this, a model made of an appropriate epoxy, is loaded and is allowed to go through a heat treatment process. At the end of the process, the loads are removed and the model is cut into thin slices for further analysis by using the principles of two-dimensional photoelasticity. The slice cut from the model will retain the same stress distribution, as it was part of the 3-D model, provided, the slicing operation is carried out carefully so that no machining stresses are introduced. This restriction is usually satisfied by using single point cutting tools and providing appropriate coolant while cutting. Although the slice has a finite thickness and will have some variation of stresses over the thickness, their effect is usually ignored for a simplified analysis. Further details on the evaluation of stress distribution can be found in the books listed at the end.

It is worthwhile to know the basic process involved in stress freezing. The model is subjected to a heat treatment process for locking the deformation of the model and hence the stresses. However it is termed as *stress freezing*, because the deformation locking phenomena can be explained similar to the case of a stretched spring kept in a water bath subjected to freezing in a refrigerator. At the end of the freezing, water turns ice and it prevents the spring to come back to its original position even when the loads are removed.

The epoxies used for making photoelastic models are composed of long-chain hydrocarbon molecules. Some of the molecular chains are well bonded but a large number of molecules are loosely bonded. These can be termed as primary and secondary bonds respectively. At the critical temperature of the polymer, the secondary bonds break and the primary bonds entirely carry the applied loads. When the temperature is lowered, keeping the loads still, the secondary bonds will re-form between the elastically deformed primary bonds and serve to lock them into their deformed positions as ice preventing the stretched spring. When the loads are removed, the primary bonds relax slightly, but a significant portion of their deformation is not recovered. Since, these deformations are locked on a molecular scale; the deformation and accompanying birefringence are maintained even in small sections cut from the original model.

1.21.2
Principle of Optical Equivalence

The methodology of integrated photoelasticity mainly depends on this most powerful and simple concept for idealising the three-dimensional model. It has been noted earlier that in view of the variation of secondary principal stresses and their orientation, one cannot relate the optical phenomena to the stresses in a simplistic manner as in two-dimensional photoelasticity. The principal of optical equivalence offers hope for relating the optical phenomena to the stress distribution along the light path.

The principle states that the optical effect introduced by a three-dimensional model to an incident light can be represented by a simpler system consisting of a *retarder* and a *rotator*. This means that for the same incident light, the light ellipse coming out of the 3-D model and its corresponding optically equivalent model are the same (Fig. 1.21). The retarder introduces the retardation equivalent to the one introduced by the 3-D model. However, the azimuth of the exit light ellipse may not be the same as that for a 3-D model. The necessary correction is done by the rotator. The parameters representing the retarder and the rotator are the experimental parameters to be determined in integrated photoelasticity. We have already seen that to specify a retarder, one has to provide the retardation introduced by it and also give the orientation of its polarising axes. The retardation is termed as *characteristic retardation* denoted by the symbol 2Δ and the orientation of the retarder is termed as the *primary characteristic direction* denoted by the symbol θ. Rotator is an idealised optical element, which simply rotates the azimuth of the light ellipse and is specified by the angular rotation γ it provides. The angle γ is referred to as *characteristic rotation*. Various methods for finding these parameters can be found in ref. [18].

One can construct a matrix representation of the optical equivalent model as follows. In Eq. (1.30), one should use the retardation as 2Δ and the

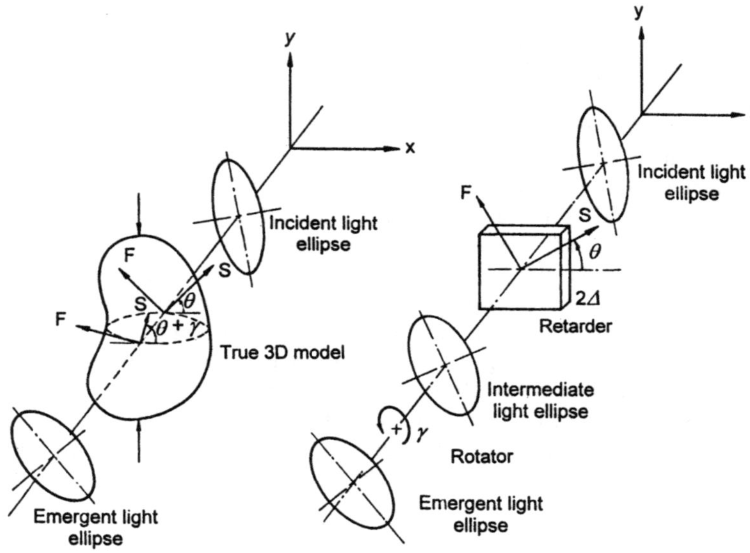

Fig. 1.21. Figure showing the principle of optical equivalence

orientation of the retarder as θ to get the matrix representation of the retarder forming the optically equivalent model. The matrix turns out to be

$$\begin{bmatrix} \cos\Delta - \mathrm{i}\sin\Delta\cos 2\theta & -\mathrm{i}\sin\Delta\sin 2\theta \\ -\mathrm{i}\sin\Delta\sin 2\theta & \cos\Delta + \mathrm{i}\sin\Delta\cos 2\theta \end{bmatrix} \tag{1.61}$$

For a rotator that provides an anti-clockwise rotation, the rotation matrix is

$$\begin{bmatrix} \cos\gamma & -\sin\gamma \\ \sin\gamma & \cos\gamma \end{bmatrix} \tag{1.62}$$

The matrix representation [*M*] of the optically equivalent model is obtained by pre-multiplying Eq. (1.61) by Eq. (1.62) as

$$[M] = \begin{bmatrix} P - \mathrm{i}Q & -R - \mathrm{i}S \\ R - \mathrm{i}S & P + \mathrm{i}Q \end{bmatrix} \tag{1.63}$$

where,

$$\begin{aligned} P &= \cos\Delta\cos\gamma \\ Q &= \sin\Delta\cos(2\theta + \gamma) \\ R &= \cos\Delta\sin\gamma \\ S &= \sin\Delta\sin(2\theta + \gamma) \end{aligned} \tag{1.64}$$

Experimentally one can find the characteristic parameters, and from these the matrix given in Eq. (1.63) can be constructed. Theoretically one can assume an appropriate stress distribution and the simplest of these is to assume them as appropriate polynomial functions. The 3-D model can be represented as a train of retarders and using Jones calculus, one can obtain the net effect of these. If the stress distribution is correctly modelled then the final matrix should be identical to the optically equivalent model. In other words, one should try to evaluate the polynomial coefficients, such that these are same. It is very easily said, but in practice it becomes very difficult to solve even simple problems. A single experimental measurement may not be sufficient to find all the polynomial coefficients and additional measurements along different light paths may have to be used. Unlike in 2-D photoelasticity, wherein the change in wavelength does not provide any new information, one does get additional equations by employing different wavelengths here. Further details can be found in ref. [11].

Integrated photoelasticity has been successful in analysing axisymmetric problems. It has also been widely used for evaluating the residual fringes in glass components.

1.22 Model to Prototype Relations

One of the first questions that come up in anybody's mind is how does one relate the results from conducting experiments on plastics to metallic prototypes? The validity can be explained from considering the nature of the equations that are necessary for solving the stress distribution. In two-dimensional problems, the equations of equilibrium together with the boundary conditions and the compatibility conditions in terms of stress components give us a system of equations that are usually sufficient for complete determination of the stress distribution. The equilibrium equations are

$$\begin{aligned} \frac{\partial \sigma_x}{\partial x} + \frac{\partial \tau_{xy}}{\partial y} + F_x &= 0 \\ \frac{\partial \tau_{xy}}{\partial x} + \frac{\partial \sigma_y}{\partial y} + F_y &= 0 \end{aligned} \tag{1.65}$$

where, F_x and F_y are the body forces in x and y directions respectively. The compatibility conditions for plane stress and plane strain situations in terms of stress components are respectively

$$\left(\frac{\partial^2}{\partial x^2}+\frac{\partial^2}{\partial y^2}\right)(\sigma_x+\sigma_y)=-(1+\nu)\left(\frac{\partial F_x}{\partial x}+\frac{\partial F_y}{\partial y}\right)$$
$$\left(\frac{\partial^2}{\partial x^2}+\frac{\partial^2}{\partial y^2}\right)(\sigma_x+\sigma_y)=-\frac{1}{(1-\nu)}\left(\frac{\partial F_x}{\partial x}+\frac{\partial F_y}{\partial y}\right) \tag{1.66}$$

It is interesting to note that in the case of constant body forces, the equations determining the stress distribution do not contain the elastic constants of the material. Thus, the stress magnitudes and their distribution are same for all isotropic materials. However, the strains and also the deformations are functions of elastic constants. Obviously, a plastic model will get deformed more than the metallic prototype. Equations (1.65) and (1.66) also indicate that the stress distribution in the elastic state is independent of the magnitude of the loads and the scale of the model. Thus for making photoelastic models, the scale of the models and the loads may be chosen as convenient.

The model results can be related to the prototypes following the laws of similarity, provided, the model and the prototypes are geometrically similar and the distribution of the loading is the same for both. The similarity relations do not apply near points of loading and the principle of Saint Venant is usually invoked for other regions.

For exact similarity, the model and prototype should be geometrically similar when deformed by their respective loads. This necessitates that the corresponding strains in the model and prototype should be equal. To achieve this, the force scale should be appropriately taken. Assuming that both the model and the prototype have the same Poisson's ratio, the stress in the model and the prototype are related as

$$\sigma_m=\sigma_p\frac{F_m}{F_p}\left(\frac{L_p}{L_m}\right)^2 \tag{1.67}$$

where L_m and L_p are characteristic lengths of the model and the prototype. The subscript m refers to the model and p refers to the prototype. Replacing stresses in terms of strains and the Young's moduli in Eq. (1.67), the force scale is obtained as

$$F_m=F_p\left(\frac{L_m}{L_p}\right)^2\frac{E_m}{E_p} \tag{1.68}$$

In three-dimensional problems, the compatibility conditions are given by Beltrami and Mitchell equations [3]. These equations contain Poisson's ratio and unlike 2-D problems, the stress distribution is dependent on Poisson's ratio even if the body force is constant. In certain special cases, in 2-D problems also, the stress distribution is dependent on Poisson's ratio. When one works on multiply connected bodies, such as, a plate with a hole having a resultant force acting at the boundary of a hole, the stress distribution is a function of the Poisson's ratio.

Thus in model studies, the mismatch of Poisson's ratio is a crucial factor in relating stresses in the model to prototypes. For room temperature studies, the mismatch of Poisson's ratio between the plastics and the metallic prototypes is usually negligible and its effect is ignored. In conventional three-dimensional photoelasticity, at the stress-freezing temperature, the model's Poisson's ratio approaches 0.5 and its influence become significant. However, experiments have shown that the mismatch of Poisson's ratio affects only the minor principal stress significantly and the major principal stress is relatively unaffected [16].

Table 1.3 summarises the properties of some of the most commonly used photoelastic model materials. For general studies, one usually uses models made of epoxies. Models made of polyurethane are used for preparing demonstration models. They are also used in dynamic studies, as the velocity of stress wave propagation is the least and the recording of the phenomena is comparatively simple. By changing the composition of the polyurethane, it is possible to simulate the soil as layered elastic media for foundation studies in Civil engineering [17]. Models made of Plexiglas are used if one is interested in recording the isoclinic information. In certain problem situations, where the loading due to self-weight is to be simulated, one selects the use of gelatin.

Table 1.3. Approximate properties of a few photoelastic materials

Material	Stress Fringe Value $F_\sigma^{\dagger}$ (N/mm/fringe)	Young's Modulus‡ E (MPa)	Proportional Limit‡ (MPa)	Poisson's Ratio‡ ν	Figure of Merit§ (1/mm)
Polycarbonate	8	2,600	3.5	0.28	325
Epoxy	12	3,300	35	0.37	275
Glass	324	70,000	60	0.25	216
Homolite 100	26	3,900	48	0.35	150
Homolite 911	17	1,700	21	0.40	100
Plexiglass	140	2,800	-	0.38	20
Polyurethane	0.2	3	0.14	0.46	15
Gelatin	0.1	0.3	-	0.50	3

§ Higher the figure of merit, the better is the material suitable for photoelastic analysis
† Stress fringe value corresponding to Sodium Vapour source (589.3 nm).
‡ Room temperature properties.

1.23
Closure

Photoelasticity is not only an experimental tool to evaluate stresses but has contributed significantly to the development of solution methods (analytical/numerical) for stress analysis. During the early stages of development of Theory of Elasticity, the solutions obtained by it were verified with the experimental results of Photoelasticity. Rapid strides in the field of Fracture Mechanics were possible only because of the experimental verification by Photoelasticity. After the advent of the numerical technique of Finite Elements it was thought that one could do away with experimental approach. It is true that a numerical approach is convenient for a detailed parametric analysis compared to an experimental approach. However, validation of numerical modelling has to be done *a priori*. In that sense, Photoelasticity is a very convenient tool. With the developments in Digital Computers, the cost of performing a 2-D analysis by FEM and Photoelasticity are comparable. However, for a true three-dimensional analysis Photoelasticity works out cost effective. In the design of crankshaft of an automobile it is mandatory to go through a 3-D photoelastic analysis. One of the reasons why Photoelasticity is not widely practised is that one needs training to interpret the fringes unlike the case of strain gauge technique where the instrumentation is developed to an extent to give directly strains. With the developments in the use of digital image processing techniques for data reduction, the training needed to interpret fringes can be made minimal and it is the purpose of this book to summarise the significant developments that have taken place in the last two decades in that direction.

Exercises

1.1 Explain what is meant by polarization of light. How linearly polarised light is achieved in commercial diffused light polariscopes?

1.2 Prove that light ellipse of any azimuth and ellipticity can be produced by a polarizer and quarter-wave plate combination.

1.3 A light ellipse of ellipticity 0.36 with an azimuth at 80° from the horizontal needs to be generated for an experiment. Determine the angular orientations of the polarizer and quarter-wave plate to generate this light ellipse.

1.4 Derive the expressions for azimuth, major and minor axes of a light ellipse generated by the superposition of two simple harmonic motions travelling in perpendicular directions with a phase difference of δ. The respective amplitudes of the waves are a_x and a_y.

1.5 Compare the passage of light through an optically isotropic material and a crystalline medium.

1.6 For the formation of isochromatic fringes, one needs two polarised beams whose axis of polarization is perpendicular to each other, travelling in the same direction but with different velocities are needed. What causes this in conventionally used model materials.

1.7 Explain the physics behind the formation of fringes in a plane polariscope.

1.8 Determine the intensity of light transmitted in a plane polariscope by Jones calculus for the case when both polarizer and analyzer are parallel to each other.

1.9 In Fig. 1.13, if the first quarter-wave plate (slow axis) is kept at 45°, find the intensity of light transmitted by Jones calculus.

1.10 The quarter-wave plates used in a circular polariscope (Fig. 1.13) are matched for a wavelength of 540 nm. If a sodium vapour source (589.3 nm) were used to illuminate the model, how would you model its influence in the analysis? Use Jones calculus to substantiate your answer.

1.11 Mention the sequence of steps in Tardy's method of compensation.

1.12 Consider the case wherein the first quarter-wave plate (Fig 1.13) is removed and the polarizer is kept at 45° to the horizontal. The analyzer is rotated until a fringe order passes through the point of interest. Determine the relationship between the analyzer angle β and the fractional retardation of the model.

1.13 A circular disc of 60 mm diameter is loaded diametrically by a load of 1200 N. If the fringe order observed at the centre is 4.25, determine the material stress fringe value of the model.

1.14 The material stress fringe value of an epoxy model is found to be 12 N/mm/fringe using a sodium vapour source (589.3 nm). A model made of the material is then viewed in Mercury green (546.1 nm). At the point of interest, the fringe order observed is 3.28. Determine the principal stress difference at that point.

1.15 In Prob. 1.14 if the orientation of isoclinic is found to be 35°, determine the in-plane shear stress and the normal stress difference at the point of interest.

1.16 In Probs. 1.14 and 1.15 if the point of interest actually lies on a free boundary, determine the state of stress at the point of interest.

1.17 For a disc of 60 mm diameter and thickness 6 mm subjected to a diametral load of 325 N, the isochromatics are shown in Fig. 1.14. Using the software *fsigma.c* determine the material stress fringe value. Verify the evaluation by using *reconstr.c*.

1.18 Figure 1.22 shows the dark-field isochromatics of a beam subjected to four-point bending. Label the fringe orders.

Fig. 1.22.

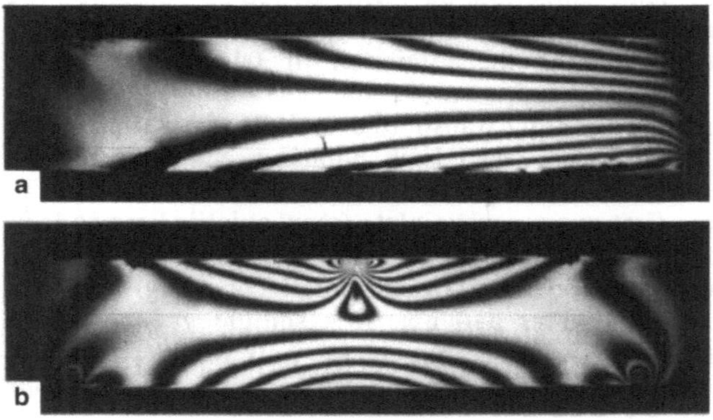

Fig. 1.23.

1.19 Figures 1.23a and 1.23b show the dark-field isochromatics of a cantilever beam and a beam subjected to three-point bending respectively. Determine the fringe orders using the principles of solid mechanics. On the basis of the experimental results, comment on the existence of neutral axis in these cases.

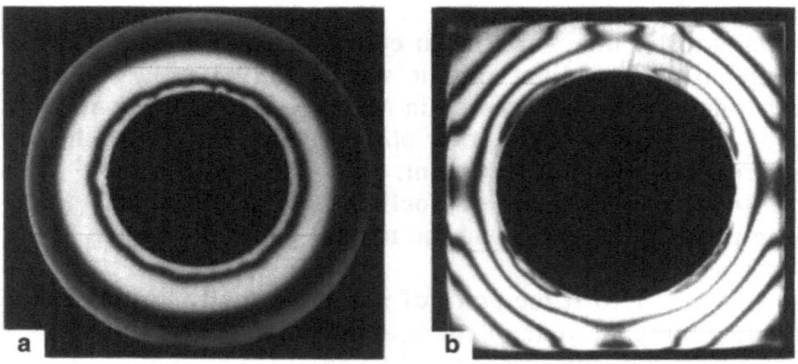

Fig. 1.24.

1.20 Figures 1.24a and 1.24b show the dark-field isochromatics of an internally pressurised thick ring and a square tube respectively. For which of these cases, can you determine the fringe orders?

1.21 In Prob. 1.19 if the material stress fringe value of the model is 12 N/mm/fringe and the model thickness is 6 mm and the beam depth is 25 mm, determine the bending moment applied. Do you have sufficient data to determine the stress tensor variation over the depth of the beam? Comment on the role of stress separation techniques for this case.

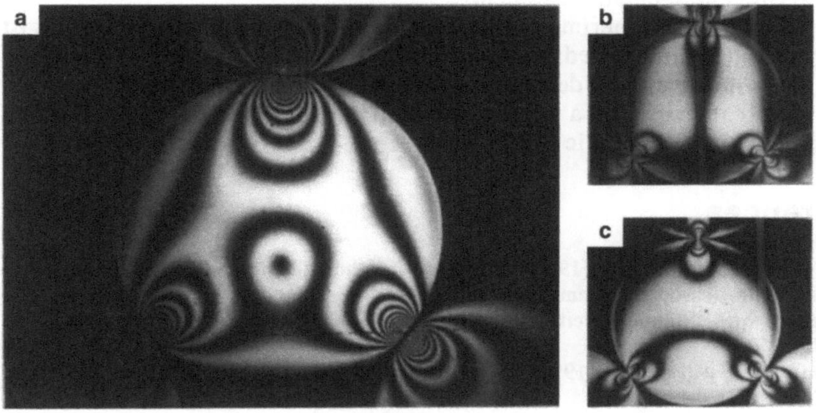

Fig. 1.25.

1.22 Figure 1.25a shows the dark-field isochromatics of a disc subjected to three-point loading. Figures 1.25b and 1.25c show the 0° isoclinic and 45° isoclinic respectively. Determine the fringe orders in the disc.

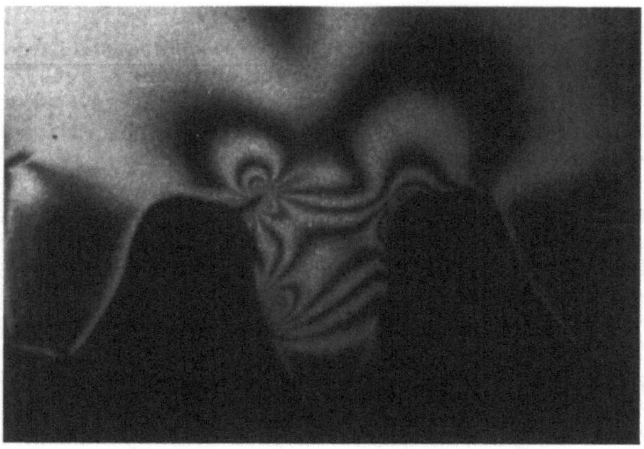

Fig. 1.26.

1.23 Figure 1.26 shows the dark-field isochromatics of a gear tooth with a crack in its tensile root fillet. Label the fringe orders in the complete domain.

1.24 Mention two methods for determining the sign of the boundary stresses.

1.25 What are secondary principal stresses? How is its variation over the light path is accommodated in the modelling of the problem in conventional 3-D photoelasticity and integrated photoelasticity.

1.26 Write a note on stress freezing.

1.27 What are the experimental parameters to be determined for a point of interest in integrated photoelasticity? If the stress field is known *a priori* how can you determine these parameters theoretically?

1.28 Justify the use of a plastic in photoelastic analysis to predict the behaviour of metallic prototypes.

References

1. Coker EG, Filon LNG (1967) A treatise on photoelasticity. First edn 1931. Revised edn Jessop HT, Cambridge University Press, London
2. Frocht MM, Photoelasticity. Vol 1 (1941), Vol 2 (1948) John Wiley and Sons, New York
3. Dally JW, Riley WF (1991) Experimental stress analysis. McGraw-Hill, New York
4. Durelli AJ, Riley WF (1965) Introduction to photomechanics. Prentice Hall International, Englewood Cliffs
5. Kuske A, Robertson G (1974) Photoelastic stress analysis. John Wiley and Sons, London
6. Srinath LS, Raghavan MR, Lingaiah K, Gargesha G, Pant B, Ramachandra K (1984) Experimental stress analysis. Tata McGraw-Hill, New Delhi
7. Theocaris PS, Gdoutos EE (1979) Matrix theory of photoelasticity. Springer-Verlag, Berlin Heidelberg
8. Oppel G (1937) The photoelastic investigation of three-dimensional stress and strain conditions. NACA TM No 824 (Translation by Vernier J)
9. Weller R, Bussey JK (1939) Photoelastic analysis of three-dimensional stress system using scattered light. National Adv Committee for Aeronautics, Tech Note 737, Washington DC
10. Srinath LS (1983) Scattered light photoelasticity. Tata McGraw-Hill, New Delhi
11. Aben HK (1979) Integrated photoelasticity. McGraw-Hill, New York
12. Brewster D (1816) On the communication of the structure of doubly refracting crystals to glass, muriate of soda, flour spar and other substances by mechanical compression and dilatation. Philosophical Transactions of the Royal Society, pp 156–178. Also in: Billings BH (ed) (1992) Applications of polarised light. SPIE Milestone series Vol MS57, pp 491–504
13. Durelli AJ, Mulzat AP (1965) Large strain analysis and stresses in linear materials. J Engng Mech ASCE 91:65–91
14. Sanford RJ (1980) Application of the least squares method to photoelastic analysis. Exp Mech 20:192–197
15. Ramesh K (1997) PHOTOSOFT_H: a comprehensive photoelasticity simulation module for teaching the technique of photoelasticity. Int J Mechanical Engineering Education 25(4):306–324
16. Srinath LS (1982) Lecturer Notes. IISc Banglore
17. Chandrasekhara K, Joseph Antony S, Santanu Manna (1994) An experimental-numerical hybrid method for elastic contact problem. Mech Struct & Mach 22(4):487–504
18. Srinath LS, Ramesh K, Ramamurti V (1988) Determination of characteristic parameters in three-dimensional photoelasticity. Opt Engng 27(3): 225–230. Also in: Applications of polarised light. SPIE Milestone series Vol MS57, pp 589–594

2 Reflection Photoelasticity

2.1 Introduction

Reflection photoelasticity is an extension of transmission photoelastic analysis for the analysis of opaque prototypes. A thin temporarily birefringent coating is pasted on the prototype with a reflective backing at the interface. The prototype is loaded appropriately and a reflection polariscope is used for collecting the optical information.

Unlike in transmission photoelasticity, the birefringent plastic is not directly loaded but the surface strains developed on the prototype are transmitted through the adhesive to the coating. In view of this, the elastic properties of both the specimen and the coating, particularly their Poisson's ratio has a strong influence on the interpretation of the optical response of the coating. Further, the optical response achievable is an order of magnitude less than what is possible in transmission photoelasticity. This has necessitated the use of white light for reflection photoelastic analysis.

Mesnager [1] was the first to study the concept of photoelastic coatings. He used segments of glass as a coating. The method was further developed by Oppel [2] in 1937. He used flat sheets of Bakelite. Both glass and Bakelite were not the ideal choice as materials for photoelastic coating. Glass has a high modulus and tends to reinforce the specimen significantly. Bakelite has a significant *time-edge* effect. Lack of proper adhesives to bond these were also a problem in the early development of photoelastic coatings. The availability of epoxy resins in the 1950s contributed significantly to the development of the technique. Zandman [3] in 1960 developed a unique procedure for preparing contourable plastic for coating complicated geometries of industrial components. Zandman et al. [4] studied the problem of reinforcement caused by the photoelastic coating and derived correction factors for interpreting the fringe patterns for engineering use.

Reflection photoelasticity is truly an engineering tool for solving practically important and complicated problems. Several approximations are made in the interpretation of the optical information recorded. Using the principles of transmission photoelasticity, the optical response of the coating is initially related to the coating stresses. Based on the principles of mechanics of solids, the specimen stresses are determined from the coating stresses. The analysis is improved by the use of appropriate correction factors. The various aspects of reflection photoelasticity are discussed here.

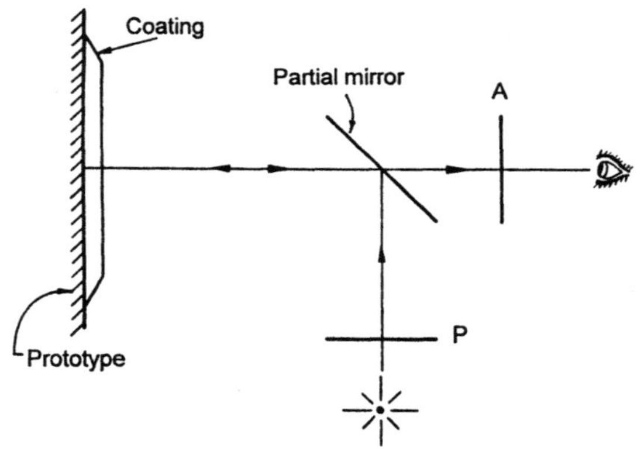

Fig. 2.1. Optical arrangement to obtain normal incidence in a reflection polariscope

2.2 Reflection Polariscope

While using photoelastic coatings, the light must be reflected from the interface between the coating and the prototype. As in transmission photoelasticity, the optical phenomenon can be easily related to the stresses, only if a normal incidence is used. Simplest optical arrangement for obtaining normal incidence is shown in Fig. 2.1. Though this arrangement is quite good for a plane polariscope, for circular polariscope one has to provide appropriate compensation as the partial mirror alters the ellipticity of polarization. Usually, polariscopes of this type are large and cumbersome for portable use. Figure 2.2 shows another arrangement, which is portable and is also commercially available. The very nature of the arrangement of the elements do not guarantee normal incidence. The angle of oblique is usually of the order of 4° whose influence on the optical pattern interpretation

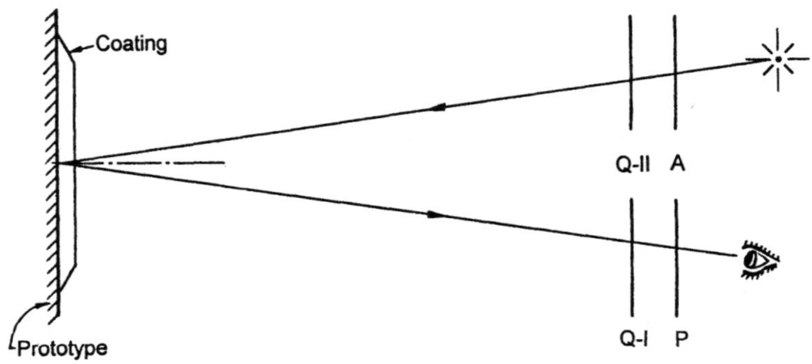

Fig. 2.2. Optical arrangement used in commercially available reflection polariscopes

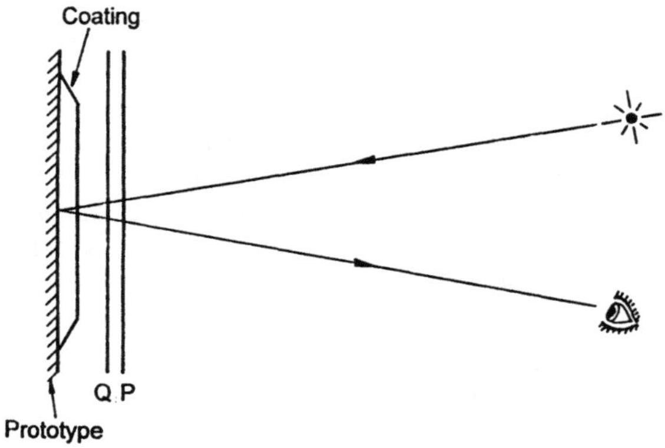

Fig. 2.3. Optical arrangement used in photoelastic strain gauges

is considered negligible. Thus, the engineering approximation used in reflection photoelasticity starts right from the data collection stage. One can also observe photoelastic fringes by directly keeping polarizer and quarter-wave plate as shown in Fig. 2.3. This arrangement is used in constructing photoelastic strain gauges wherein the circular polarizer comes integrated with the birefringent material.

2.3 Stress and Strain-Optic Relations for Coatings

Extension of the concepts of two-dimensional transmission photoelastic analyses to three-dimensional stress analyses through stress freezing and slicing has made the photoelastic analysis useful from the industrial point of view. A slice of finite thickness is used in 3-D photoelasticity and the optical information is interpreted assuming that the stress variation over the thickness is negligible. Due to the nature of application of the loads on photoelastic coatings, the stresses in general vary over the thickness for a finite thickness coating. However, the concepts of two-dimensional transmission photoelastic analyses can be extended for reflection photoelasticity too assuming that this variation is negligible. The analysis is improved by the use of appropriate correction factors.

In a plane polariscope arrangement, one can observe the superposition of isoclinics over isochromatics and in a circular polariscope arrangement, one will observe only isochromatics. The isochromatics can be related to the difference in principal stresses or strains and isoclinics correspond to the orientation of principal stress/strain direction.

Both the incident and reflected light contribute to photoelastic effect and for a coating of thickness h_c, the stress-optic law can be written as

$$\sigma_1^c - \sigma_2^c = \frac{NF_\sigma}{2h_c} \tag{2.1}$$

where N is the fringe order, F_σ is the material stress fringe value and $(\sigma_1^c - \sigma_2^c)$ represents the principal stress difference in the coating.

The birefringence in the coating is introduced through the surface deformations at the interface and it is useful to represent the photoelastic phenomena to the strains developed. The strain-optic law is

$$\varepsilon_1^c - \varepsilon_1^c = \frac{NF_\varepsilon}{2h_c} \tag{2.2}$$

where F_ε is the strain-optic coefficient and $(\varepsilon_1^c - \varepsilon_2^c)$ gives the principal strain difference in the coating. It is well known that the retardation introduced by the birefringent material is a function of the wavelength λ and the strain-optic coefficient is usually expressed as

$$F_\varepsilon = \frac{\lambda}{K} \tag{2.3}$$

where K is the strain coefficient, supplied by the manufacturer or to be determined by calibrating the coating. For a perfectly linear elastic photoelastic material, one can find an inter relationship between the parameters F_σ and F_ε. From stress-strain relations, the principal strain difference in the coating can be related to the principal stress difference as

$$\varepsilon_1^c - \varepsilon_2^c = \frac{1+\nu_c}{E_c}(\sigma_1^c - \sigma_2^c) \tag{2.4}$$

Using Eqs. (2.4) and (2.1), the relationship between F_σ and F_ε can be obtained:

$$F_\varepsilon = \frac{1+\nu_c}{E_c} F_\sigma \tag{2.5}$$

For a given photoelastic coating, inspection of Eqs. (2.1) and (2.2) shows that the optical response will increase if the coating thickness is increased.

2.4 Coating and Specimen Stresses

The ultimate objective of reflection photoelasticity is to find the surface stresses of the prototype on which the photoelastic coating is pasted. An ideal situation is that the coating is firmly pasted on the surface. Further, its thickness be very small compared to the specimen thickness, yet produces a sufficient amount of birefringence for data acquisition.

It is seen in Sect. 2.3, that the optical response can be increased if the coating thickness is increased. Use of coatings of 3-mm thickness is not uncommon. In principle, coatings of finite thickness serve as a three-dimensional extension of the part and serve to share some of the external load thereby reinforcing the part. Since the loading of the coatings is through surface deformations at the interface, the mismatch of Poisson's ratio between the coating and the specimen also significantly influences the optical response of finite thickness coatings. Further, when the specimen is subjected to bending or torsional loads, the strain introduced will not be uniform throughout the coating thickness, affecting the interpretation of the optical response to even the coating stresses.

The approach adopted in reflection photoelasticity is as follows. Initially, expressions relating the coating stresses to specimen stresses are obtained based on the following assumptions:

1. The thickness of the coating is very small.
2. Both the specimen and the coating have the same Poisson's ratio.
3. The specimen and the coating are in a state of plane stress.

Later, correction factors are used to account for the role of a finite thickness coating.

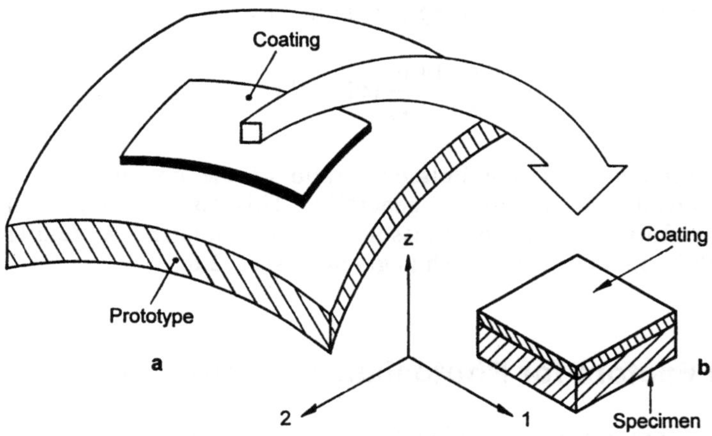

Fig. 2.4. Figure showing the reference axes to analyse an elemental coating-specimen interface

Let a small element of a coating and specimen, far removed from the edges be as shown in Fig. 2.4b. The reference axes are taken as the principal stress/strain axes. Let the surface strains of the specimen are transmitted to the coating through the adhesive without loss or amplification. Thus,

$$\varepsilon_1^c(x,y)=\varepsilon_1^s(x,y) \text{ and } \varepsilon_2^c(x,y)=\varepsilon_2^s(x,y) \tag{2.6}$$

Since it is assumed that both the coating and the specimen are in a state of plane stress, the stress-strain relations give:

$$\begin{aligned} \varepsilon_1^s &= \frac{1}{E_s}(\sigma_1^s - \nu_s\sigma_2^s), \quad \varepsilon_1^c = \frac{1}{E_c}(\sigma_1^c - \nu_c\sigma_2^c) \\ \varepsilon_2^s &= \frac{1}{E_s}(\sigma_2^s - \nu_s\sigma_1^s), \quad \varepsilon_2^c = \frac{1}{E_c}(\sigma_2^c - \nu_c\sigma_1^c) \end{aligned} \tag{2.7}$$

where the superscript/subscript c refers to the coating and s refers to the specimen. From Eqs. (2.6) and (2.7), one can find the coating stresses as

$$\begin{aligned} \sigma_1^c &= \frac{E_c}{E_s(1-\nu_c^2)}\left[(1-\nu_c\nu_s)\sigma_1^s + (\nu_c - \nu_s)\sigma_2^s\right] \\ \sigma_2^c &= \frac{E_c}{E_s(1-\nu_c^2)}\left[(1-\nu_c\nu_s)\sigma_2^s + (\nu_c - \nu_s)\sigma_1^s\right] \end{aligned} \tag{2.8}$$

From Eq. (2.8) the inter relationship between the difference in principal stresses of the specimen and the coating is obtained as

$$\sigma_1^s - \sigma_2^s = \frac{E_s}{E_c}\frac{1+\nu_c}{1+\nu_s}(\sigma_1^c - \sigma_2^c) \tag{2.9}$$

In Eqs. (2.7) through (2.9), the Poisson's ratio of the coating and the specimen are indicated separately for generality. This does not mean that these equations fully account for Poisson's ratio mismatch. A detailed discussion on Poisson's ratio mismatch is given in Sect. 2.6.

2.5 Correction Factors for Photoelastic Coatings [4]

If the coating is thin and is also subjected to a state of plane stress, then using Eq. (2.9) and Eqs. (2.2)–(2.4), one can directly determine the principal stress difference in the specimen as

$$\sigma_1^s - \sigma_2^s = \frac{E_s}{1+\nu_s}\frac{N}{2h_c}\frac{\lambda}{K} \tag{2.10}$$

It is interesting to observe that the elastic properties of the coating do not appear in Eq. (2.10). This is because, rather than using F_σ, as in transmission photoelasticity, here K is used for relating the optical information to the stress difference. Thus, strain coefficient K is the fundamental optical property of the coating needed for data reduction.

If a finite thickness coating is used, it is necessary the following points be considered:

1. The coating reinforces the specimen.
2. A strain variation exists over the thickness of the coating and
3. Mismatch of Poisson's ratio between the coating and the specimen needs to be accounted for.

For various types of loading situations, correction factors have been developed to account for the factors (1) and (2) considering that there is no mismatch of Poisson's ratio. The Eq. (2.10) needs to be rewritten as

$$\sigma_1^s - \sigma_2^s = R_f \frac{E_s}{1+\nu_s} \frac{N}{2h_c} \frac{\lambda}{K} \tag{2.11}$$

where R_f is the reinforcement factor which is a function of the mechanical properties of the coating and the specimen, as well as the geometry of the specimen and loading.

For specimens subjected to plane stress, the correction factor is due to the reinforcement effect of the coating only and the strain does not vary over the thickness of the coating. The correction factor R_f^a is obtained as

$$R_f^a = 1 + \frac{h_c E_c (1+\nu_s)}{h_s E_s (1+\nu_c)} \tag{2.12}$$

The determination of correction factor for flexure of beams or plates is little more involved, because there is reinforcement effect as well as distortion effect due to the strain gradient in a direction perpendicular to the surface. The fringes observed are proportional to the average strain through the thickness and not to the surface strain.

The expressions for the correction factor could be greatly simplified by representing the ratio of coating and specimen properties as

$$e = \frac{E_c}{E_s}, \quad g = \frac{h_c}{h_s}, \quad \text{and } m = \frac{1-\nu_s^2}{1-\nu_c^2} \tag{2.13}$$

The correction factor for a beam in bending is given as follows

$$R_f^b = \frac{(1+eg)}{(1+g)} \left[4(1+eg^3) - \frac{3(1-eg^2)^2}{(1+eg)} \right] \left[\frac{1+\nu_s}{1+\nu_c} \right] \tag{2.14}$$

For the bending of thin or medium thick plates, in regions of uniform stress, the correction factor is

$$R_f^{bp} = \frac{(1+emg)}{(1+g)}\left[4(1+emg^3) - \frac{3(1-emg^2)^2}{(1+emg)}\right] \tag{2.15}$$

The Eq. (2.15) is applicable for common geometrical discontinuities such as a hole or a series of holes in a plate under flexure. The correction factor is not valid for cases wherein the plate has an abrupt change in thickness. For the restricted case of pure-bending loads, the correction factor may be applied for thick plates too. For $v_s = v_c$ Eq. (2.15) is identical to Eq. (2.14). Even for different values of Poisson's ratios, the numerical values of R_f^b and R_f^{bp} for most engineering materials are the same.

Frequently, a structural part is subjected to a combination of plane and flexural loads. For thick coatings, proper weighting of the correction factors is important. One approach could be to find a coating thickness such that the factors R_f^a, R_f^{bp} are equal. If done, one will observe that the thickness of the coating is larger than the thickness of the structure. Alternately for analysing thin plates, a coating of convenient thickness be first applied to examine the general distribution of strains and to locate high strain regions. Then the coating is stripped off and replaced in the critical regions of high strain by a thin coating.

For torsion of circular shafts, the correction factor is due to the reinforcement effect as well as the distortion effect due to the strain gradient through the coating. The correction factor is given by

$$R_f^t = \frac{2}{(1+c)}\left[1 + \frac{G_c(c^4-1)}{G_s(1-a^4)}\right] \tag{2.16}$$

where G is the shear modulus, $a = r_i/r_o$ and $c = r_c/r_o$ (Fig. 2.5). The correction factor is not applicable for the regions such as keyways and abrupt

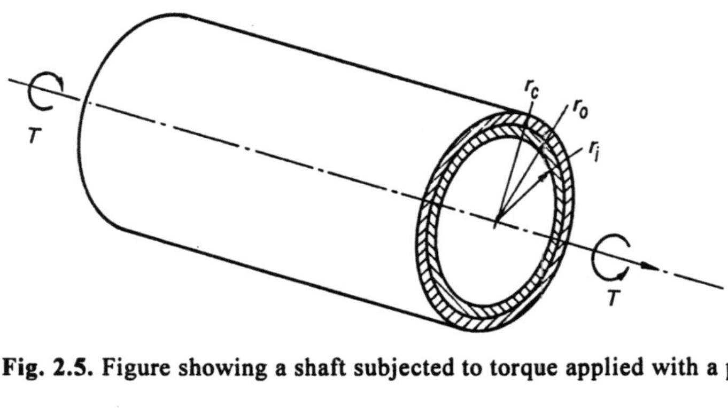

Fig. 2.5. Figure showing a shaft subjected to torque applied with a photoelastic coating

changes in diameter. In such regions, thin coatings should be employed and they will reveal sufficient birefringence due to higher levels of stress and no correction factor need to be applied.

A birefringent coating reinforces a pressure vessel in two ways simultaneously; viz., it restrains radial displacements of the vessel and it restrains longitudinal deformation of the member (Fig. 2.6). Unlike the earlier mentioned cases, the restraints are not equal for the transverse and longitudinal directions and the evaluation of correction factor is more complicated.

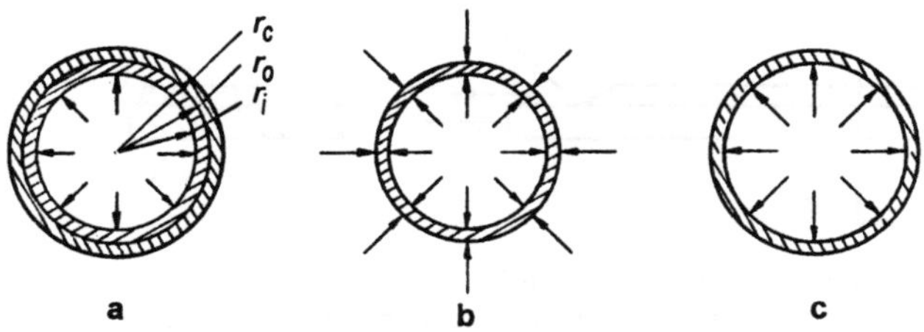

Fig. 2.6. Reinforcement effect of the coating applied to a pressurised cylinder: **a** cylinder with the coating **b** loading of the pressurised cylinder **c** load sharing of the coating

In the case of a long cylindrical pressure vessel with closed ends, the correction factor (R_f^p), assuming that the Poisson's ratio of the coating and the specimen is the same, is given by

$$\frac{1}{R_f^p} = \left[\frac{2(1-2\nu+c)(1-\nu)}{(1-2\nu+c^2)+P(1-2\nu+a^2)} - \frac{(1-2\nu)}{(1+P)}\right]$$
$$P = \frac{em(c^2-1)}{(1-a^2)} \tag{2.17}$$

The applicability of Eq. (2.17) is very restrictive and it is valid only for regions in continuous vessels remote from closures, discontinuities, nozzles, etc. In most practical problems, the coating thickness will be small in relation to the outside radius of the vessel and the correction factor will be close to unity. For the analysis of small diameter tubing the role of correction factor becomes significant.

The correction factors reported in this section are not valid if the part is subjected to strains beyond the elastic limit. In such cases, one should use thin coatings to avoid the need of a correction factor.

2.6 Poisson's Ratio Mismatch

Consider a simple tension specimen coated with a finite thickness photoelastic coating (Fig. 2.7). The specimen is in a state of uniaxial stress but the state of strain is biaxial due to the influence of Poisson's ratio. At the interface, the coating faithfully follows the specimen and the strains developed in the coating and the specimen are identical.

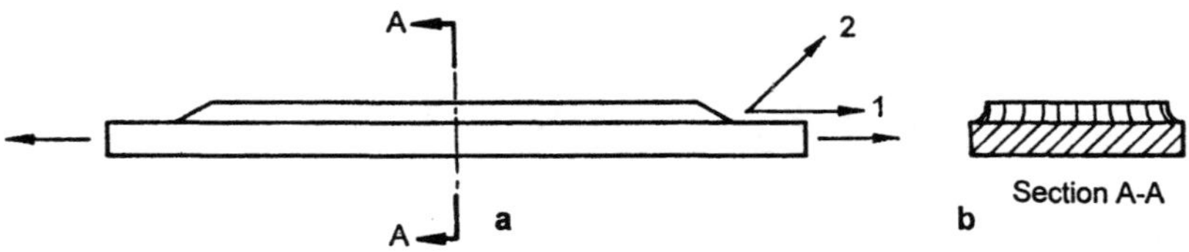

Fig. 2.7. a Tension plate with photoelastic coating. **b** The vertical lines in the centre of the coating are straight but the ones at the free end are bent due to Poisson's ratio mismatch

The longitudinal strain is governed by the load applied but the lateral strain is a function of both the load applied and the Poisson's ratio of the specimen. Thus, one has

$$\varepsilon_1^c = \varepsilon_1^s, \quad \varepsilon_2^c = \varepsilon_2^s \quad \text{and} \quad \varepsilon_2^s = -\nu_s \varepsilon_1^s \tag{2.18}$$

However, on the free end of the coating the lateral strain is governed by the Poisson's ratio of the coating as

$$\varepsilon_2^c = -\nu_c \varepsilon_1^s \tag{2.19}$$

This introduces a strain variation through the thickness of the coating. Considering that the Poisson's ratio of the coating is larger than the specimen, the fringe order N observable lies in the range:

$$\frac{2h_c(1+\nu_s)\varepsilon_1^s}{F_\varepsilon} < N < \frac{2h_c(1+\nu_c)\varepsilon_1^s}{F_\varepsilon} \tag{2.20}$$

Experiments have been conducted [5] to study the influence of Poisson's ratio mismatch using glass fibre epoxy tension specimens where the specimen Poisson's ratio is varied between 0.097 to 0.35. For a fixed coating Poisson's ratio of 0.36, the boundary fringe order is always found to be higher than the interior. These are governed by the equations:

$$
\begin{aligned}
N_{\text{boundary}} &= \frac{2h_c(1+\nu_c)\varepsilon_1^s}{F_\varepsilon} \\
N_{\text{interior}} &= \frac{2h_c(1+\nu_s)\varepsilon_1^s}{F_\varepsilon}
\end{aligned}
\tag{2.21}
$$

A transition zone exists near the boundary where the fringe order lies between these two extremes. This can be expressed as

$$
N_{\text{transition}} = 2h_c\left[1+\nu_s + C_\nu(\nu_c - \nu_s)\right]\frac{\varepsilon_1^s}{F_\varepsilon}
\tag{2.22}
$$

where C_ν is a correction factor accounting for the mismatch. The length of the transition zone is found to be four times the thickness of the coating. For metallic specimens, $(\nu_c - \nu_s)$ is usually less than 0.06. Hence, the effect of Poisson's ratio mismatch is often neglected in the analysis of most metallic components.

The expected change in the fringe order due to Poisson's ratio mismatch has been discussed above. However, from the stress analysis point of view, one needs to do the inverse problem. A case in point is the problem of evaluation of stress concentration factor using photoelastic coatings. In theory of elasticity, the stress concentration factor (SCF) is defined as the maximum stress divided by the far field or average stress. Let us consider the problem of a hole in a plate subjected to tensile loading. Let the maximum fringe order observed at the boundary of the hole be $N_{\max}$ and let the average fringe order be $N_{\text{far-field}}$. The ratio of these does not directly give the SCF but one has to evaluate SCF by

$$
\text{SCF} = \frac{N_{\max}}{N_{\text{far-field}}}\frac{(1+\nu_s)}{(1+\nu_c)}
\tag{2.23}
$$

2.7 Coating Materials

An ideal photoelastic coating should have the following properties:

1. High strain coefficient K so that small coating thickness is sufficient to give enough optical information.
2. Low Young's modulus so that even a thicker coating does not reinforce the specimen.
3. Linear stress-strain and strain-fringe relations.
4. Easy bondability to various specimen materials.
5. Possession of good machinability.
6. Sufficient pliability to permit use on curved surfaces of intricate components.

In most instances, the selection of a coating material involves a compromise since no material exhibits all these characteristics. Further, the choice of the material depends on the type of specimen material and the level of strain anticipated in the experiment. Specific coating materials have been developed for pasting on high modulus (basically metallic specimens), medium modulus (non-metallic specimens) and low modulus (rubber like materials). Usually photoelastic materials are available as flat sheets of standard dimensions with thickness varying in the range of 0.25 mm to 3 mm. For contourable plastics, the material is supplied as two component liquids, which needs to be mixed appropriately to get a pliable coating. The strain coefficient K for contourable plastics is generally lower than that of flat sheets. Contourable plastics in liquid form have a shelf life of about six months. Preparation of pliable plastic from the two component liquids requires skill. However, to simplify the industrial application of photoelastic coatings, contourable plastics in sheet form are available at a

Table 2.1. Properties of photoelastic coating materials

Material	E_c (GPa)	ν_c	K	Strain Limit %	Max. usable Temp. °C	Suitability
Coatings Suitable for High Modulus Materials (Metallic Specimens)						
Polycarbonate	2.21		0.16			Flat
PS-1†	2.50	0.38	0.15	10	150	Flat
PS-2†	3.10	0.36	0.13	3	260	Flat
PS-8†	3.10	0.36	0.09	3 to 5	200	Flat
PL-1 liquid†	2.90	0.36	0.10	3 to 5	230	Contourable
PL-8 liquid†	2.90	0.36	0.08	3 to 5	200	Contourable
Polyester	3.86		0.04	1.5		Flat
Epoxy with anhydride	3.28		0.12	2.0		Flat / Contourable
Coatings Suitable for Medium Modulus Materials (Non-metallic Specimens)						
PS-3†	0.21	0.42	0.02	30	200	Flat
PL-2 liquid†	0.21	0.42	0.02	50	200	Contourable
Coatings Suitable for Low Modulus Materials (Rubber)						
Polyurethane	0.004		0.008	15		Flat
PS-4†	0.004	0.50	0.009	>50	175	Flat
PL-3 liquid†	0.014	0.42	0.006	>50	150	Contourable

† Courtesy Measurements Group, Inc., Raleigh, NC, USA.

premium price [6]. In view of the on-going polymerisation in these, the sheets are to be used immediately, say, within 24 hrs of its supply. Table 2.1 summarises the properties of some of the photoelastic coating materials.

2.8 Bonding the Coating

As in other coating techniques, such as, strain gauges and brittle coatings, the successful use of photoelastic coating depends on how well the coating is bonded to the specimen surface. For each category of coating materials, such as high, medium and low modulus, one has to use appropriate adhesives. The adhesives used in conjunction with high modulus materials have epoxy filled with aluminium powder for better shear strength. Usually, the supplier of these materials gives detailed instructions on specimen preparation and adhesive usage. For the bonding to be good, the basic requirement is that the surface of the component should be smooth and free of all foreign material. This can be easily obtained by sanding followed by cleaning with chemical agents and solvents.

For plane surfaces, the coating sheet needs to be initially machined to size and cleaned. Adhesive is applied evenly over its surface and the coating is aligned to the specimen and positioned at one end. It is slowly rotated and pressed into position to remove air bubbles entrapped in the adhesive. Usually it takes about 24 hrs for the adhesive to cure and no external pressure is applied to the coating to avoid the introduction of residual stresses. The final layer of adhesive for a perfect bond can be of the order of 0.08 to 0.25 mm. Coating is normally pasted on a portion of the specimen surface. Care should be taken to bevel the edges while finishing the ends such that no abrupt change in the thickness occurs on the coated part. Alternately, this can be achieved by filling the ends with adhesive (Fig. 2.8).

For complicated geometry, one should prepare initially a pliable plastic using the two-component coating material. A flat mould is used for this purpose. Normally, when the two component materials are mixed as per the manufacturers recommendations, the polymerisation starts and it may take about 16 to 24 hrs to cure. However to get the pliable plastic, the sheet is not allowed to complete the polymerisation on the moulding plate and is stripped from the plate at an appropriate time. The sheet thus removed is highly pliable. The strain coefficient is very low and large deformations do not cause significant fringe formation. This sheet is contoured on the component to be coated and allowed to cure over it. At the end of the process, one gets a shell of the birefringent coating, which is to be pasted, similar to the flat sheets. For contouring large structures, a refrigerated environment may be used to delay the polymerisation of the pliable plastic.

The time to decide the removal of the sheet from the moulding plate for contouring has to be decided based on experience. This will be of the order of two to six hours depending on the thickness of the sheet, the ambient

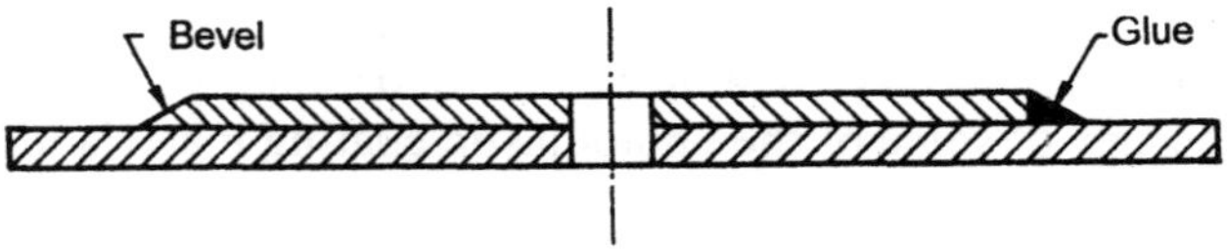

Fig. 2.8. Schemes to facilitate smooth transfer of load from the prototype to the coating

temperature and the type of coating material. Skill is required to prepare the pliable plastic and contour it. In view of contouring it by hand, the thickness of the coating may not be uniform. Since thickness is an important factor for data reduction, it has to be measured preferably before pasting the coating. However, techniques based on ultrasonic have been developed to find the thickness even after pasting it on the component surface [6]. Since there is a possibility of thickness variation, to simplify data reduction, the thickness of the coating should be selected such that any small variation does not affect the data interpretation adversely.

2.9 Selection of the Coating Thickness

Ideally, the coating thickness should be as small as possible so that the interpretation of the coating stresses to specimen stresses is simple and direct. However, the chosen coating thickness should be sufficient to produce a meaningful number of fringes for easy measurement. In principle, one can increase the optical response by increasing the applied load. However, for elastic stress analysis, the loading on the specimen cannot be increased indefinitely. If the specimen principal stresses are of opposite sign and if it follows Tresca yield criteria, then for a yield strength S_y of the material, the maximum value of $(\overset{s}{\sigma_1} - \overset{s}{\sigma_2})$ is only S_y. Substituting this in Eq. (2.10), the maximum fringe order N_{max} obtainable is

$$N_{max} = \frac{1+\nu_s}{E_s} \frac{2h_c K}{\lambda} S_y \tag{2.24}$$

Equation (2.24) shows that the maximum fringe order obtainable is linearly related to the thickness of the coating, strain coefficient of the coating and the yield strength of the specimen material. Table 2.2 summarises the range of maximum fringe order obtainable for various specimen materials with a coating thickness of 1 mm having a strain coefficient of $K = 0.15$ for a light source of wavelength 577 nm. The table shows that the maximum fringe order obtainable is quite low for many materials. This necessitates the use of higher thickness coatings for better data reduction.

Table 2.2. Maximum fringe order obtainable for various specimen materials

Material	S_y (MPa)	E_s (GPa)	ν_s	N_{max}
Steel				
HR 1020	240	207	0.292	0.78
CD 1020	310	207	0.292	1.00
HT 1040	550	207	0.292	1.78
HT 4140	900	207	0.292	2.92
Maraging	1720	207	0.292	5.58
Aluminium				
1100 H16	140	71	0.334	1.37
3004 H34	200	71	0.334	1.95
2024 T3	345	71	0.334	3.37
7075 T6	500	71	0.334	4.88
Phosphor Bronze	515	111	0.349	3.25
Beryllium Copper	480	124	0.285	2.59
Glass	21	46.2	0.245	0.29
Concrete (Compression)	28	18	0.170	0.95

If thicker coatings are employed, then appropriate correction factors are needed for data reduction. To simplify data reduction, while using flat sheets, it is generally recommended to determine the coating thickness such that the correction factor is unity. While using contourable plastics, it was mentioned earlier that it is difficult to maintain the thickness of the coating over the component surface. In such applications, the coating thickness should be found such that small variations in coating thickness do not unduly change the correction factor. The first solution can be easily obtained by equating the solution for R_f to unity. The second solution is obtained by differentiating R_f with respect to g. In order to ensure that the correction factor calculated is a minimum, the second derivative is to be computed and its sign checked. These two approaches are illustrated for the case of bending of plates.

The governing equation for evaluating the thickness, for the case wherein the correction factor is unity, is given by

$$f(g) = \frac{(1+emg)}{(1+g)}\left[4(1+emg^3) - \frac{3(1-emg^2)^2}{(1+emg)}\right] - 1 = 0 \qquad (2.25)$$

If the numerical value of em is substituted in Eq. (2.25) then one gets a non-linear equation in terms of g. The value of g has to be calculated iteratively by the Newton-Raphson method. The governing equation for least variant correction factor is obtained by differentiating Eq. (2.15) with respect to g as

$$f'(g) = 4em(1+emg^3)(1+g) + 12g^2(1+emg)(1+g)(1+g) - 4(1+emg)(1+emg^3) + 12emg(1+g)(1-emg^2) + 3(1-emg^2)^2 = 0 \tag{2.26}$$

Here again the equation is to be solved iteratively for g. To verify that the calculated thickness indeed is correct, the second derivative is to be calculated and checked for its sign.

One can also determine the coating thickness based on the precision with which the strain measurement needs to be made. The precision in strain measurement depends on the precision with which fringe order can be measured. Equation (2.2) can be rewritten as

$$\Delta(\varepsilon_1^c - \varepsilon_2^c) = \frac{\Delta N F_\varepsilon}{2h_c} \tag{2.27}$$

It is more common to talk of relative precision of measurement, which is defined as

$$\xi = \frac{\Delta(\varepsilon_1^c - \varepsilon_2^c)}{(\varepsilon_1^c - \varepsilon_2^c)_{\max}} \tag{2.28}$$

Substituting Eq. (2.28) in Eq. (2.27) and rewriting one gets

$$h_c = \frac{\Delta N F_\varepsilon}{2\xi(\varepsilon_1^c - \varepsilon_2^c)_{\max}} \tag{2.29}$$

The importance of this equation is that if the accuracy in ΔN can be improved, the thickness of the coating can be decreased. Full field manual observations of coloured fringes give $\Delta N = \pm 0.1$ fringe whereas point by point compensation techniques can give $\Delta N = \pm 0.01$ fringe. With the developments in digital image processing techniques, the accuracy of ΔN evaluation can be enhanced over the field, thereby, reducing the coating thickness and thus improving the data interpretation.

2.10 Calibration of the Coating Material

Evaluation of strain coefficient K for a photoelastic coating is known as calibration of the coating material. This is a crucial parameter and needs to be evaluated as accurately as possible. To find K, the coating has to be pasted on a specimen and subjected to a known state of stress. The circular disc specimen that is widely used for calibration of transmission photoelastic analysis is not suitable for calibrating a photoelastic coating. A cantilever beam in bending is recommended for calibration purposes. The

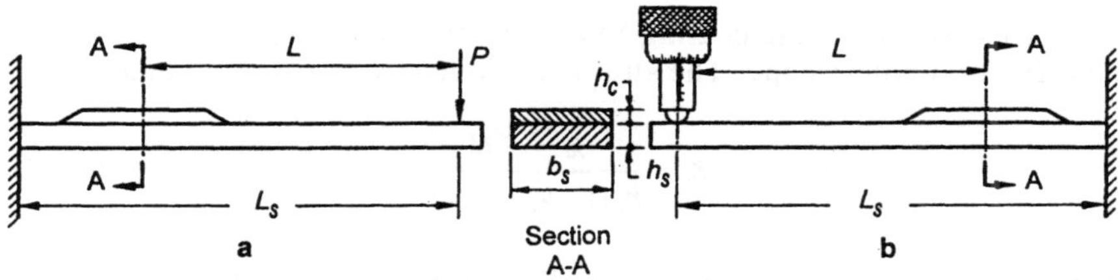

Fig. 2.9. Calibration arrangements: **a** load controlled and **b** displacement controlled

beam is under a uniaxial state of stress and using flexure formula, the state of stress at the point of interest can be determined. The experiment can be either load-controlled (Fig. 2.9a) or displacement-controlled (Fig. 2.9b). The calibration apparatus is compact for the displacement-controlled arrangement. For calibration purposes, only a small portion of the beam is to be coated with the coating. This saves the precious coating material as well as its influence on the deflection of the beam at the free end ($y_o = PL_s^3 / 3\, E_s\, I_{zz}$) will be minimal.

Due to the bending load, there exists a strain variation over the thickness of the coating and the bending correction factor R_f^b is to be used for data interpretation. For displacement-controlled experiment, the strain coefficient K is obtained as

$$K = R_f^b \frac{E_s}{1+\nu_s} \frac{b_s h_s^2 \lambda}{12 L h_c} \frac{N}{P} \tag{2.30}$$

For load-controlled experiment, the strain coefficient K is obtained as

$$K = \frac{R_f^b}{1+\nu_s} \frac{L_s^3 \lambda}{3 L h_s h_c} \frac{N}{y_o} \tag{2.31}$$

In order to determine K accurately, the fringe order N has to be measured accurately. Further, the experiment is to be repeated for various loads or displacements as the case may be and a graph is plotted between N versus P or N versus y_o. A best-fit straight line is constructed through the data points. The slope of the graph gives the ratio N/P or N/y_o which is to be used in Eq. (2.30) or Eq. (2.31) for determining K. Table 2.1 shows that for various materials, K has a wide range from 0.16 to 0.009. For materials having higher values of K the above mentioned approach is feasible but for lower values of K, one needs to apply a large strain of the order of 10 to 15%.

For materials, which have poor optical response, it is generally recommended to perform the calibration test directly on the material itself by making a tension member out of the material. Since large strains are involved, the axial strain measurement can be done by an extensometer. If the coating is transparent, the fringe order can be measured in a transmission polariscope and if the coating has a reflective backing, then, one has to use a reflection polariscope. The strain coefficient K can be obtained as

$$K = \frac{\lambda}{h_L(1+\nu_c)} \frac{N}{\varepsilon_1^c} \tag{2.32}$$

where h_L is the length of the light path. For a transmission arrangement, it is h_c and for a reflection arrangement, it is $2h_c$.

2.11 Data Collection and Analysis

Like in transmission photoelasticity, the basic data that is to be collected in photoelastic coating analysis is the fringe order N and the isoclinic parameter θ. As noted before, use of white light is widely prevalent in reflection photoelasticity. While using white light, the tint of passage corresponding to integral fringe orders is idealised to occur at a wavelength of 577 nm. For determining isochromatic fringe order, one can use the colour code (Table 1.2) or use a compensation technique such as Tardy method of compensation or an external compensator such as Babinet-Soleil compensator. Isoclinic fringe order can be measured by rotating the polarizer and analyzer in crossed combination until a dark fringe passes through the point of interest and noting the orientation of the polarizer with respect to a reference. With these data one can find the principal stress difference $(\sigma_1^s - \sigma_2^s)$ using Eq. (2.11) and the shear stress τ_{xy}^s by

$$\tau_{xy}^s = R_f \frac{G_s N}{2h_c} \frac{\lambda}{K} \sin 2\theta \tag{2.33}$$

If the coating is sufficiently thin, then the need of R_f is not required and it will make the analysis more robust. The list of reinforcement factors given in Sect. 2.5, is not exhaustive and for specific applications, one has to determine R_f from first principles. The important decision that the experimentalist has to make is whether for a given application the thickness effects are important. This can be decided experimentally by applying initially a coating of convenient thickness and later increase it by applying another layer of the same coating and recording the change in the fringe order. If the fringe order variation is linear in conjunction with the coating thickness, then the thickness effect does not play a role. Alternately, one can experimentally establish for a given application the limiting coating thickness within which R_f need not be considered.

In certain class of problems, one would like to evaluate the individual stress components and for this, auxiliary methods are needed which are discussed in Chapter 9.

2.12 Application of Photoelastic Coatings [7, 8]

Over the years, photoelastic coatings have found application in various fields ranging from aerospace to biomechanics. The technique has also been applied to analyse stress fields in a variety of specimen materials such as aluminium alloys, concrete, ceramic, bone, composite materials and rubber. Extensive tests on aircraft wing, aircraft window and landing gears have been reported. Even in the design of modern aircraft's such as A340 and Boeing 777, the weight of the landing gears are reduced by the use of photoelastic coatings. For this, 70% scaled models of landing gears are made of epoxy and coated with photoelastic coatings using a specialised process of paint-on technique using spray guns. After a detailed coating analysis, the models are stress frozen for conducting a three-dimensional study using the slicing procedure discussed in Chapter 1.

The technique has also been found to be useful for detecting assembly stresses. In view of their complexity, no mathematical modelling is easy for these classes of problems. Use of photoelastic coatings comes in handy for such difficult applications. Further, for the measurement of residual stresses too the technique has been used with success. It is mentioned earlier that due to the nature of loading, the optical response of photoelastic coatings is an order of magnitude less than what is seen in transmission photoelasticity. Hence, if one observes nice fringe patterns while using coatings, it is to be understood as a warning signal that the component is highly stressed.

2.13 Closure

The technique of photoelastic coatings has many advantages over other methods of experimental stress analysis. It provides full-field data that enables the investigator to visualise the complete distribution of surface strains/stresses. Since the coatings can be applied directly to the prototype, the need for models is eliminated. Through proper selection of coating materials, the method can be made applicable over a wide range of strain. The method is also very useful in converting complex non-linear stress analysis problems in the prototype into relatively simple linear elastic problems in the coating. For instance, plastic and viscoelastic response of a prototype can be measured in terms of the elastic response of the birefringent coating. Similarly, the anisotropic response of composite materials can be examined in terms of an isotropic response of the coating.

Exercises

2.1 The use of strain-optic law is desirable in photoelastic coating analysis. Why?

2.2 Summarise the approximations made in photoelastic coating data collection and interpretation.

2.3 Discuss the validity of Eq. (2.5) for a finite thickness photoelastic coating.

2.4 In a photoelastic coating experiment, the fringe order observed in white light at a point of interest is 1.6. The coating is of 1 mm thick and the strain optic coefficient K is 0.15. Determine the principal stress difference in the specimen for the specimen material listed in Table 2.2. Comment on the state of stress of the specimen material.

2.5 Determine the reinforcement factors R_f^a, R_f^b and R_f^{bp} for the following coating material combination.

(a) PS-1 on steels HR 1020, CD 1020, HT 1040 and HT 4140

(b) PS-1 on aluminium (c) PS-1 on rubber

(d) PS-4 on rubber

Tabulate these for various values of h_c / h_s starting from 0.1–1 in steps of 0.1.

2.6 The maximum fringe order observed in a plate with a hole subjected to tension is 2.6. The average fringe order observed is 0.9. The coating material used is PS1 sheet of 3-mm thickness. Determine the SCF if the specimen material is (a) HT1040 steel (b) 2024 Aluminium (c) Glass fibre reinforced plastic having a Poisson's ratio of 0.15.

2.7 Select a suitable coating material for analysing the following structural components.

(a) High strength steel pressure vessel. (b) Aircraft landing gear.

(c) Fibre glass boat. (d) Gas turbine blade.

2.8 Design a portable reflection polariscope equipped with a Handycam for image recording. In the design, provide a provision for independently aligning each of the optical elements.

References

1. Mesnager M (1930) Sur la determination optique des tensions interieures dans les solides a trois dimensions. Comptes Rendus, Paris, 190:1249
2. Oppel G (1937) Das polarisationsoptische schichtverfaren zur messung der oberflachenspannungen am beanspruchten bauteil ohne model. Zeitschrift des Vereines Deutscher Ingenieure 81:638
3. Zandman F (1960) Stress analysis of a guided missile tail section with the photoelastic coating method. Proc SESA 17(2):135–150
4. Zandman F, Redner S, Riegner EI (1962) Reinforcing effect of birefringent coatings. Exp Mech 2(2):55–64
5. Dally JW, Alfirevich I (1969) Application of birefringent coatings to glass-fibre-reinforced plastics. Exp Mech 9(3):97–102
6. Strainoptic™ contourable photoelastic coating sheets. Bulletin FS-100, Strain Optic Technologies, Inc, 108 West Motgomery Avenue, North Wales, Pennsylvania 19454, USA
7. Zandman F, Redner S, Dally JW (1977) Photoelastic coatings. SESA Monograph No.3, Society for Experimental Stress Analysis, USA
8. Reflections – The photostress® Newsletter. Measurements Group, Inc, P.O Box 27777, Rayleigh, NC 27611, USA

3 Digital Image Processing

3.1 Introduction

The term digital image processing (DIP) generally refers to the processing of a two-dimensional picture by a digital computer. A digital image is an array of real numbers represented by a finite number of bits. An image given in the form of a photograph or a slide is first digitised and stored as a matrix of binary digits in computer memory. This digitised image can then be processed and/or displayed on a high-resolution monitor. Early systems of image processing were configured around big computers such as a PDP 11 system. Recent advances in computer technology have brought in the development of plug-in cards, which can make a conventional PC into an image processing station. These cards are known as frame grabbers. Monochrome and colour frame grabbers are available in the market.

Digital image processing has a broad spectrum of applications, such as remote sensing via satellites, automated inspection of industrial parts, fringe pattern analysis in photomechanics, robotics, medical image processing etc.

In this chapter, the relevant concepts that are needed for understanding the role of digital image processing in digital photoelasticity are discussed. Not only the concepts but also time-tested softwares written in Microsoft C to execute these concepts are also given.

3.2 Image Sampling and Quantization

The most basic requirement for computer processing of images is that the images be available in digital form.

3.2.1 Pictures as Functions

A picture is a flat object whose brightness may vary from point to point. This variation can be shades of grey for a black and white picture. Such picture can be represented by a single real-valued function, say $g(x, y)$. It is customary to assume that these functions are analytically well behaved such that the image functions are integrable, differentiable etc.

3.2.2 Uniform Sampling and Quantization

To be suitable for computer processing, an image function $g(x, y)$ must be digitised both spatially and in amplitude. Digitisation of the spatial co-

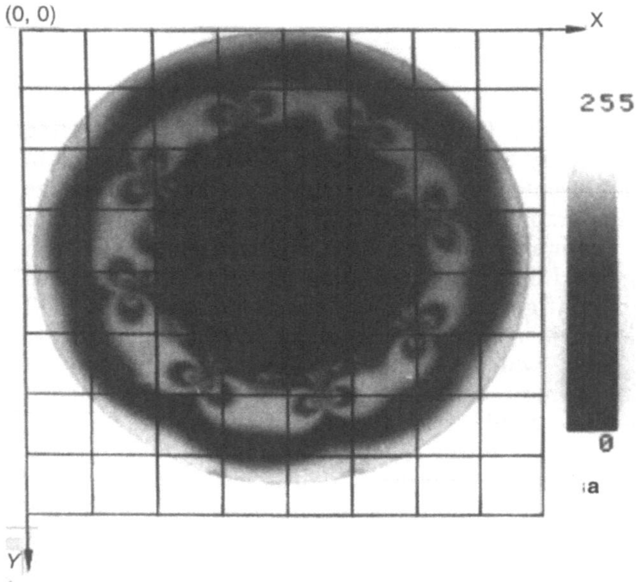

a

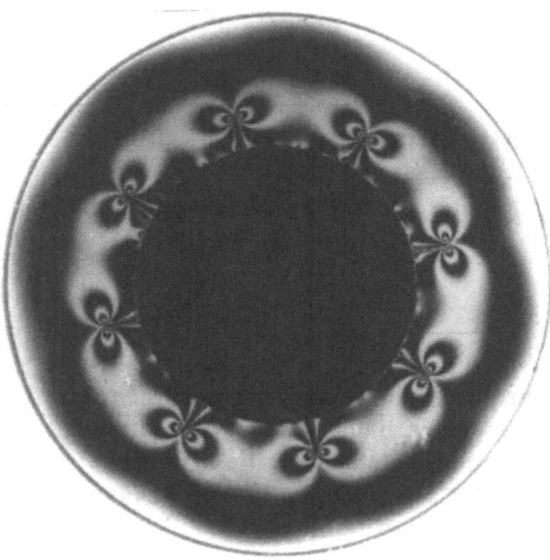

b

Fig. 3.1 **a** Digitised image of a photoelastic fringe pattern quantized using 256 grey level shades. The figure is superimposed with a grid. A single element of it contains 2500 pixels. **b** Optical image of the fringe patterns ordinates (x, y) is called image sampling, and amplitude digitisation is called grey-level quantization. The most common method of picture sampling is to use a regularly spaced square array of points, i.e., points (md, nd) whose co-ordinates are multiples of some unit distance d. The value of $g(x, y)$ at a point will be called the grey-level or brightness of the picture at that point.

Let a continuous image $g(x, y)$ be approximated by equally spaced samples arranged in the form of an $M \times N$ array given as

$$g(x,y) \approx \begin{bmatrix} g(0,0) & g(0,1) & \cdots & g(0,N-1) \\ g(1,0) & g(1,1) & \cdots & g(1,N-1) \\ \vdots & \vdots & \vdots & \vdots \\ g(M-1,0) & g(M-1,1) & \cdots & g(M-1,N-1) \end{bmatrix} \tag{3.1}$$

The right-hand side of Eq. (3.1) represents what is commonly called a digital image. Each element of the array is a discrete quantity and represents the grey level value of a picture element, or pixel. It is usual to regard these values as non-negative and bounded, i.e., $0 \leq g(x, y) \leq K$ where K is the number of discrete grey levels allowed for each pixel.

The digitisation process requires decisions about values for M, N and K. The resolution or the degree of discernible detail of an image depends strongly on these parameters. The more these parameters are increased, the closer the digitised array approximates the original image. Common practice in digital image processing is to let these quantities be integer powers of two. Studies have been conducted on how to select these values so that the image quality is not sacrificed [1].

A typical image size, comparable in quality to a monochrome TV image, should have a 512 × 512 array with 128 (7-bits) grey levels. In digital photoelasticity 8-bit grey level quantization is quite common which allows 256 grey level shades. In such a case, the value 0 indicates pitch black and 255 indicates pure white. Figure 3.1a shows a photoelastic fringe pattern of a thick ring subjected to internal pressure digitised using 256 grey level shades. The image is superimposed with a grid. An element of this contains 2,500 pixels. The figure also indicates the origin and x and y co-ordinates in a digital picture. Figure 3.1b shows the same fringe patterns recorded optically. A comparison of these two images indicate that the digitisation is reasonably good except at zones of high fringe density.

3.3 Video Standards

Video standards were originally developed for television applications. The purpose of a standard is to ensure that the devices offered by different

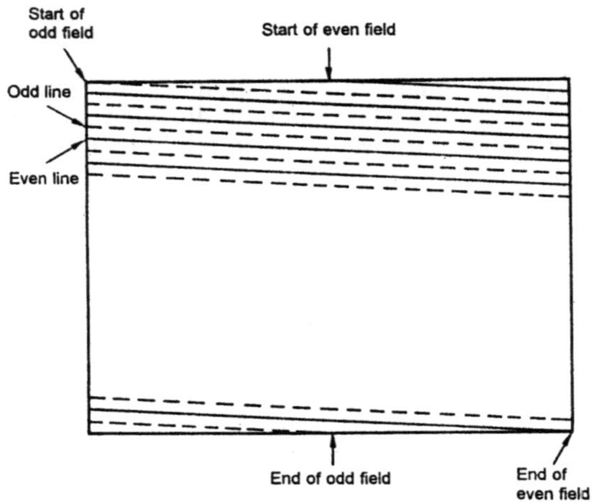

Fig. 3.2. Typical interlaced scanning of an analog video monitor

manufacturers will operate together. The standard monochrome video formats are EIA 170 (originally called RS 170) used in USA and CCIR used in Europe. The colour television standards are the NTSC, SECAM and PAL. The NTSC (national television systems committee) is used in North America and Japan. The Sequential Couleur à Mémoire (SECAM, or sequential chrominance signal and memory) is used in France, Eastern Europe, and the Soviet Union; and the phase alternating line (PAL) is used in India, parts of Asia, West Germany, United Kingdom, parts of Europe, South America, and Africa. Some television engineers have been known to refer to these standards as *never twice same colour* (NTSC), *something essentially contradictory to the American method* (SECAM), and *peace at last* (PAL). Table 3.1 summarises the salient characteristics of these standards. Each image has two fields namely odd and even field, which

Table 3.1. Characteristics of video standards

Characteristics	CCIR (B & W) PAL/SECAM (Colour)	EIA 170 (B & W) NTSC (Colour)
Format	Analog signal 2 interlaced fields per frame 768 pixels × 572 lines	Analog signal 2 interlaced fields per frame 640 pixels × 480 lines
Frame rate	25 frames/s	30 frames/s (B & W) 29.97 frames/s (Colour)
Vertical Resolution	625 lines per frame 575 lines are visible	525 lines per frame 485 lines are visible
Bandwidth	5.5 MHz	4.2 MHz

Fig. 3.3. Opto-mechanical drum scanner

are interwoven on the display (Fig. 3.2). This type of display is also termed as interlaced display.

The sampling rate (number of pixels per unit area) is determined by the bandwidth of the video standard. In these standards, the type of interlaces and scan times are selected to minimise the electronic bandwidth and still provide a cosmetically pleasing image to the viewer with no flicker effects.

Non-standard video formats include digital video and high-resolution video. Digital video is a digitised waveform of any format of video signal such as NTSC, PAL etc. High-resolution video includes a spatial resolution of 1024 pixels × 1024 lines and higher. This requires increased sampling rate and thus increases the bandwidth requirements.

3.4 Image Sensors

The image digitisation in early systems was achieved through opto-mechanical drum scanners (Fig. 3.3), which are essentially slow devices. A variety of electronic image sensors is available today. These have made the possibility of digitising the image at video rates namely 25 frames/s for PAL systems and 30 frames/s for NTSC systems. Electronic imagers based on vidicon tubes were initially used for digital photoelasticity. In these, an object film or transparency is continuously illuminated to form an electron image on a photosensitive plate, called the target, which is photoconductive. A finite-aperture electron beam scans the target and generates current, which is proportional to the light intensity falling on the target. Though these systems have reached a high state of development since their introduction more than 50 years ago, their importance to typical imaging applications has decreased dramatically in favour of solid state imagers.

Charge-coupled devices (CCD), charge injection devices (CID) and metal oxide silicon capacitors (CMOS) convert light into an electronic signal. Of these, the CCD's have found the most widespread use.

Charge-coupled devices (CCD's) were invented by Boyle and Smith [2] in 1970. The heart of a solid state camera is a CCD array. It provides the conversion of light intensity to measurable voltage signals. Figure 3.4 shows a typical CCD camera. The lens assembly images the light onto the detector array. Its output is a series of analog pulses that represent the image intensity at a series of discrete locations. When the array output is amplified and formatted into a standard video format, a CCD camera is created. Most standard cameras provide only a video output. However in specialised cameras, digital output is possible which could be directly fed to an image processor. While using an analog output camera, the video signal must pass through an external analog to digital converter (ADC). These multiple conversions may modify the imagery. Thus, staying within the digital domain ensures maximum image fidelity.

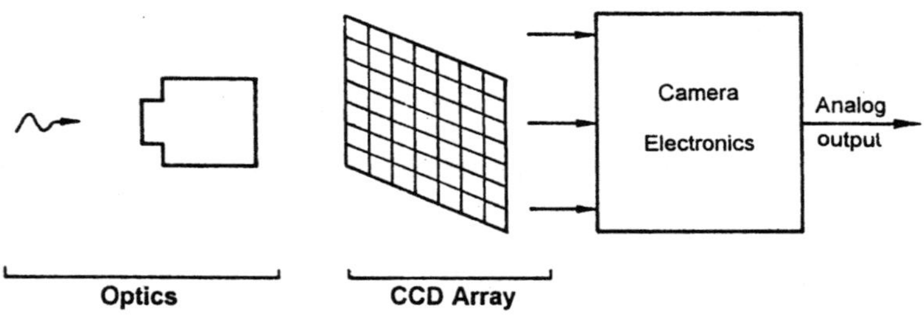

Fig. 3.4. Elements of a CCD camera

3.5 Image Display

Though for image capture, solid state devices are used, the image display is still based on tube based system. Figure 3.5 illustrates a typical monochrome cathode ray display. The digital to analog converter (DAC) provides the analog voltages that modulate the electron beam. The electron beam paints an image on the CRT phosphor in the scan format (Fig. 3.2). Electron beam density variations create variation in the visible image intensity. The output brightness of a CRT depends on the grid voltage and follows a power law relationship.

$$L_{display} = kV^{\gamma}_{grid} \tag{3.2}$$

On a log-log scale, the slope of Eq. (3.2) is the display Gamma. Gamma can range from one to five. For design purposes, NTSC has standardised Gamma at 2.2. PAL and SECAM use a value of 2.8.

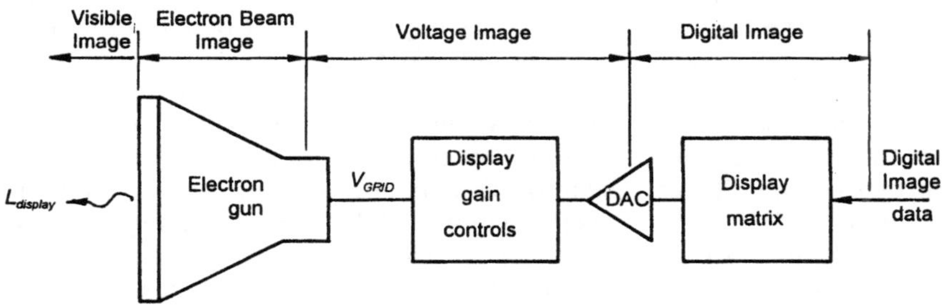

Fig. 3.5. Monochrome cathode ray tube (CRT) display

Printing image display devices are useful primarily for low-resolution image processing work. Common means of recording an image directly on paper include laser printers, heat-sensitive paper devices, and ink-jet systems.

With developments in printing technology, photo quality image outputs are possible. This is usually achieved by a halftone display [3]. In this, only black or white dots are written. By making the dot size much smaller than the pixel size, white or black dots are dispersed pseudo randomly such that the average number of dots per pixel area is equal to the pixel grey level. Due to spatial integration performed by the eye, such a black and white display renders the perception of a grey-level image. Newspapers, magazines, several printer/plotters, graphic displays, and facsimile machines use the halftone method of display.

3.6 Image Perception

In qualitative applications such as TV transmission, videophones etc., the final image has to be perceived by a human observer. Understanding of the visual perception process is important for developing measures of image fidelity, which aids in the design and evaluation of image processing algorithms and imaging systems. Such studies are needed to decide what compromises can be made for sending images through videophone etc., where the bandwidth of the signal has to be limited or in studies involving image compression and image restoration. However, for scientific applications like machine vision or photomechanics no alteration of the original signal is desirable. Hence, the studies on perception by a human observer play a secondary role in such applications.

3.7
Image Storage

The smallest and most fundamental unit of computer data is a *bit*. The word "bit" is the contraction of the term binary digit. The most important collection of bits is a *byte*. The *byte* is eight bits taken together as a single unit.

The amount of data associated with visual information is quite large. Storage is measured in Kbytes (meaning kilo or one thousand bytes), Mbytes (meaning mega or one million bytes). Gbytes (meaning giga, or one billion bytes), and Tbytes (meaning tera, or one trillion, bytes). An 8-bit image of size 512 × 512 pixels requires 0.27 MB of storage. Thus, providing adequate storage is usually a challenge in the design of image processing systems. Digital storage for image processing applications falls into three principal categories: (1) short term storage for use during processing, (2) on-line storage for relatively fast recall, and (3) archival storage, characterised by infrequent access.

One method of providing short-term storage is computer memory. Another is by specialised boards, called frame buffers, that store one or more images and can be accessed rapidly, usually at video rates of 25 (PAL) or 30 (NTSC) complete images per second. The amount of storage on a frame buffer card is limited to the physical size of the card and by the storage density of the memory chips used. In PC based frame grabbers, one MB of storage is quite common. On-line storage generally takes the form of magnetic disks. Winchester disks with few Gbytes are common. Magnetic tapes and optical disks are the usual media for archival applications.

3.8
Some Basic Relationships and Mathematical Operations Between Pixels

3.8.1
Neighbours of a Pixel

A pixel p at co-ordinates (x, y) has two horizontal, two vertical and four diagonal neighbours. The horizontal and vertical neighbours together are called the 4-neighbours of p. The four diagonal neighbours of p together with the 4-neighbours, are called the 8-neighbours of p (Fig. 3.6). Some of the neighbours of p lie outside the digital image if (x, y) is on the border of the image.

	(x, y-1)	
(x-1, y)	(x, y)	(x+1, y)
	(x, y+1)	

a

(x-1, y-1)	(x, y-1)	(x+1, y-1)
(x-1, y)	(x, y)	(x+1, y)
(x-1, y+1)	(x, y+1)	(x+1, y+1)

b

Fig. 3.6. Neighbours of a pixel with their co-ordinates: **a** four neighbours **b** eight neighbours

3.8.2 Arithmetic and Logic Operations

Arithmetic and logic operations between pixels are used extensively in most branches of image processing. The basic arithmetic operations are addition, subtraction, multiplication and division. These operations have specific use in image processing. They can be performed on grey level images and these are to be carried out pixel by pixel. Since they involve only one spatial pixel location at a time, the result of performing an arithmetic operation at location (x, y) can be stored in that location itself in one of the existing images. This is known as processing *in-place.*

The principal use of image addition is for image averaging to reduce noise. Image subtraction can be used to remove static background information. One of the principal uses of image multiplication (or division) is to correct grey level shading resulting from non uniformity's in illumination or in the sensor used to acquire the image.

The principal logic operations used in image processing are AND, OR and COMPLEMENT. These operations are functionally complete as they can be combined to form any other logic operation. Logic operations apply only to binary images. As logic operations involve only one pixel location at a time, they can also be done *in-place*, as in the case of arithmetic operations.

Logic operations are basic tools in binary image processing, where they are used for tasks such as masking, feature detection, and shape analysis.

3.8.3 Neighbourhood Oriented Operations

In addition to pixel-by-pixel processing on entire images, arithmetic and logic operations are used in neighbourhood-oriented operations. In neighbourhood processing, the value assigned to a pixel is a function of its grey level and the grey level of its neighbours. For instance, a simple neighbourhood operation is to replace the value of $g(x, y)$ with the average value of the pixels in a 3×3 region centred at the pixel. This is done by an

$w_{-1,-1}$	$w_{0,-1}$	$w_{1,-1}$
$w_{-1,0}$	$w_{0,0}$	$w_{1,0}$
$w_{-1,1}$	$w_{0,1}$	$w_{1,1}$

Fig. 3.7. 3 × 3 Mask with the weightings

arithmetic operation of the form:

$$g^R(x,y) = \frac{1}{9}\sum_{i=-1}^{1}\sum_{j=-1}^{1} g(x+i, y+j) \tag{3.3}$$

and assigning to $g(x, y)$ the value of $g^R(x, y)$.

A generic way of representing such neighbourhood operations is by a mask (Fig. 3.7). Mask is also referred to as a filter or a template. The size of the mask represents the number of pixels involved in the neighbourhood operation. The elements of the mask represent the weightings that have to be assigned to each pixel under the mask for calculation purposes. The mask centre is to be moved to the pixel of interest and the result is calculated as follows:

$$g^R(x,y) = \sum_{i=-1}^{1}\sum_{j=-1}^{1} w_{ij}\, g(x+i, y+j) \tag{3.4}$$

For scanning the whole image, the mask is then moved to the next pixel location in the image and the process is repeated. This is continued until all pixel locations have been covered. For pixels that are located in the border of the image, the value of $g^R(x, y)$ is computed by using partial neighbourhoods. In addition, the usual practice is to store the value of $g^R(x, y)$ in a different frame buffer, forming a new image, instead of changing pixel values *in-place*. This practice avoids using grey levels in Eq. (3.4) that have been altered as a result of an earlier application of this equation.

Equation (3.4) is used widely in image processing. Proper selection of the weightings and application of the mask at each pixel position in an image makes possible a variety of useful image operations, such as noise reduction, edge detection etc. In fact, the operation described in Eq. (3.4) performs noise reduction if we let $w_{ij} = 1/9$, and is known as spatial averaging. This eliminates the presence of salt and pepper noise.

Applying a mask at each pixel location in an image is computationally an expensive task. For example, applying a 3 × 3 mask to a 512 × 512 image requires nine multiplications and eight additions at each pixel location. This adds up to a total of 2,359,296 multiplications and 2,097,152 additions. Masks of sizes 5 × 5 or 7 × 7 are also in common use.

3.9 Basic Steps in Image Processing

The basic steps in digital image processing are image acquisition, pre-processing, segmentation, analysis and display.

The first step in the process is image acquisition, which is to acquire a digital image. This requires an imaging sensor and the capability to digitise the signal produced by the sensor. The sensor could be a monochrome or colour TV camera that produces an entire image of the problem domain every 1/25 (PAL) or 1/30-sec (NTSC). The imaging sensor could also be a line-scan camera that produces a single image line at a time. After a digital image has been obtained, the next step deals with pre-processing that image. The role of pre-processing is to improve the image in ways that increase the chances for success of the other processes. Pre-processing is application dependent and one has to establish the necessary steps by trial and error.

The next stage deals with segmentation. Broadly defined, segmentation partitions an input image into its constituent parts of objects. In general, autonomous segmentation is one of the most difficult tasks in digital image processing. Once the image is segmented, depending on the application, the image needs to be analysed for extracting useful information. This may be identification of a fingerprint in forensic analysis or extraction of fringe orders in fringe pattern analysis.

The various steps indicated are interdependent and may require input from the user. The level of sophistication accomplished is inversely related to the level of user interaction in the image processing process. Total elimination of user interaction is usually difficult to achieve and thus, most image processing applications are usually interactive in nature.

3.10 Typical Image Processing Systems for Digital Photoelasticity

PC based image processing systems are quite popular in digital photo-elasticity in view of their low cost and easy integration with the polariscope. Figure 3.8a shows a CCD camera replacing a human observer in the polariscope. The camera is connected to a PC containing an appropriate frame grabber. Figure 3.8b shows the high-resolution monitor for image display and the PC monitor for graphic output. Frame grabbers of a wide variety such as PIP, MVP_AT (Matrox Corporation, Canada) and DT285 (Data Translation of USA) have been used extensively by the researchers. With the introduction of PCI bus in Pentium based computers, specialised cards compatible to PCI architecture are also available. They are PULSAR, GENESIS and METEOR of Matrox Corporation. In view of their recent introduction, not much literature is available on their use. In principle, these cards can also be used for digital photoelasticity.

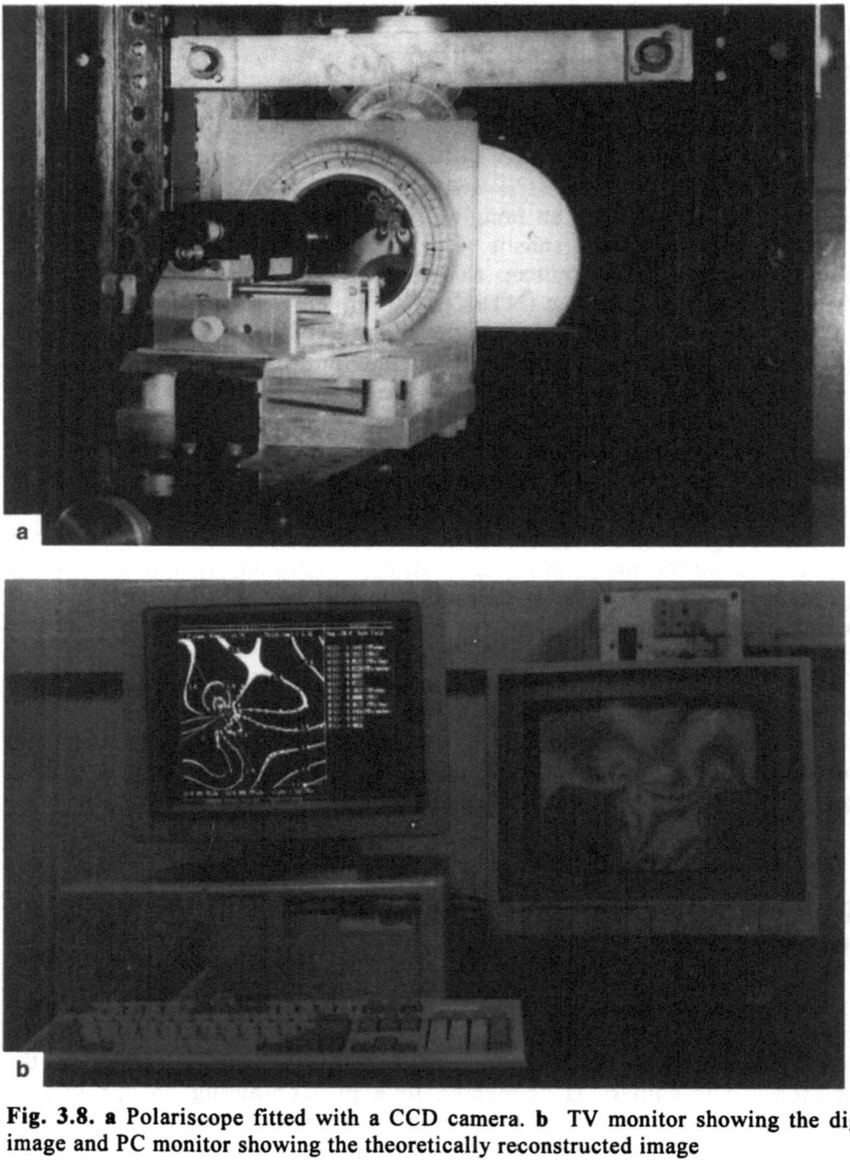

Fig. 3.8. **a** Polariscope fitted with a CCD camera. **b** TV monitor showing the digital image and PC monitor showing the theoretically reconstructed image

The role of a frame grabber is to understand the signal from the camera, facilitate image-processing operations for data extraction, and finally send an appropriate signal for image display. Frame grabber cards usually come with on-board memory called frame buffers, which provide quick access for data analysis and are quite useful for storing intermediate results of an image processing sequence. To reduce cost, some frame grabbers do not have on-board memory for display. For digital photoelasticity, it is desirable to select only those boards that have on-board memory as it greatly simplifies the structure of the software.

3.11 Software Structure and Design

Frame grabber cards usually come with basic software support to perform standard imaging and graphic operations. These commands can be directly used in what is called as an *interpreter* mode or can be called from a higher level language such as "C". Thus, a specific image-processing task along with data processing can be effectively done by writing a "C" code by invoking the library of imaging functions supplied with the frame grabber card. The *interpreter* mode of the card is useful, as the specific operation of each command in the imaging library can be individually executed and its operation can be understood for its later use in a "C" code. If a problem requires only the use of functions from the imaging library, then the sequence of commands needed be stored in the interpreter environment as a macro for later use.

In this book, for Black & White image processing, the software is reported for a Matrox PIP 1024B (PAL compatible) board in view of its simple architecture. PIP 1024B has four frame buffers which can be configured using *quadm* (quadrant mode) command to either a single screen of size 1024 × 1024 or four quadrants as shown in Fig. 3.9. Usually the four-quadrant mode is selected for digital photoelasticity. A high-resolution monitor is required for image display. Images can be either grabbed to full quadrant of size 512 × 512 or in zoom mode to a size of 256 × 256. The imaging operations can be effected within the user-defined window using the *select window* command. Portions of the image can be copied/moved to any quadrant through *cpws* (copy workspace) command. The intensity of pixel can be read by *pixelr* (pixel read) command and it can be changed by *pixelw* (pixel write) command. In the C-code the PIP 1024B based commands appear with a prefix *fg_*. Sample software is presented

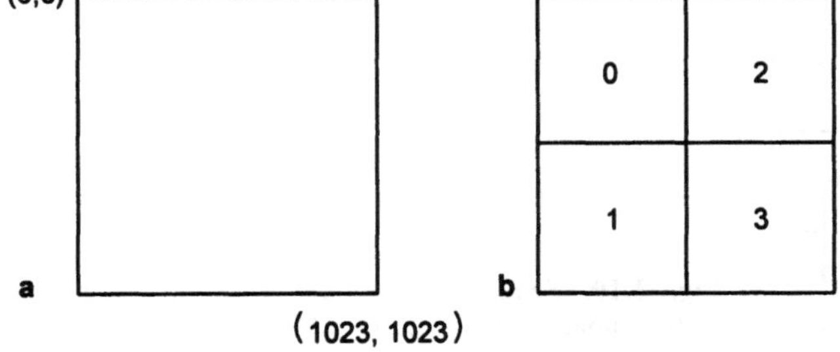

Fig. 3.9. Workspace configurations in a PIP system: **a** single quadrant mode **b** four quadrant mode and their order

in the next section. Source code for most of the algorithms discussed are enclosed in the accompanying CD. The program names of the codes are indicated in the text for easy usage of the CD. A detailed command list is also given in the CD, which can be used for decoding the softwares, reported in this book, to other boards.

A modular approach to software development is adopted in this book. Software segments that will be useful for more than one imaging application is written as a header file. For example, all the operations needed for cursor operations are written in a file called *curupip.h* (CD) and it can be called in any application where cursor operations are needed by just including this header file in the beginning of the "C" code. The softwares reported in this book are time-tested and compiled in a Microsoft-C compiler version 5.1 and linker version 3.65. The softwares given in various chapters can be assembled to form an image-processing module for digital photoelasticity. The global variables needed for all the softwares are put in a definition file called *pipdef.h* (CD). This file also has the routine for initialising the board for image processing operations.

3.12 Image Acquisition

Using a Matrox PIP board, images of size 512 × 512 or 256 × 256 can be captured from a standard camera. To capture images of 256 × 256, the board has to be specifically initialised in the *initpip* command (*snapzoom.c*, CD). Image capturing is a crucial step for digital photoelasticity. If the captured image is good, then it greatly simplifies the analysis procedure. While using a diffused light polariscope, it should be ensured that the intensity of light is uniform over the field. The model should be adjusted to the centre of the field and the camera should be kept perpendicular to the model. To minimise the influence of electrical noise, it is recommended to capture as many images as possible and then take an average of these images for further processing. This process of frame averaging is known as time averaging in image processing parlance.

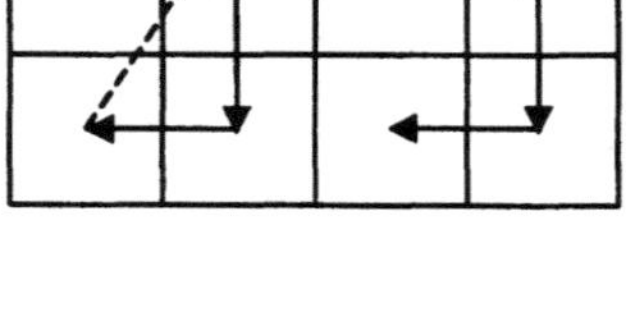

Fig. 3.10. Sequence of image flow for capturing sixteen images of size 256 × 256

While capturing images of size 512 × 512 up to four images can be captured and for images of size 256 × 256 up to sixteen images are possible using the *pan* and *scroll* commands. The sequence of image flow while capturing 256 × 256 image is shown in Fig. 3.10. Since the CCD array can record images at video rates (i.e., 25 frames/s in PAL and 30 frames/s in NTSC) the whole process of digitisation gets over in a fraction of a second. The program code for image capturing and time averaging of images of sizes 512 × 512 is as follows.

Program 3.1. Image capture and time averaging for an image of size 512 × 512

```
//snap512.c
// This program is for time averaging over a maximum of four 512 × 512
// pictures and uses full resolution without zoom while snapping.

#include "pipdef.h"
void take512(n,quad_no)
{
 int l1 = 0, h1 = 255, l2 = 20, h2 = 235, y_cood, x_cood, i;
 char c,gen_buf[512], lut_buf[256], buf1[512];

 quad_no %=4;   // use modulo 4 of the given value
 n %= 5;        // use modulo 5 of the given value to avoid trouble
 printf("quad_no = %2d.  # of quads = %2d \n", quad_no, n);
 fg_quadm(1); fg_setwin(0,0,511,511); fg_chan(2);
 fg_sync(1); fg_dquad(quad_no); fg_snap(1);
 fg_autoset(); // setup the gain and offset to the correct value
 fg_zoom(0);      // do not use zoom
 fg_sbuf(0);   // enable input from camera to the screen

 fg_scaling(l1,l2,h1,h2,lut_buf); //  rescale all the LUTs so that 0 to l1
 // levels are at l2 for the picture snapped and h1 to 255 are at h2
 // this enhances the contrast. Further the levels 0 thro l2 are needed
 // for implementing a menu driven pseudo color imaging.

 fg_lutm(lut_used); fg_lutd(lut_used,all_colours,0,256,lut_buf);
 // a pause for object adjustment
 printf("Press enter to start snapping\n"); c = getch();
 fg_sbuf(1); fg_pan(0); fg_scroll(0);

 if (n<=1) fg_snap(1);

else {      // start grabbing the picture
   for (i=0; i < n; i++)
    { fg_dquad(i);     // choose quadrant i
    fg_snap(1); }  // A 512 x 512 image  is snapped
 // start averaging now and put the contents in quad quad_no specified.
    for (y_cood=0; y_cood< 512; y_cood++)
     { for (x_cood=0; x_cood < 512; x_cood++) gen_buf[x_cood] = 0;
       for (i=0; i<n; i++)
         { fg_rowr(y_cood, i, buf1);
```

```
      for (x_cood=0; x_cood< 512 ; x_cood++)
       gen_buf[x_cood] += buf1[x_cood] / n;
 // averaged over all the quadrants 0 thro n for a given (x,y)
        }
    fg_dquad(quad_no); fg_roww(y_cood, quad_no, gen_buf);
    }
  printf("Averaging completed.\n");
    }
}

main()
{
 int i=1, j=0;
 initialise_pip();
 printf("# of quads over which I average = ");
 scanf("%1d", &i);
 printf("Quad into which I put the output image = ");
 scanf("%1d", &j);
 take512(i,j);
 fg_exit();
 }
```

In this program the image is captured and digitised using the *snap* command. While performing digitisation, the first 16 grey levels are kept reserved for pseudo colouring using LUT (Look up Table) operations.

3.13 Tools for Image Understanding

3.13.1 Pseudo Colouring

Pseudo colouring implies that artificial colouring can be done to the image for its understanding. This is quite useful to verify whether the illumination is uniform over the field. One can also find the intensity range of fringe areas to implement image segmentation. The pseudo colouring is done through output LUT operations and a program code is given in the enclosed CD (*pseudcol.c*).

By pseudo colouring, the original image is not altered and only the display is affected. Figure. 3.11 shows the principle of LUT operation for mapping the input grey levels to a different set of grey levels for display. In pseudo-colouring, appropriate colours are generated by using the LUT for *R*, *G* and *B* planes.

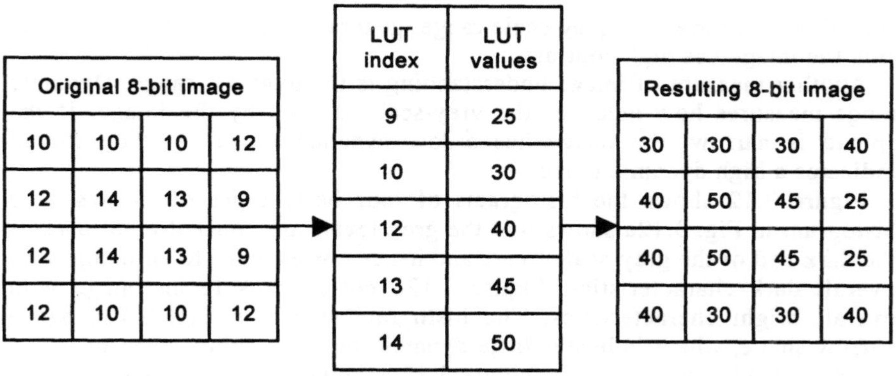

Fig. 3.11. Principle of LUT operation

3.13.2 Histogram

A histogram is a graph, showing the distribution of pixels as a function of intensity values. By examining the histogram, one can determine what contrast enhancements will improve the quality of the image. Contrast, measures the brightness and darkness in an image. If the spread of the histogram is narrow, the image has low contrast. On the other hand, if the

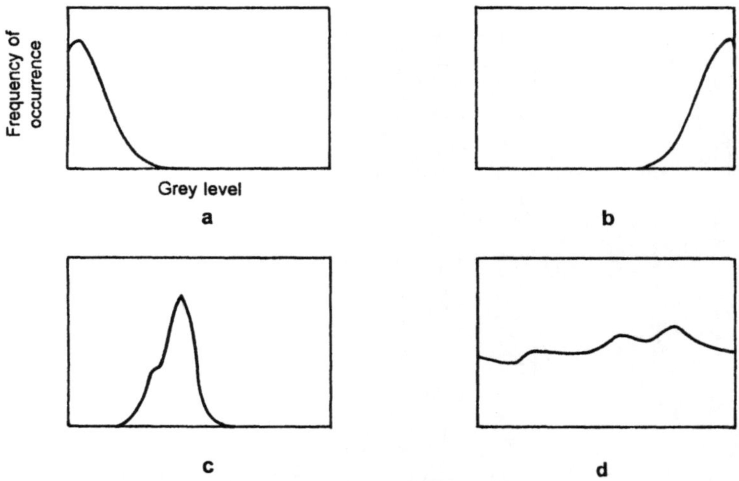

Fig. 3.12. Histograms for typical image types: **a** dark image **b** bright image **c** low contrast image **d** high contrast image

spread is over the entire grey-scale range, generally with two peaks at either end, the image has high contrast.

Another measure of image understanding is the dynamic range. Dynamic range measures how much of the grey-scale is used by the image. If the spread is narrow, the image has a low dynamic range. A wide spread indicates a high dynamic range.

Figure 3.12 shows the histograms of four basic types of images. The histogram in Fig. 3.12a shows that the grey levels are concentrated towards the dark end of the grey-scale range and hence corresponds to an image with overall dark characteristics. Figure 3.12b corresponds to an image with overall bright characteristics. The histogram shown in Fig. 3.12c has a narrow shape, which indicates little dynamic range and thus corresponds to an image having low contrast. Figure 3.12d shows a histogram with significant spread, corresponding to an image with high contrast.

The properties discussed above are global descriptions and do not say anything specific about image content.

3.13.3 Two-Dimensional and Three-Dimensional Intensity Plots

Two-dimensional intensity plot provides intensity variation along a line. Normally, the eye is not sensitive enough to see the difference in grey levels over a fringe. Figure 3.13 shows the intensity variations across different lines of a ring under diametral compression. The figure clearly brings out

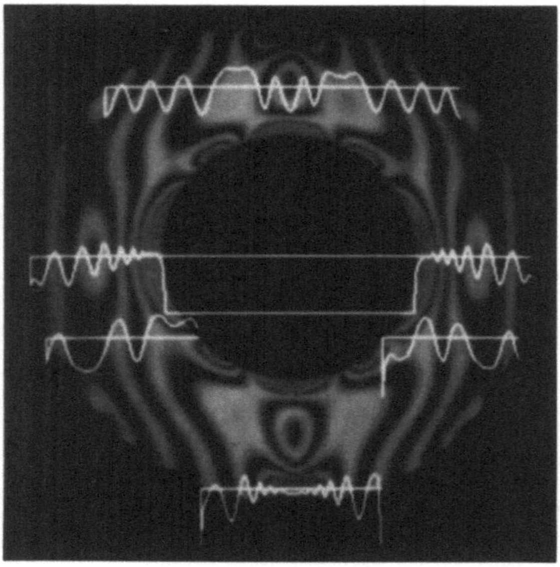

Fig. 3.13. Two dimensional plot of intensity variations for various lines in the dark field isochromatics of a ring under diametral compression

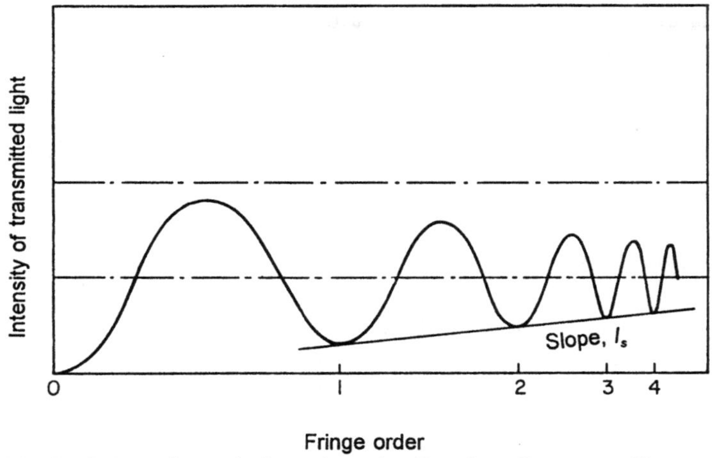

Fig. 3.14. Intensity variations along the direction of stress gradient

the fact that the CCD camera (TM 560 PULNiX) used is good enough to record intensity variations over the fringe thickness. An interesting observation is that in high stress gradient zones, the intensity variation is as shown in Fig. 3.14. Further details about this behaviour are discussed in Chapter 6.

For obtaining 3-D intensity plot, co-ordinate transformation needs to be applied for image data. The z co-ordinate of the pixel is its intensity value. For a pixel located at (x, y, z), initially the origin of the window is translated to the screen origin (Fig. 3.15) as

$$\begin{Bmatrix} x_1 \\ y_1 \\ z_1 \end{Bmatrix} = \begin{Bmatrix} x+O_x \\ y+O_y \\ z \end{Bmatrix} \tag{3.5}$$

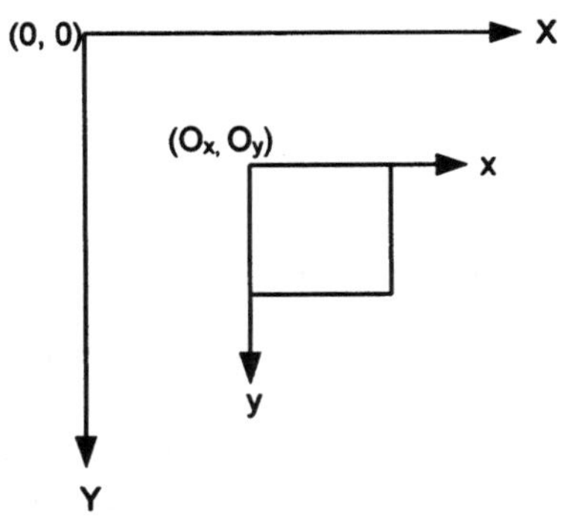

Fig. 3.15. Screen and image co-ordinates

Then rotation about x, y and z-axes are to be performed sequentially. The rotation about x axis (θ_x) is effected by

$$\begin{Bmatrix} x_2 \\ y_2 \\ z_2 \end{Bmatrix} = \begin{bmatrix} 1 & 0 & 0 \\ 0 & \cos\theta_x & \sin\theta_x \\ 0 & -\sin\theta_x & \cos\theta_x \end{bmatrix} \begin{Bmatrix} x_1 \\ y_1 \\ z_1 \end{Bmatrix} \tag{3.6}$$

Rotation about y-axis is obtained by

$$\begin{Bmatrix} x_3 \\ y_3 \\ z_3 \end{Bmatrix} = \begin{bmatrix} \cos\theta_y & 0 & \sin\theta_y \\ 0 & 1 & 0 \\ -\sin\theta_y & 0 & \cos\theta_y \end{bmatrix} \begin{Bmatrix} x_2 \\ y_2 \\ z_2 \end{Bmatrix} \tag{3.7}$$

Rotation about z-axis is then done by

$$\begin{Bmatrix} x_4 \\ y_4 \\ z_4 \end{Bmatrix} = \begin{bmatrix} \cos\theta_z & \sin\theta_z & 0 \\ -\sin\theta_z & \cos\theta_z & 0 \\ 0 & 0 & 1 \end{bmatrix} \begin{Bmatrix} x_3 \\ y_3 \\ z_3 \end{Bmatrix} \tag{3.8}$$

Finally, the picture is transformed to the window origin as

$$\begin{Bmatrix} x_5 \\ y_5 \\ z_5 \end{Bmatrix} = \begin{Bmatrix} x_4 - O_x \\ y_4 - O_y \\ z_4 \end{Bmatrix} \tag{3.9}$$

The transformed values of x_5, y_5 and z_5 are plotted with suitable scaling for the intensity value (z_5) for getting a 3-D picture of the fringe patterns. Figure 3.16 shows the 3-D plot of intensities for the entire ring, *source, sink* and a *saddle point*. The values of θ_x, θ_y and θ_z are to be chosen such that the resulting image is informative as well as aesthetically pleasing. In Fig. 3.16, the values used are 10°, -10° and 45° respectively.

3.14 Filtering in Spatial Domain

The use of spatial masks for image processing is usually called spatial filtering, and the masks themselves are called spatial filters. Filters in general can be classified as low pass; high pass and band pass filters. They can be further classified as linear and non-linear filters.

Slowly varying characteristics of an image, such as overall contrast and average intensity, constitute low frequency component of an image. The

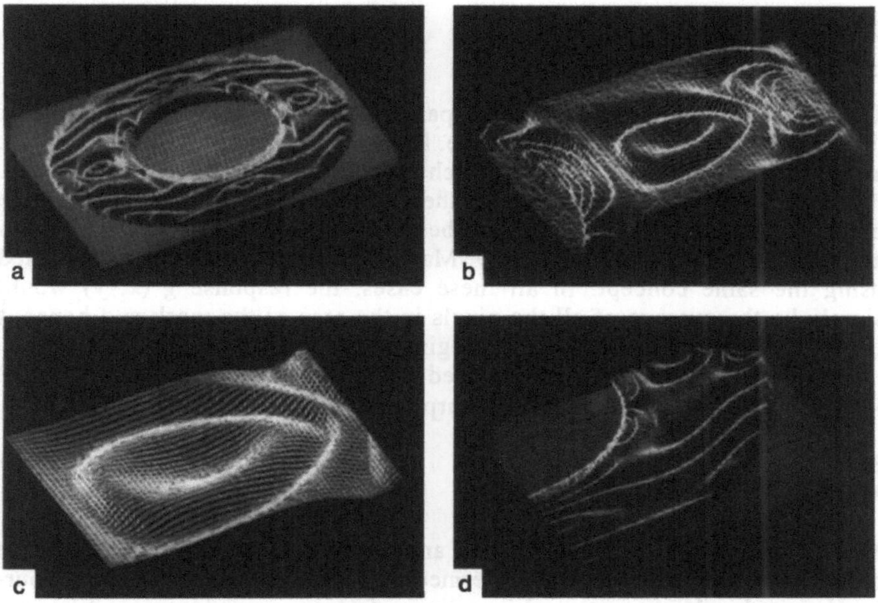

Fig. 3.16. Three-dimensional plots of intensity variations for various isochromatic fringe features of a ring under diametral compression: **a** whole ring **b** source **c** sink **d** saddle point

edges and other sharp details in an image constitute high frequency components of an image.

Low pass filters attenuate or eliminate high-frequency components. The net effect of low pass filtering is image blurring. Similarly, high pass filters attenuate or eliminate low-frequency components. High pass filtering introduces apparent sharpening of edges and other sharp details but reduces image contrast and average intensity. A third type of filtering, called band pass filtering, removes selected frequency regions between low and high frequencies which are used for image restoration.

The linear filters are termed as convolution filters and non-linear filters are called non-convolution filters. Convolution filters use the neighbourhood values in a linear fashion as in Eq. (3.4) whereas; non-convolution filters use the neighbourhood values differently, which is discussed in Sect. 3.14.2. High pass filtering is discussed later under the heading "Edge Detection".

The weightings of the masks used in convolution operation are also termed as kernels in image processing literature. Several commercial softwares allow the specification of the kernel by the user.

3.14.1 Low Pass Spatial Filtering

The essential requirement for a low pass filter is that all the weightings of the mask (Fig. 3.7) should be positive. For a 3 × 3 spatial filter, the simplest arrangement would be a mask in which all coefficients have a value of one. From Eq. (3.4), the response would then be the sum of grey levels for nine pixels, which could cause $g^R(x, y)$ to be out of the valid grey-level range. To avoid this, one has to divide it by 9. Masks of larger size can be constructed using the same concept. In all these cases, the response $g^R(x, y)$ would simply be the average of all the pixels in the area of the mask and hence it is referred to as neighbourhood averaging.

Low pass filtering is usually used to remove salt and pepper noise. However, it blurs edges and other sharp details.

3.14.2 Median Filtering

Median filter is a non-linear filter and it can be used to achieve noise reduction without blurring. It is termed as non-linear since the neighbourhood grey levels are not used in a linear fashion as in low pass filtering. The grey level of each pixel under consideration is replaced by the median of the grey levels in the neighbourhood of that pixel. It is quite useful when the noise pattern consists of strong, spike like components and the characteristic to be preserved is edge sharpness.

The median of a set of values is such that half the values in the set are less than the median value and half are greater than the median value. Thus, in median filtering, points with distinct intensities are forced to be more like their neighbours. This eliminates intensity spikes that appear isolated in the area of the filter mask.

3.15 Image Enhancement

In image enhancement, the goal is to accentuate certain image features for subsequent analysis or for image display. The enhancement process itself does not increase the inherent information content in the data. It simply emphasises certain specified image characteristics. Examples include contrast and edge enhancement, noise filtering, sharpening, and magnifying. Image enhancement is useful in feature extraction, image analysis, and visual information display. Enhancement algorithms are generally interactive and application-dependent.

Image enhancement techniques fall into two broad categories: spatial domain methods and frequency domain methods. The spatial domain refers to the image plane itself, and approaches in this category are based on direct manipulation of the pixels in an image. Frequency domain processing techniques are based on modifying the Fourier transform of an image.

Image enhancement functions in the spatial domain, map each grey level into another grey level by a predetermined transformation and can be expressed as

$$g^R(x,y) = T_R\{g(x,y)\} \tag{3.10}$$

where $g(x, y)$ is the input image, $g^R(x, y)$ is the processed image, and T_R is an operator on g, defined over some neighbourhood of (x, y).

3.15.1 Contrast Stretching

Poor illumination, lack of dynamic range in the imaging sensor, or wrong setting of a lens aperture during image acquisition can result in a low-contrast image. The idea behind contrast stretching is to increase the dynamic range of the grey levels in the image being processed. Figure 3.17 shows a typical transformation used for contrast stretching. The locations of points (m_1, n_1) and (m_2, n_2) control the shape of the transformation function. In general, m_i and n_i are selected so that the function is single-valued and changing monotonically. This is necessary to preserve the order of grey levels, thus preventing the creation of intensity artefacts in the processed image.

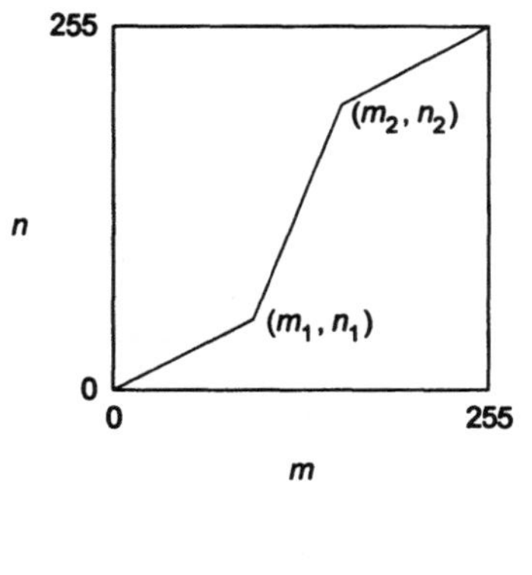

Fig. 3.17. Generic grey level transformation for contrast stretching

If $m_1 = m_2$, $n_1 = 0$ and $n_2 = 255$, the transformation becomes a thresholding function (Fig. 3.18a) that creates a binary image. Other values of (m_1, n_1) and (m_2, n_2) can be used to produce various degrees of spread in the grey levels of the output image (Figs. 3.18b and 3.18c), thus affecting its contrast. To get image negatives, the order from black to white has to be reversed so that the intensity of the output image decreases as the intensity of the input increases (Fig 3.18d). Some fringe multiplication techniques in digital photoelasticity use image negatives as an intermediate step.

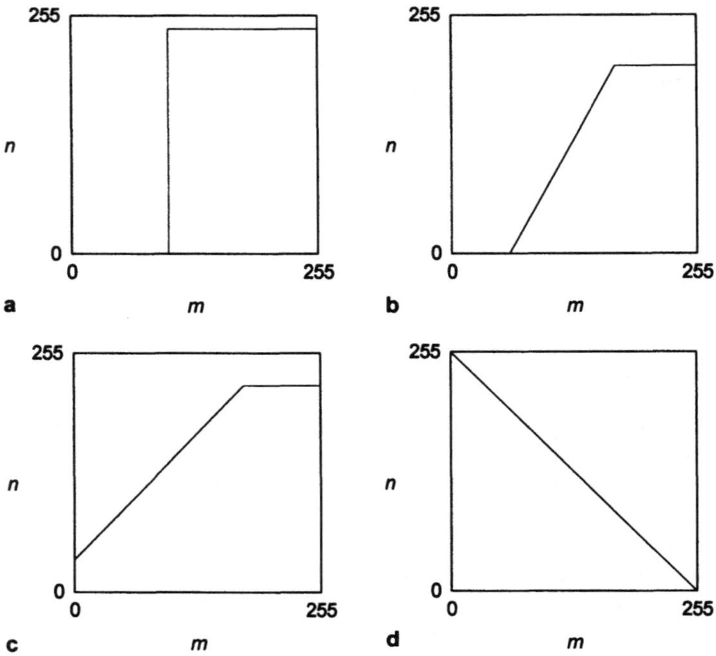

Fig. 3.18. Grey level transformations for typical operations: **a** thresholding **b** contrast stretching-I **c** contrast stretching-II **d** image negative

3.15.2 Histogram Equalisation

Histogram equalisation alters the dynamic range and contrast of an image. To perform histogram equalisation, one has to select a function representing a curve to modify the histogram. Most of the commercial frame grabbers have in-built routines to perform histogram equalisation. The available curves include *Linear, Bell, Cube, Log* and *Exponential*. This operation has the effect of lightening or darkening an overall image, depending on the curve chosen.

The *Linear* equalisation curve moves the indexes so that the average index density across the grey-scale is even. This equalisation produces a high contrast image with the highest possible dynamic range. The *Bell* equalisation curve moves the indexes in a similar manner to the *Linear* equalisation curve, but groups the indexes more closely to the centre of the scale. This equalisation produces a high contrast image with not quite as high a dynamic range as would be achieved with a *Linear* equalisation curve.

The *Cube* equalisation curve concentrates the range of indexes at the low end of the scale, making the overall image darker. This equalisation produces a high contrast image with a low dynamic range. The *Log* equalisation curve concentrates the range of indexes at the low end of the scale, making the overall image darker. This equalisation also produces a high contrast image with a low dynamic range. The *Exponential* equalisation curve concentrates the range of indexes at the high end of the scale, making the overall image lighter. The image modified will have a high contrast but with a low dynamic range.

3.16 Image Segmentation

The first step in image analysis generally is to segment the image. Segmentation subdivides an image into its constituent parts or objects. The level to which this subdivision is to be carried depends on the problem being solved. That is, segmentation should stop when the objects of interest in an application have been isolated. In digital photoelasticity, the model has to be segmented from the background and within the model the fringe areas need to be identified for which segmentation techniques can be used.

3.16.1 Thresholding

Thresholding is one of the most important approaches to image segmentation. The grey level histogram shown in Fig. 3.19 corresponds to Fig. 3.13. In Fig. 3.13, the background is dark but within the model domain, one has dark fringes in a white background, which are grouped into two dominant modes. One obvious way to extract the objects from the background is to select a threshold T that separates these modes. Then, any point (x, y) for which $g(x, y) > T$ is called an object point; otherwise, the point is called a background point.

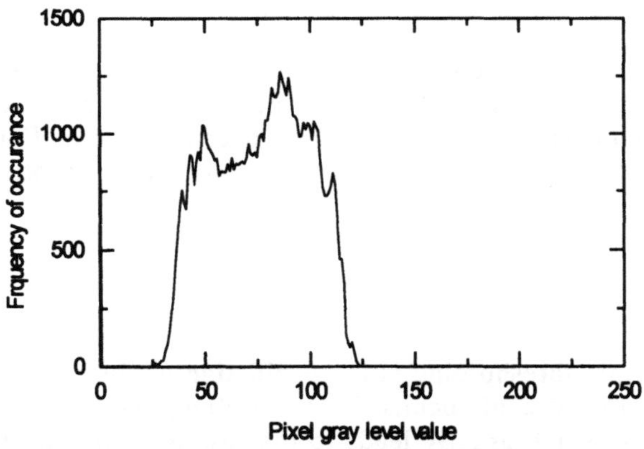

Fig. 3.19. Histogram of the RING image showing two peaks

In a generic sense, thresholding may be viewed as an operation that involves test against a function T of the form

$$T = f\{x, y, p(x,y), g(x,y)\} \tag{3.11}$$

where $g(x, y)$ is the grey level of point (x, y) and $p(x, y)$ denotes some local property of this point, for example, the average grey level of a neighbourhood centred on (x, y). A thresholded image $b(x, y)$ is defined as

$$b(x,y) = \begin{cases} 1 & \text{if } g(x,y) > T \\ 0 & \text{if } g(x,y) \le T \end{cases} \tag{3.12}$$

When T depends only on $g(x, y)$, the threshold is called global. If T depends on both $g(x, y)$ and $p(x, y)$, the threshold is called local. If, in addition, T depends on the spatial co-ordinates x and y, the threshold is called dynamic.

3.16.1.1 Global Thresholding

The simplest of all thresholding techniques is to partition the image histogram by using a single threshold. Segmentation is then accomplished by scanning the image pixel by pixel and labelling each pixel as object or background, depending on whether the grey level of that pixel is greater or less than the value of T. As indicated earlier, the success of this method depends entirely on how well the histogram can be partitioned.

In practice, global thresholding can be expected to be successful in highly controlled environments. One of the areas in which this is possible is where illumination control is usually feasible which is the case in transmission photoelasticity.

3.16.1.2 Semi Thresholding

In semi-thresholding, the image is not fully converted as a binary image. Certain grey levels, e.g., those above a threshold, become 255, but the remaining grey levels remain unchanged, rather than becoming zero. For determining fringe areas, this approach is quite useful in digital photoelasticity.

3.16.1.3 Dynamic Thresholding

As mentioned earlier, dynamic thresholding is the evaluation of the threshold of each pixel locally. Because of localised thresholding, the contrast variation within the image – which usually leads to the loss of information in global thresholding – could be overcome largely. Based on the histogram of the image, maximum threshold level, T_{max} and minimum threshold level, T_{min} are chosen. The intensity of each pixel is compared with these two. If it is more than T_{max}, the pixel is made white and if it is less than T_{min}, it is made black. For pixels with intensity values between T_{max} and T_{min}, another decision criterion is applied. Two quantities, namely the edge detection operator D_c and the ambient grey level B_o are calculated [4] based on the pixel's neighbours as

$$D_c = \left[\frac{g(i,j)-g(i-1,j)}{2}\right] + \left[\frac{g(i,j)-g(i,j-1)}{2}\right] \tag{3.13}$$

$$B_o = \left[\frac{g(i-1,j)-g(i,j-1)}{2}\right] \tag{3.14}$$

where $g(i\text{-}1, j)$ and $g(i, j\text{-}1)$ are the grey levels of the pixels directly above in the previous row and in the last column of the same row respectively. A threshold T is calculated based on the following equation

$$T = -D_c + \alpha B_o \tag{3.15}$$

where α is a parameter, which has to be chosen carefully for any particular image to control the optimum information of the image. A header file to implement semi-thresholding both globally and dynamically is enclosed in the CD (*threspip.h*).

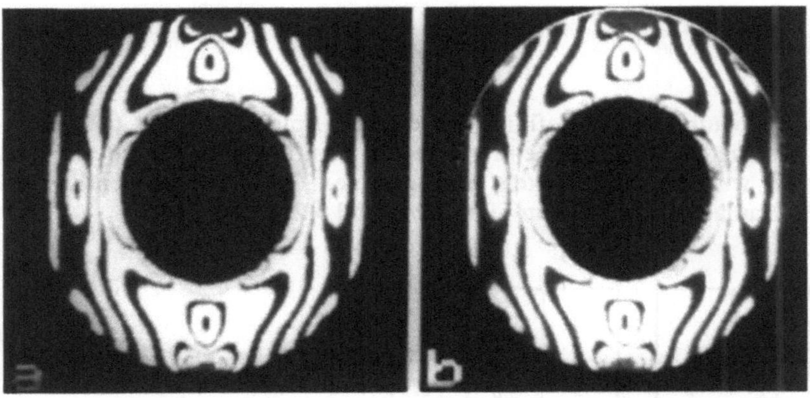

Fig. 3.20. Fringe areas of the RING image obtained by two schemes of thresholding: **a** global thresholding and **b** dynamic thresholding

Figure 3.20a shows the globally thresholded image with a threshold value of 80. In Fig. 3.20b dynamically semi-thresholded image is given. The parameters for dynamic thresholding are; top threshold is 84 and the parameter α is 2. Dynamic thresholding is only marginally better than global thresholding. The fringe areas near the load application points are better identified by dynamic thresholding.

3.16.2
Edge Detection

An edge is the boundary between two regions with relatively distinct grey-level properties. Edges in digital images are generally slightly blurred as a result of sampling. It is reasonable to assume for the purpose of analysis that the regions in question are sufficiently homogeneous so that the transition between the two regions can be determined based on grey-level discontinuities alone.

Averaging of pixels over a region tends to blur the details in an image. As averaging is analogous to integration, differentiation can be expected to have the opposite effect and thus sharpen the image. The idea underlying the most edge-detection techniques is the computation of a local derivative operator.

3.16.2.1
Edge Detection by Convolution Filters

Edge detection can be achieved through the use of convolution and non-convolution filters. The convolution filters are Laplacian, horizontal edge and vertical edge. Laplacian filter is the digital implementation of performing the Laplace operator over the image function $g(x, y)$ as

$$\nabla^2 g = \frac{\partial^2 g}{\partial x^2} + \frac{\partial^2 g}{\partial y^2} \tag{3.16}$$

The basic requirement in defining the digital Laplacian is that the coefficients associated with the centre pixel is positive and the coefficients associated with the outer pixels are negative. The most frequently used forms of Laplacian are given in Figs. 3.21a and 3.21b.

	-1	
-1	+4	-1
	-1	

a

-1	-1	-1
-1	+8	-1
-1	-1	-1

b

Fig. 3.21. Masks for various convolution filters: **a** Laplacian 1 **b** Laplacian 2

-2	0	2
-2	0	2
-2	0	2

c

-2	-2	-2
0	0	0
2	2	2

d

Fig. 3.21. Masks for various convolution filters: **c** Vertical edge **d** Horizontal edge

In general, Laplacian removes the low-contrast components and leaves only the edges. The horizontal edge filter accents the horizontal edges and removes all other parts of the image. Similarly, the vertical edge filter accents the vertical edges and removes all other parts of the image. The convolution kernels for these filters are given in Figs. 3.21c and d.

It is interesting to note that in all these filters, the sum of the coefficients is zero and hence the response is zero whenever the point in question and its neighbours have the same value.

3.16.2.2
Edge Detection by Non-Convolution Filters

Edge detection can also be achieved by gradient operators. The gradient of image function $g(x, y)$ at a point is given as

$$\nabla g = \begin{Bmatrix} G_x \\ G_y \end{Bmatrix} = \begin{Bmatrix} \dfrac{\partial g}{\partial x} \\ \dfrac{\partial g}{\partial y} \end{Bmatrix} \tag{3.17}$$

The magnitude of this vector is

$$|\nabla g| = \left[G_x^2 + G_y^2\right]^{\frac{1}{2}} \tag{3.18}$$

The filters Roberts, Prewitt and Sobel are various digital implementation of the Eq. (3.18). In order to simplify the computations, the gradient is approximated as the sum of absolute values as

$$\nabla g \approx |G_x| + |G_y| \tag{3.19}$$

Robert filter uses a 2 × 2 mask to calculate G_x and G_y. It is also termed as a cross-gradient operator and is used to detect high frequency edges. Masks

1	0
0	-1

a

0	1
-1	0

b

Roberts

-1	-1	-1
0	0	0
1	1	1

a

-1	0	1
-1	0	1
-1	0	1

b

Prewitt

-1	-2	-1
0	0	0
1	2	1

a

-1	0	1
-2	0	2
-1	0	1

b

Sobel

Fig. 3.22. Masks for various non-convolution filters. The masks shown in **a** and **b** are to be used to calculate G_y and G_x respectively

of even sizes are awkward to implement. Prewitt and Sobel use a 3 × 3 mask to calculate G_x and G_y. The respective kernels for these operations are shown in Fig. 3.22.

While the effect of each filter can be described in general, there is no way to predict the exact result a particular filter will have on a particular image. In other words, filtering is a trial and error process. Thus, one has to experiment with different types of filters until one identifies a filter that produces those effects, which are desirable for a particular application.

3.16.2.3
Edge Detection by Thresholding

Thresholding can also be used to extract the edges of objects, rather than the objects themselves. This is possible in situations where the objects are dark and the background is light, and intermediate grey levels occur only on the borders between objects and background. This is quite so for fringe regions in digital photoelasticity. If the intermediate levels are mapped into 255 and all other levels into zero, one obtains the outlines of the fringes.

Figure 3.23 gives the edge detected by various filters for the image shown in Fig. 3.13. It clearly brings out the advantage of using thresholding for edge detection in digital photoelasticity.

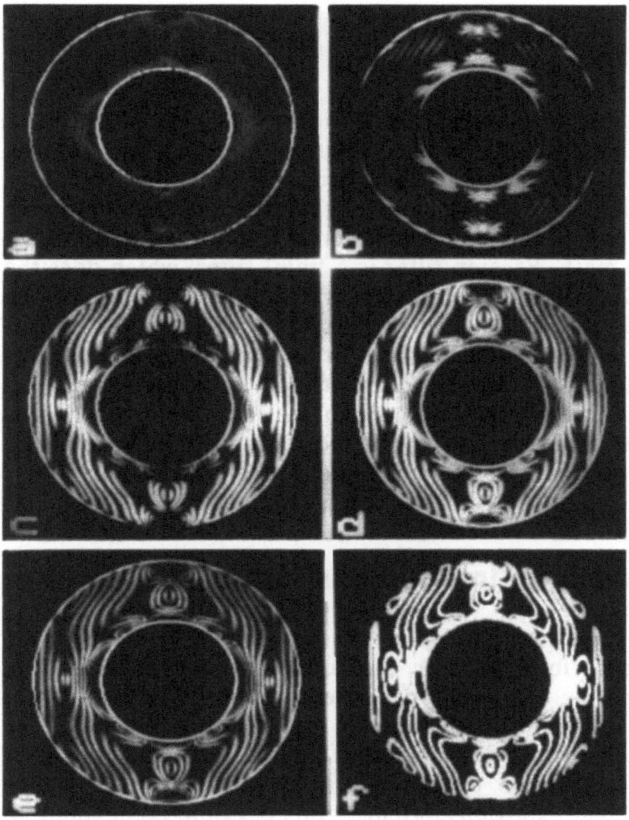

Fig. 3.23. Edge detected by various methods for the RING image: **a** Laplacian-2 **b** Horizontal filter **c** Vertical filter **d** Prewitt filter **e** Sobel filter **f** Global theresholding

A difficulty with edge detection as an approach to picture segmentation is that the detected edges often have gaps in them, at places where the transitions between regions are not abrupt enough. Moreover, edges may be detected at points that are not part of region boundaries, if the given picture is noisy. Thus, the detected edges will not necessarily form a set of closed, connected curves that surround closed, connected regions. One way to overcome these problems is to use tracking techniques to follow the edges around the regions. Such techniques can be designed to tolerate gaps in edges, which do lie on a curve, and to reject as noise edges, which do not. Another possibility is to apply curve detection operations to the edge detector output. This rejects edge points that do not lie on curves, and can be designed to fill gaps in edges that do.

3.17 Morphological Filters

Morphological filters change the shape of an image. They are non-linear (non-convolution) which include erosion, dilation, opening and closing filters. Erosion filter erodes the edges and generally darkens the image. Dilation filter dilates the edges and generally lightens the image. Dilation is a 'max' filter, which changes the pixel under consideration to the maximum grey level value in the mask. Erosion filter is a 'min' filter, which does the opposite operation. Opening filter consists of two operations involving erosion first followed by dilation. This is used to open bright images containing dark areas close to their borders or to separate bright objects joined by narrow bridges. These are quite useful in medical imaging. Closing filter performs dilation followed by an erosion operation. This is used to close small dark holes in a bright image. This filter is quite useful for improving discontinuous fringe skeletons in digital photoelasticity.

3.18 Further Discussions on Image Sensors

The developments in electronic imaging are very rapid. However, the potential uses of sensors introduced in the future should be assessable, given a basic understanding of the interdependence between current sensor architecture and their possible application to digital photoelasticity. The optical and electronic characteristics of sensors and the accompanying error sources have a direct influence on the technical possibilities of image acquisition in digital photoelasticity. Hence, this section is devoted to describing the operation and characteristics of electronic sensors, particularly, the charge coupled devices.

3.18.1 Operation of CCD arrays

The basic building block of the CCD is the metal oxide semiconductor (MOS) capacitor. MOS gate (transfer site) for p-type silicon is shown in Fig. 3.24a. The number of transfer sites (gates) per detector area of the image sensor generally varies from two to four. Applying a positive voltage to the gate causes the mobile positive holes in the p-type silicon to migrate towards the ground electrode. A region devoid of positive charge is created which is termed as a depletion region. When a photon is absorbed in the depletion region, it produces an electron-hole pair. The electron stays within the depletion region whereas, the hole moves to the ground electrode. The number of electrons that can be collected is proportional to the applied voltage, oxide thickness and gate electrode area.

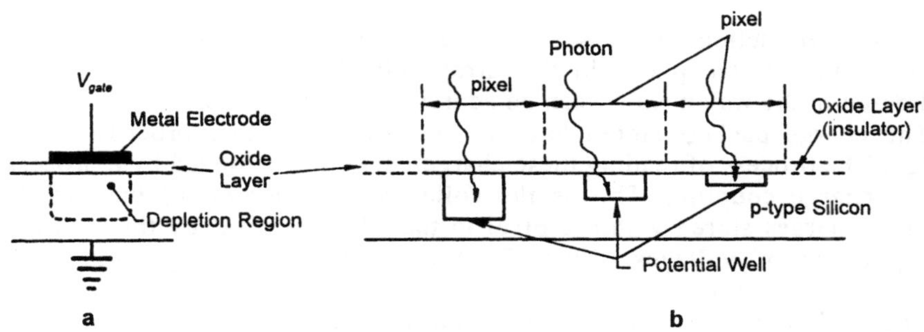

Fig. 3.24. **a** CMOS capacitor. **b** Linear array of CCD with potential well of different sizes proportional to the incident light intensity

When a CCD sensor is specified, one refers to an array of many individual CCD's either in the form of a line (Fig. 3.24b) or arranged in a rectangular array (Fig. 3.25). The individual CCD element in the sensor is a picture element abbreviated as pixel. When light falls on to this, a linearly proportional number of free electrons as mentioned before are generated due to photoelectric effect. Electrons continue to accumulate for the duration of the pixel's exposure to light. The pixel's storage capacity is limited and is described by its full-well capacity, which is measured in electrons per pixel. Typically, this ranges from 10, 000 to 100, 000 electrons per pixel. Modern day CCD cameras have a protective circuitry called anti-blooming which protects the CCD array due to accidental over exposure of light [5]. The number of free electrons generated (charge packet) in each site of a CCD array will be directly proportional to the incident radiation on that specific pixel and the charge packets created, in finite pixel array, will be a

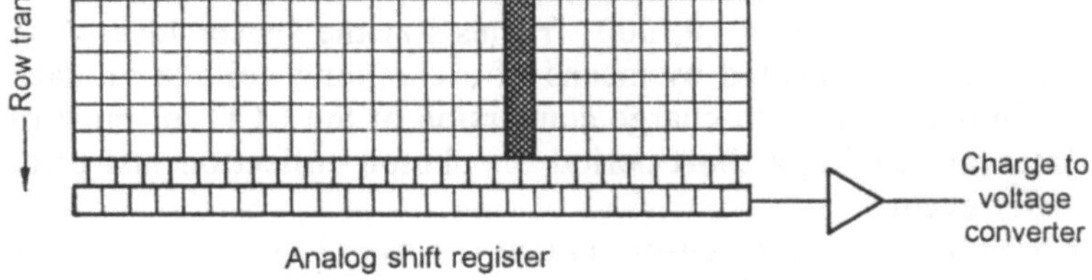

Fig. 3.25. Rectangular array of CCD showing the readout sequence

faithful reproduction of the scene projected on its surface. It has to be pointed out that the pixels in this array (Fig. 3.25) cannot be randomly addressed in the way memory can be addressed in a computer.

The charge packets stored have to be read for further processing the image. The process of reading these charge-packets is achieved by a process called charge coupling. This is the collective transfer of all the mobile electric charges stored within a pixel to the adjacent pixel by the external manipulation of voltages. Thus, CCD cameras are actually solid state devices with CCD readout.

The reading of accumulated charges has to be done sequentially in a two step process. First, it has to be shifted vertically one row at a time into a masked off analog shift register on the lower edge of the sensor's active area (Fig. 3.25). Each row in the analog shift register is then clocked, pixel by pixel, through a charge to voltage converter and thereby providing one voltage per pixel. The stream of pixel voltages along with a variety of synchronisation pulses composes the actual analog signal. The most common read-out systems consist of reading all the odd rows first and then the even rows are accessed. This conforms to standard interlaced video format. Other formats for reading exists which will be discussed in Chapter 11.

One of the significant advantages of CCD image sensors over vacuum tube sensors is the precise knowledge of the photo sensor locations with respect to one another. In a camera tube, the video image is read from a photosensitive material by a scanning electron beam. The position of the beam is never precisely known because of the uncertainty in the sweep circuits resulting from random electrical noise.

3.18.2 Interline Transfer CCD

One of the most commonly available CCD sensors is the interline transfer CCD. This consists of additional vertical transfer registers located between each pixel (Fig. 3.26). Charge accumulated in the active area of the pixel can be readily transferred into the storage area in about 1 μs. Most transfer register designs are based upon standard video timing. Each field is read every other 1/50 sec (PAL) or 1/60 sec (NTSC) and thus these cameras are able to digitise the image at video rates.

The fast charge-dumping feature also opens the possibility of full electronic shuttering of the order of 10, 000 frames/s at the sensor level. The electronic shutter is implemented by means of a clamping voltage on each pixel which inhibits the photo-to-charge conversion of the CCD for most of the framing interval; leaving a short period for photon collection just prior to charge transfer event.

The major drawback of these sensors is that it leaves less area for active sensors. This is assessed by a factor called fill factor, which is the active area of the pixel to the total pixel area. The fill factor may be as low as

20%. However, the fill factor can be improved by the use of micro lenses in conjunction with the array.

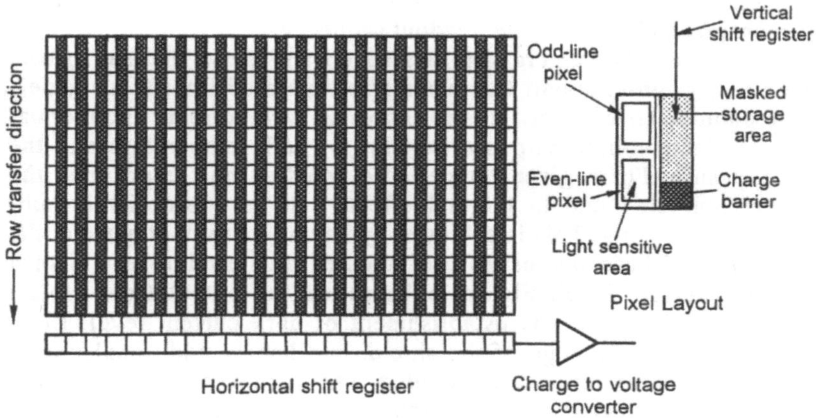

Fig. 3.26. Interline transfer readout of a CCD array. Note the role of vertical transport registers in reducing the light sensitive area

3.18.3 Linearity and Dynamic Range

Each electron captured in the potential well adds linearly to the collected charge and hence, the output signal voltage for the individual pixel is directly proportional to the collected charge. Over exposure or poorly designed output amplifiers can result in non-linearities of the output. With careful design, linearities with deviations of less than 1% are possible.

Dynamic range is the maximum illumination level divided by the minimum illumination level. It depends on the lens *f*-number, aperture setting and integration time. The spectral content of the source and the array spectral responsivity also affect the camera output voltage. Thus, the camera dynamic range can be quite variable depending on the test conditions. However, in general, the dynamic range of a CCD increases as the temperature is lowered. Standard video sensors operating at room temperature typically have a dynamic range of about 100–200 grey level (8-bits). Once digitised, the useful signal is 7–8 bits in depth. It is to be noted that this itself exceeds that of human vision. By careful camera design with appropriate cooling, a dynamic range of 65,536 grey levels (16-bits per pixel) is possible.

3.18.4
Sources of Noise

One can identify two main sources of noise; one due to the thermal effects and the other introduced during the readout sequence.

During operation, the camera gets heated up and these thermal effects also generate electron-hole pair that cannot be separated from those generated by the photoelectric effect. This is known as dark current. The thermal generation of electron-hole pair is a function of both temperature and the length of exposure. For a given temperature and for short exposures, the dark current generated due to thermal effects can be appropriately subtracted before processing. This is achieved by having extra photosites at the end of the array. These photosites are shielded and used to establish a reference dark current level. Figure 3.27a shows the output from typical pixel and Fig. 3.27b shows the establishment of dark current level. Light leakage may occur at the edge of the shield, and hence, additional photosites known as isolation pixels are included. The average value of the dark current pixels is subtracted from the active pixels leaving only the light generated signal.

For long exposures such as that used in astronomy, the CCD array needs to be cooled and temperature maintained to minimise dark current noise. Such considerations may be necessary for digital scattered light photoelasticity where long exposures are needed due to weak intensities of the scattered light.

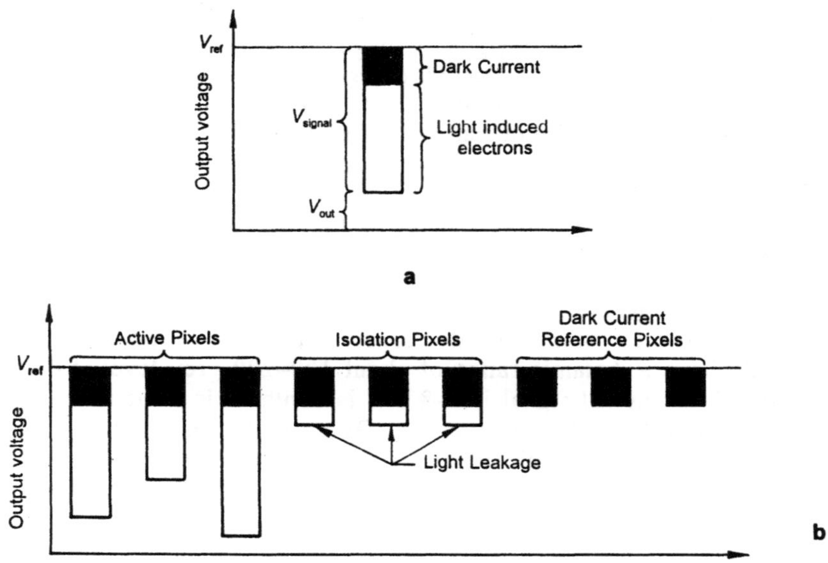

Fig. 3.27. **a** Typical output from a pixel. **b** Establishment of dark current in a CCD array

Noise may be introduced in the readout sequence of converting charge to voltage. This is called read noise or shot noise. This noise is directly related to the readout frequency and thus for scientific applications, *slow-scan* cameras may be desirable.

3.19 Digitisation of the Camera Video Signal

The CCD camera, in general, sends out a video signal as per one of the standard video formats. This video signal has to be digitised for further analysis. To produce a digital image, a sampling rate called the *pixel clock* is needed which determines how many pixels will be extracted from the active portion of the analog signal. In other words, the pixel clock is used to divide the incoming horizontal line of video into separate pixel by specifying the exact location, in time, of each pixel. The pixel clock time can come from one of the three timing mechanisms; a crystal oscillator; a phase locked loop (PLL) or an external clock.

A crystal oscillator provides an accurate and stable frequency at standard rates. A phase locked loop creates a pixel clock by locking itself to the horizontal sync of the video signal. This ensures that the number of pixels in each line remains constant. Depending on the design of PLL, a certain degree of error called *pixel jitter* occurs. It is measured in nanoseconds. If the value of the jitter is an appreciable proportion of a single pixel time, the quality of the captured data will be reduced. In properly designed PLL circuits, the pixel jitter is under 10 nanoseconds, which is about 1/10 th of the size of the pixel. The pixel clock can be generated by the frame grabber and fed to the camera or vice versa. In general, external pixel clock can support very high digitisation rates and allow for exact pixel location.

3.20 Resolution of an Image Processing System

It is to be noted that the resolution of the image processing system depends on the detector array, characteristics of the frame grabber, bandwidth of the video standard and the display. Each video standard can support a specific number of pixels both horizontally and vertically. This is decided by the bandwidth of the video standard. To reduce cost, a camera may contain fewer detector elements and may not fully take advantage of the capacity of the video bandwidth. The imagery of these cameras will be poorer compared to those cameras that fully use the available bandwidth. Thus, in general, monitor pixels need not be the same as the camera pixels. However, current camera design philosophy matches the number of detectors to the number of pixels supported by the video bandwidth.

3.21 Gamma Compensation

In Sect. 3.5 it has been mentioned that in CRT displays, the output brightness is related to the grid voltage through a power law, which is characterised by a parameter called gamma. In view of the use of CRT displays, if the camera has the inverse gamma then the radiometric fidelity is preserved at the display. Hence, the camera manufacturer usually integrates a gamma correction circuitry within the camera. In view of this, the output from the camera is non-linear for image processing purposes. However, this may be quite desirable for a consumer electronics camera where the camera output is directly connected to a CRT display. The inverse gamma is fixed as 0.45 for all the TV standards although it is exact only for the NTSC system. With PAL and SECAM, the luminance still has a residual gamma of $0.45 \times 2.8 = 1.27$ which provides some contrast enhancement.

For scientific applications, the role of gamma needs to be accounted for. This can be done in two ways: one, in which the inverse gamma is not activated in the camera and the other, to do appropriate correction before image processing. Though the camera has a circuit for inverse gamma, in common parlance, it is simply referred as gamma. Scientific cameras may have a continuously variable inverse gamma from 0.4 to 1 or may be pre selectable to 0.45 or 1. By setting inverse gamma as one in the camera, the gamma compensation can be effected for image processing purposes. If this is not possible, then before any image processing, gamma has to be corrected digitally. This is usually done by LUT's. The concern for gamma is expected to go when flat panel digital displays are used in place of CRT displays.

Exercises

3.1 Explain the process of translating a visual picture into a form suitable for processing by a digital computer.

3.2 What are video standards? Mention their role in digital image processing.

3.3 Compare the performance of a solid state camera and a vidicon tube based camera.

3.4 In DIP studies, considerable effort is aimed at understanding human visual response. Is it relevant for digital photoelasticity? Explain.

3.5 Why masks of odd sizes are preferred? Compare the computational requirements for performing spatial averaging using masks of sizes of 3×3, 5×5 and 7×7.

3.6 What are the image characteristics one can understand by inspecting a histogram?

3.7 A digital photoelastic fringe pattern is supplied to you. How would you check the quality of illumination based on this?

3.8 What are the differences between convolution and non-convolution filters?
3.9 What are the image characteristics that can be altered by filtering?
3.10 Figure 3.13 indicates that in high fringe density zones, the image is brighter. Write a software code to employ different local thresholding for different regions (tiles) of an image.
3.11 Mention the various techniques for image segmentation. Among these, what technique would you choose for identifying fringe areas? Why?
3.12 What is a CCD? How does it generate voltage proportional to the intensity of light falling on it?
3.13 What is electronic shuttering? In what class of CCD's is this possible?
3.14 What is pixel jitter? Compare the performance of a CCD camera with a digital output and an analog output.
3.15 What are the sources of noise in a CCD camera? How can the dark current be compensated?
3.16 What parameters influence the dynamic range of the CCD camera?
3.17 Is the monitor pixel in a CRT display same as the camera sensor pixel? What is the current design philosophy?
3.18 What do you understand by Gamma compensation?

References

1. Gonzalez RC, Woods RE (1993) Digital image processing. Addison-Wesley Publishing company, California
2. Boyle WS, Smith GE (1970) Charge coupled semiconductor devices. Tech J Bell Systems 49:587–593
3. Ulicheny RA (1988) Dithering with Blue Noise. Proceedings of the IEEE 76(1):56–79
4. Morrin TH (1974) A black & white representation of a grey scale picture. IEEE Trans – Computers 23:184–186
5. Holst GC (1996) CCD arrays, cameras and displays. SPIE Optical Engineering Press, Bellingham Washington

4 Fringe Multiplication, Fringe Thinning and Fringe Clustering

4.1 Introduction

Isochromatics and isoclinics are the two fringe contours obtained in a photoelastic experiment. Isochromatics are contours of constant principal stress difference and isoclinics are the loci of points along which the principal stress orientation (with respect to a reference axis) is a constant. For stress analysis at any point, the photoelastic data required are,

1. The isochromatic fringe order and
2. The isoclinic parameter at that point.

The positional co-ordinates of the point under consideration with respect to any chosen point (origin) in the model is also required for quantitative analysis. Generally, the fringes do not appear as thin lines, but they appear as broad bands. So the first step in fringe processing is to identify the actual fringe in this dark band.

Early methods to measure the fringe data were manual. Photographs of the fringe pattern were magnified for facilitating better accuracy and the fringe edges were traced; midpoints between the edges along the length of the fringe bands were joined to produce the fringe lines. In order to improve this process, a high-contrast film was used while photographing the fringes. The use of high-contrast film helped in reducing the fringe width. Efforts were also made to reduce the width of the isochromatic fringes optically. Post [1] reported a method using partial mirrors. The partial mirrors, which are parallel to each other, were inserted into the field of the polariscope on both sides of the model. The effect of the partial mirrors was to cause the light to propagate back and forth through the model and a portion of it was transmitted at each reflection point. Hence, the intensity of the ray passing back and forth through the model progressively decreased and caused fringe sharpening.

With the advent of the electronic revolution, photoelectric devices [2] were used to measure the darkest points (or the minimum intensity positions) those form the actual fringe contour. Brown and Hickson [3] and Frocht et al. [4] first fitted a photoelectric photometer to their polariscopes in the 1950s to facilitate the location of the skeleton of isochromatic and isoclinic fringes. The detector was mounted behind an aluminium plate with a central hole into which aperture plates of various sizes could be fitted. By moving either this structure or the projected photoelastic image along parallel planes, it was possible for them to examine any part of the fringe pattern using the photometer to a fineness limited by the size of the aperture. Now a distinction is to be made between the fringe *central line* and the *fringe skeleton*. When intensity is taken into consideration, one locates the minimum intensity points forming the fringe band, which are referred to as fringe skeleton points. This is the correct way of identifying the fringe contour in a mathematical sense.

Though photoelasticity is a whole field technique, in early methods, data was collected only at fringe locations. Thus, the quantum of data that can be collected is limited, if only a few fringes are seen. Hence, efforts were aimed at schemes to obtain fringe multiplication. Talukdar and Ghosh [5] achieved fringe multiplication by superposition of negatives. Post [6] achieved fringe multiplication by modifying the fringe sharpening apparatus such that one of the partial mirrors is slightly inclined with respect to the other and the model is immersed in a liquid bath having the same refractive index as that of the model. In general, one requires a high intensity light source for obtaining fringe sharpening and fringe multiplication optically.

In the early development of digital photoelasticity, DIP systems were essentially used to automate the procedures that were done either optically or manually. For example, techniques for fringe multiplication used the DIP hardware as a paperless camera and in fringe thinning methodologies fringe skeletonization is effected by using the DIP hardware as a sensor in identifying minimum intensity points. In this chapter, the techniques for fringe multiplication are discussed followed by algorithms for fringe thinning. Fringe clustering has two steps namely fringe tracing and fringe ordering. The digital techniques needed for these are discussed next.

4.2 Fringe Multiplication

It is seen in Chapter 1, that in a circular-polariscope dark-field arrangement, one observes integral fringe orders. The governing intensity equation is

$$g_d(x,y) \approx I_{\mathrm{d}} = I_{\mathrm{a}} \sin^2\frac{\delta}{2} \tag{4.1}$$

For bright-field arrangement, the fringe orders are observed in steps of 0.5 starting from 0.5 fringe order. The governing intensity equation is

$$g_\ell(x,y) \approx I_\ell = I_{\mathrm{a}} \cos^2\frac{\delta}{2} \tag{4.2}$$

If the loading is small or the optical response of the model material is very low, then one may not have sufficient information if only dark and bright-field images are recorded.

In fringe multiplication techniques, the intensity variation of the digitised image is appropriately changed to provide light extinction to occur at fractional fringe orders. Simplest of these techniques is by Toh et al. [7] who proposed that a simple digital subtraction of bright and dark-field images could result in fringe multiplication by an order of two (*frnmul.c*, CD). Subtracting Eq. (4.1) from Eq. (4.2), one gets

$$g(x,y) \approx I_a \cos\delta \tag{4.3}$$

In Eq. (4.3), the extinction of light will occur when $\delta = (2n+1)\pi/2$ i.e., fringe order N obtainable from the resultant image is equal to $(2n+1)/4$ where $n = 0, 1, 2,\ldots$. The resultant image is termed as a *mixed image* and the fringe orders are $N = 0.25, 0.75, 1.25,\ldots$. Figure 4.1 shows the fringe multiplication obtained for the problem of a disc under diametral compression. The *mixed image* does not have the original image characteristics of grey level feature [8] i.e., the *mixed image* has intensity variation as a function of $\cos\delta$ and not as $\cos^2\delta/2$.

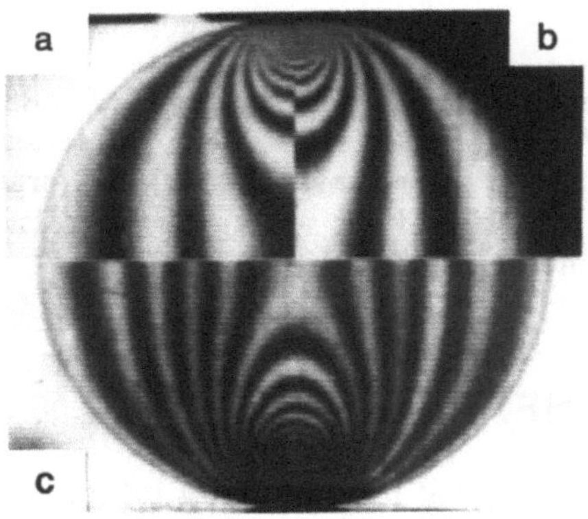

Fig. 4.1. Fringe multiplication by image subtraction of bright and dark-field images: **a** bright-field image **b** dark-field image and **c** mixed-image

In 1994, Chen [9] coined another approach for fringe multiplication, which used only the bright-field image. Referring to Eq. (4.2), if the amplitude of the incident light is known then by dividing it on both sides of the Eq. (4.2), the grey level value corresponding to the pixel can be thought of to represent $\cos^2\delta/2$. For digital implementation, the unloaded model is also to be digitised and image division of the loaded model is to be performed by the unloaded model, pixel by pixel. The normalised intensity of the fringe patterns is related to the retardation as

$$g_n(x,y) = \cos^2\frac{\delta}{2} \tag{4.4}$$

It is to be noted, that in view of the normalisation, the range of $g_n(x, y)$ is from 0 to 1. To obtain fringe multiplication, a grey level transformation is introduced, in which, the intensity varies as $\cos^2 m\delta/2$ where m represents the order of fringe multiplication. Since each pixel intensity represents $\cos^2 \delta/2$, the grey level transformation is effected by using the trigonometric identities. For fringe multiplication by two times, the transformed intensity $g_t(x, y)$ is obtained as

$$g_t(x,y) = \cos^2\delta = 4g_n^2(x,y) - 4g_n(x,y) + 1 \tag{4.5}$$

The extinction of intensity $[g_t(x, y)]$ occurs when $\delta = (2n+1)\pi/2$ for $n = 0, 1, 2$ .... The respective fringe orders observed will be 0.25, 0.75, 1.25 etc. The value of $g_t(x, y)$ evaluated by Eq. (4.5) also varies from 0 to 1 and this has to be mapped from 0 to 255 for generating a visual picture. For multiplication by three times, one has to introduce a grey level transformation as

$$g_t(x,y) = \cos^2\left(3\frac{\delta}{2}\right) = 16g_n^3(x,y) - 24g_n^2(x,y) + 9g_n(x,y) \tag{4.6}$$

With this transformation, one will have fringe orders in steps of 0.333 starting from a fringe order of 0.167. Fringe multiplication up to sixteen times has been reported in the literature. To extend this approach for three-dimensional stress frozen specimens, the bright-field setting of the polariscope without the slice is digitised for normalisation [10].

4.3 Half Fringe Photoelasticity (HFP)

Historically, one of the earliest approaches that used intensity information for fringe ordering was by Voloshin and Burger [11]. In this, the hardware feature of the B & W image processing system is effectively used to directly find the fractional fringe orders between 0 to 0.5. In view of this, the technique came to be known as half fringe photoelasticity (HFP). Since, for an 8-bit system, 256 grey level shades exists between pitch black and pure white, the method can be thought of to be a fringe multiplication technique with a factor of 512.

One-to-one correspondence between the digitised value and the fractional fringe order is established by properly calibrating the polariscope-camera system. Voloshin and Burger used a tube-based camera and hence, they proposed that the digitised value is non-linearly related to the brightness as

$$g(x,y) \propto I_d^{\gamma} \tag{4.7}$$

where γ is the slope of the vidicon tube sensitivity line on a log I_d vs. log $g(x, y)$ plot, I_d is the dark-field intensity of the image, $g(x, y)$ is the grey level value of the pixel.

Combining Eq. (4.7) and Eq. (4.1) one gets the fringe order N which is $\delta/2\pi$ as

$$N = \frac{1}{\pi}\sin^{-1}\left[kg(x,y)^{1/2\gamma}\right] \tag{4.8}$$

where, k accounts for the proportionality constant for Eq. (4.7) and I_a of Eq. (4.1).

The values of k and γ are found by measuring the fractional fringe order N and the grey level value $g(x, y)$ for any two points in the optical field. To find the fractional fringe order, Tardy method of compensation is recommended. To improve accuracy, the compensation is to be done digitally using the DIP system to identify the minimum intensity position.

While using a CCD camera, whose gamma is compensated for image processing, one can evaluate the fractional fringe orders, by performing image division that was done for fringe multiplication (Sect. 4.2). Replacing δ in terms of the total fringe order N in Eq. (4.4), one gets

$$g_n(x,y) = \cos^2 N\pi \tag{4.9}$$

By using inverse trigonometric operation in Eq. (4.9), it is possible to get fractional fringe orders between 0 and 0.5. By interactive ordering of full fringe orders, the technique can be made whole field [12]. This can be mathematically expressed as

$$N = \begin{cases} n + \dfrac{1}{\pi}\cos^{-1}(g_n(x,y))^{\frac{1}{2}} & n = 0,1,2,\ldots \\ n - \dfrac{1}{\pi}\cos^{-1}(g_n(x,y))^{\frac{1}{2}} & n = 1,2,3,\ldots \end{cases} \tag{4.10}$$

In view of its simplicity and ease of implementation, HFP has found wide acceptance and a wealth of literature is available for solving various problems [13–15].

4.4 DIP Methods for Fringe Thinning

Due to the nature of the formation of fringes, the intensity of pixels in the fringe band (dark/bright) does not remain constant but a variation does exist. Though the variation is not resolvable to a fine degree by a human eye, the present day CCD cameras can easily recognise it and quantize it for further processing. This has been brought out in Sect. 3.13.3.

Although the use of intensity information leads to a better approach for fringe thinning, the early methods were confined to treating the fringe field as a binary image [16–18]. The algorithms developed for optical character recognition (OCR) [19] were directly used for fringe processing. Later intensity based methods were developed [20–25]. These techniques can be further classified into *mask-based methods* and *global methods*.

4.5 Algorithms Based on Considering the Fringe Patterns as a Binary Image

In this approach, the fringe areas are initially identified by applying a suitable global threshold. The fringe centre lines are determined by a process of erosion, wherein the outer layers forming the fringe are progressively removed until only the fringe centre line is left. In view of progressive thinning, the approach is basically iterative in nature. It is to be noted that, in broad fringes the centre lines may not be the actual fringe location.

One of the most elegant methods for fringe thinning in this category is due to Chen and Taylor [18]. The binary image obtained after thresholding is scanned left to right, right to left, top to bottom and bottom to top sequentially to eliminate border pixels forming the fringe band. During each such scan, for every pixel with a grey-level-value below the threshold (a point on a fringe), a 3 × 3 pixel mask is considered (Fig. 4.2a) to eliminate the border pixel. For example, when the image is scanned from left to right, it is checked whether the point $g(0, 0)$ is an edge point. This is so when $g(-1, 0)$ is not a fringe point and $g(1, 0)$ is a fringe point. If $g(0, 0)$ is an edge point, then in a 3 × 3 mask there can be four combinations of pixels above and below it. The four combinations are

1. $g(0, 1)$ and $g(0, -1)$ are both fringe points,
2. $g(0, -1)$ is a fringe point and $g(0, 1)$ is not a fringe point,
3. $g(0, -1)$ is not a fringe point and $g(0, 1)$ is a fringe point or
4. Both $g(0, 1)$ and $g(0, -1)$ are not fringe points.

The point $g(0, 0)$ is eliminated if any of the first three conditions are satisfied. The fourth condition indicates that the point g(0, 0) forms a skeleton point in the horizontal direction and hence it has to be retained.

The eliminative conditions for scanning the image left to right, right to left, top to bottom and bottom to top are pictorially shown in Fig. 4.2b. In the pictorial representation, the filled circles are fringe pixels, the open circles are non-fringe pixels, and the filled triangle at the centre is the fringe pixel that is considered for elimination. The elimination of each border pixel makes the fringe a little bit thinner. This process is continued until no more fringes are eligible for elimination. A program segment

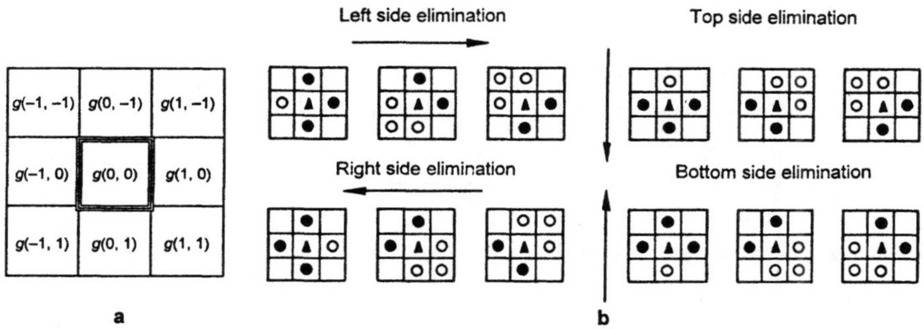

Fig. 4.2. **a** 3 × 3 pixel mask. **b** Pictorial representation of the eliminative conditions needed for scanning the image left to right, right to left, top to bottom and bottom to top. In the figure, filled circles are fringe pixels, the circles are non-fringe pixels and the filled triangles are the pixels that are considered for elimination

implementing the algorithm is included in the header file *thinpip.h* enclosed in the CD.

4.6 Mask-Based Algorithms for Skeleton Extraction Using Intensity Variation within a Fringe

Mask based methods basically use non-convolution type filters. The criteria for identifying minimum intensity, for a chosen mask, are specified appropriately for each technique. The mask is then moved over the complete image pixel by pixel to identify the fringe skeleton points.

One of the simplest approaches [20] is to use a 5 × 5 pixel mask (Fig. 4.3a) to detect fringe minima locally. With respect to the four directions shown in Fig. 4.3b, the minima conditions are defined for two typical directions 1 and 4, respectively, as

$$\begin{aligned} &g_{00} + g_{0-1} + g_{01} < g_{-21} + g_{-20} + g_{-2-1} \quad \text{and} \\ &g_{00} + g_{0-1} + g_{01} < g_{21} + g_{20} + g_{2-1} \end{aligned} \tag{4.11}$$

$$\begin{aligned} &g_{00} + g_{-11} + g_{1-1} < g_{22} + g_{21} + g_{12} \quad \text{and} \\ &g_{00} + g_{-11} + g_{1-1} < g_{-2-2} + g_{-2-1} + g_{-1-2} \end{aligned} \tag{4.12}$$

Figure 4.3c shows the pixels involved for each scan direction. The pictorial representation clearly brings out the meaning of the conditions represented by Eqs. (4.11) and (4.12) better and is self-explanatory. Similar to Eqs. (4.11) and (4.12) minima conditions can be easily written for directions 2 and 3.

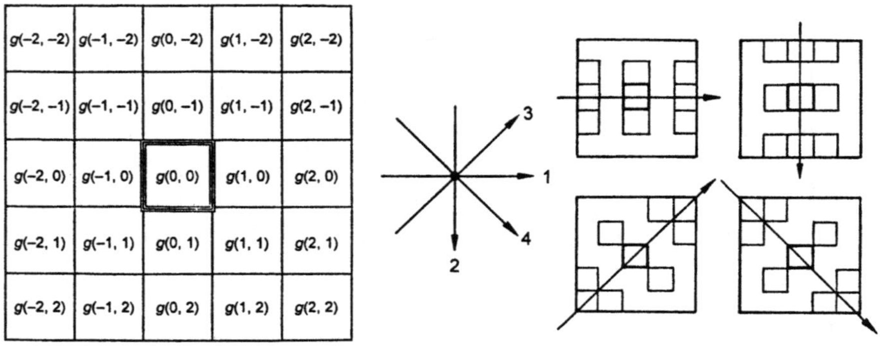

a b c

Fig. 4.3. **a** 5 × 5 pixel mask. **b** Scan directions needed for fringe minima detection in mask-based algorithm-I. **c** Pixels involved in each scan direction

When the minima conditions are satisfied for any of the two or more directions, the object point $g(0, 0)$ is recognised as a point on the fringe skeleton. The fringe skeleton thus obtained is, in general, not of one pixel width and in order to thin it further, it is suggested to use a binary based fringe thinning algorithm to iteratively determine the fringe skeleton. A program segment implementing the above algorithm is included in the header file *thinpip.h* (CD). In the implementation of the algorithm, the loop for detecting minimum intensity point is terminated if the conditions are satisfied for any two directions. The additional refinement by a binary-based algorithm is not incorporated.

A more exhaustive scan of the 5 × 5 pixel mask (Fig. 4.3a) is considered by Umezaki et al. [22]. In all, the intensity distribution for eight directions is obtained (Fig. 4.4a). For directions 1, 2, 3 and 4, this is the grey level value of the individual pixels in these directions. For directions 5, 6, 7 and 8, the grey level of the individual pixels does not directly give the intensity distribution. Average of neighbouring pixel values those are hatched in Fig. 4.4a also has to be computed. For a typical direction 6, the intensity distribution is

$$(g_{-2-1}, (g_{-10} + g_{-1-1})/2, g_{00}, (g_{11} + g_{10})/2, g_{21}) \tag{4.13}$$

Point $g(0, 0)$ (Fig. 4.3a) is selected as fringe point and made white if it is a point of minimum intensity for at least two or more of the scan directions. To decide whether the point $g(0, 0)$ is of minimum intensity for a particular scan direction, one has to check whether it satisfies any one of the five conditions shown in Fig. 4.4b. A program segment for implementing this algorithm is included in the header file *thinpip.h* (CD). In the implementation of the algorithm, the loop for detecting minimum intensity point is terminated if the conditions are satisfied for any two directions.

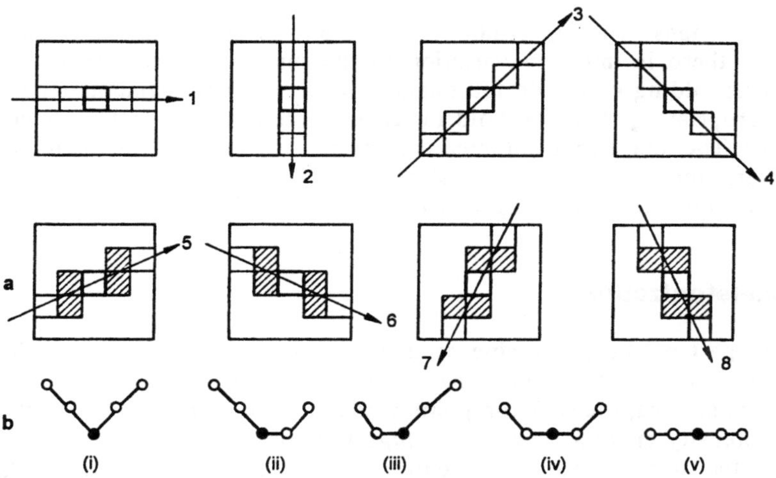

Fig. 4.4. a Scanning scheme used in the mask-based algorithm-II. **b** The five possible intensity variations along each scan direction

Wei-Chung Wang [25] reported a technique called ERBP (extraction of relative bright point) which is similar to the one reported by Umezaki et al. but determines the brightest point rather than the pixel minima. Gillies [21] reported a *differential zero crossing* algorithm which used very complex edge detection operators for processing photoelastic fringes. The technique is computationally highly demanding as masks of size 11×11 and more are used and further, the results obtained are poorer in comparison to the algorithms discussed here.

4.7 Global Identification of Fringe Skeletons Based on Intensity Variation

In general, photoelastic images have high contrast and the edges of the fringes can be extracted easily. Further, the use of edge information to identify the fringe skeletons eliminates the occurrence of spurious fringes. Thus in global methods, fringe thinning involves the two steps of edge detection and fringe skeletonization.

4.7.1 Edge Detection

Edge detection is an important step. The better is the edge detection algorithm; the better will be the skeleton identification [26]. It is seen in the previous chapter, that edge detection by thresholding works very well for photoelastic images. The algorithm is simple and is also computationally very fast. If there is intensity variation in the image one can use either dynamic thresholding or divide the image into convenient segments (tiles) and use appropriate global threshold for each of the segment [27] to detect fringe edges. The image thus obtained is to be stored in one of the available frame buffers and labelled as the *edge image* for further processing. In practice, the *edge image* can simply be a semi-thresholded image.

4.7.2
Fringe Skeletonization

The second part of the processing sequence is the skeletonization of the fringes. For skeletonization, the fringe areas are to be identified first and within the fringe area, the skeleton point needs to be identified based on the minimum intensity criteria. The *edge image* previously stored can be used to identify the fringe areas. To identify the skeleton point, the pixels between the edges of the fringe need to be scanned appropriately. One of the simplest scanning schemes is row-wise scanning of the image. Let us look at the basic steps needed for employing this approach.

4.7.2.1
Row-Wise Scanning Algorithm

This algorithm is applicable for fringes that are vertical or almost vertical (Fig. 4.5a). The image is scanned row-wise. Briefly, the steps involved are as follows:

1. Scan the edge image (stored after the edge detection) row-wise from the first pixel to identify the fringe edge
2. The scanning is carried out from one edge of the fringe to the other edge. Store the grey level values of the pixels within the boundaries of the fringe. Identify the minimum intensity point and mark it as a fringe skeleton pixel.
3. Steps (1) and (2) are repeated until all the pixels in the image are scanned and at the end of the process, the thinned image is obtained.

A program segment implementing the algorithm is as follows:

Program 4.1. Row-wise thinning algorithm

```
void rowthin(r,s,w,t)
int r,s,t,w;
{
int i,j,k,c,cext,count,min_ind,getmin,xcord=0;
 for (j=qd2.y+winxy[1]+1;j<=qd2.y+winxy[3]-1;++j)
  { count = 0;  getmin = 0;
   for (i=qd2.x+winxy[0]+1;i<=qd2.x+winxy[2]-1;++i)
    { c = fg_pixr(i,j); cext = fg_pixr(((r*512)+i),((s*512)+j));
```

```
        if ((c >= 250) && (count > 0)) getmin = 1;
        if (c < 250) { inten[count]=cext; x[count]=i; count++;
                    if (i == qd2.x+winxy[2]-1) getmin = 1; }
        if (getmin == 1) { xcord=x[0]; min_ind=inten[0];
          for (k = 0; k < count ; ++k)
          { if ( min_ind > inten[k] ) { min_ind=inten[k]; xcord=x[k];}}
          if ((w==0) && (t==0)) min_ind = 250;
          fg_pixw(((w*512)+xcord),((t*512)+j),250);
          getmin = 0; count = 0;}
      } // pixels in a row
    } // next row
}
```

4.7.2.2 Algorithm for Fringe Skeleton Extraction for Arbitrarily Shaped Fringes

If the scanning is made column-wise, the algorithm can be applied to extract fringe skeletons which are primarily horizontal (Fig. 4.5b) such as those occur in beams subjected to pure bending. Figure 4.5c shows the geometry of an isochromatic fringe pattern that is observed near a crack tip. To process the fringe pattern in Fig. 4.5c, one requires a more refined scanning method taking into account the curvature of the fringes. This requires the evaluation of the edge normal from geometric considerations of the fringe. Though the idea is conceptually simple, it is very difficult to implement from the software point of view. Even if it is implemented, one has to do an enormous amount of computations to extract the fringe skeletons.

Fringe skeletons that are continuous and free of noise can be easily extracted by appropriately processing the results of fringe skeletons obtained by a finite number of orthogonal scans. Once the edges are deter-

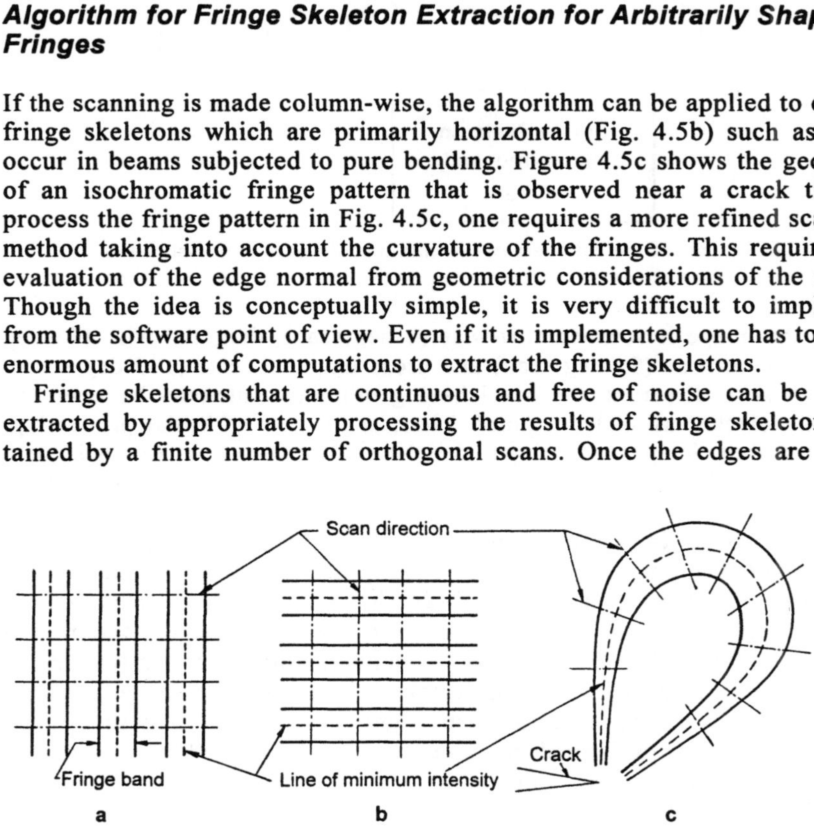

Fig. 4.5. Scan directions needed to identify fringe minima for various fringe types: **a** fringes that are nearly vertical **b** fringes that are nearly horizontal and **c** fringes that have a general shape with fringe thickness varying over it

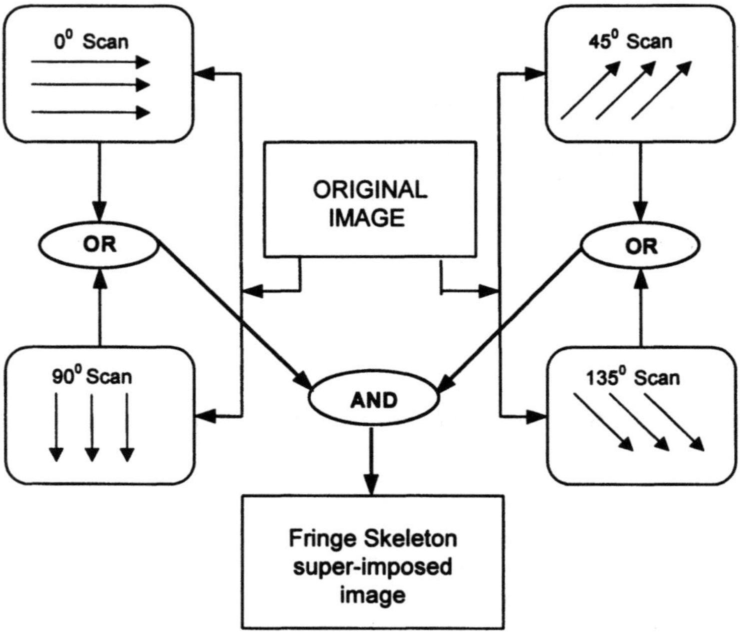

Fig. 4.6. The scheme of logical operators to obtain continuous fringe-skeleton

mined, the image is scanned row-wise (0-deg scan), diagonal-wise (45-deg scan), column-wise (90-deg scan) and cross diagonal wise (135-deg scan). For each scan direction, the pixel having minimum intensity between the edges is selected as skeleton point. Thus, at the end of initial processing, one gets four images of fringe skeletons corresponding to each scan direction. It is to be noted that these scans are done globally and are not for each individual mask as seen earlier. Noise points which are scan direction dependent, appear in each of the above scans. Further, the fringe skeleton is not continuous in one scan. The discontinuity is also scan direction dependent. Ramesh and his co-workers [23–24] have shown that the use of the logical operators shown in Fig. 4.6 can eliminate the fringe discontinuity and noise. The logical "OR" operation between the orthogonally scanned images helps to get a continuous fringe skeleton but the image contains noise. The noise is scan direction dependent and is removed by the logical "AND" operation performed between the logically "OR"ed images. A header file incorporating this scheme is given in the enclosed CD (*orthopip.h*).

In digital image processing, it is usually difficult to foresee completely the performance of a proposed algorithm and one gets a grasp of its success or bottlenecks only by actually testing it on sample images. The algorithm has to be evolved by solving a set of intelligently chosen test images.

In order to test the above algorithm, let us take a computer generated grey level image as shown in Fig. 4.7. The image can be generated by the code *annul.c* enclosed in the CD and let the image be called as ANNUL.

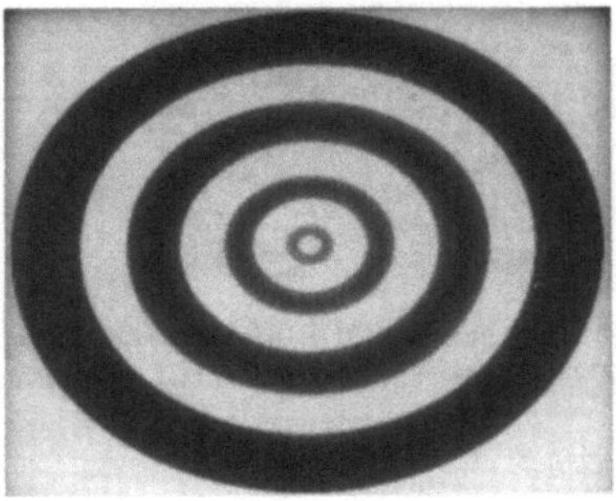

Fig. 4.7. Test image ANNUL showing fringes of different thickness

The test image has concentric circles (appear as ellipses due to the aspect ratio of the monitor) and width of each circle varies from 10 pixels (inner

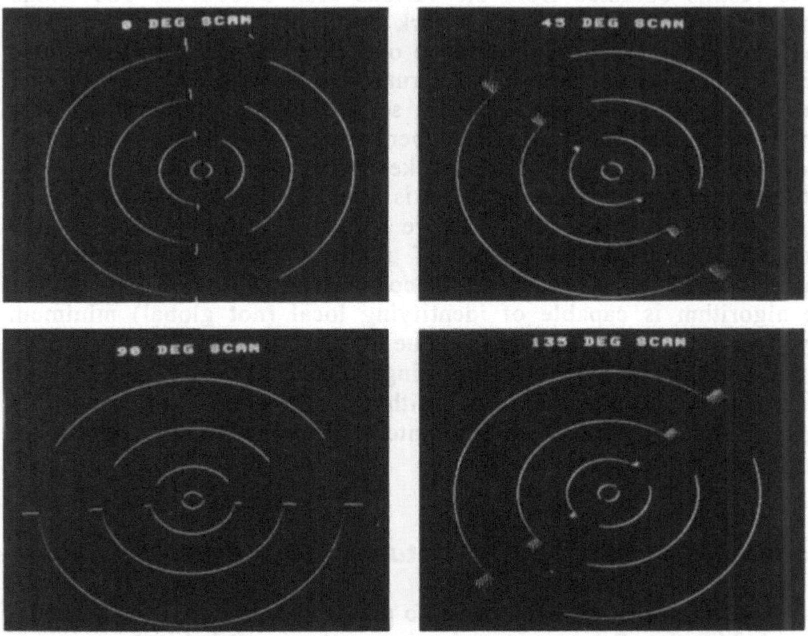

Fig. 4.8. Fringe skeletons that are obtained by 0-, 45-, 90-, and 135-deg scans. The discontinuities and noise are scan direction dependent

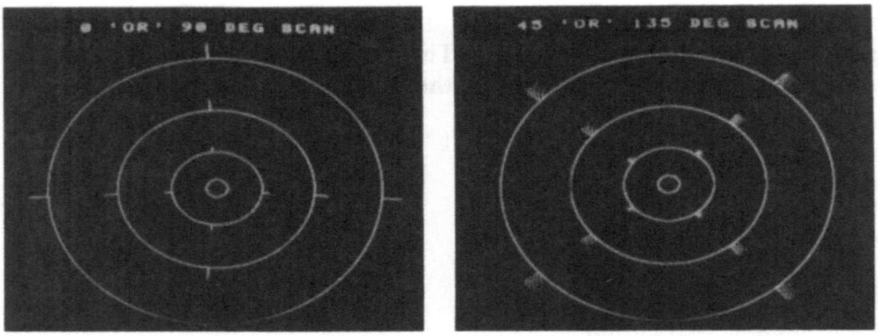

Fig. 4.9. Fringe skeleton obtained by **a** 0-deg OR 90-deg scans and **b** 45-deg OR 135-deg scans. In each case, the fringe skeleton is complete, however noise is still present. The noise is different in **a** and **b**

most) to 50 pixels (outermost). Further, the minimum intensity is not the same for each circle and varies from a grey level value 75 to a grey level value 95 from the outermost to the innermost circle in steps of 5 grey level values. Due to this variation, fringe skeletons cannot be extracted by specifying a single value of the minimum intensity globally. Figure 4.8 shows the results obtained from the various scan directions. The fringe skeletons are shown white against a dark background. Each of the scans show that the fringe is not continuous in one scan and also spurious noise exists in each of these scans. A close scrutiny of these images reveals that both the discontinuity and the noise are scan direction dependent. Figure 4.9 shows that logical "OR" operation performed between two orthogonal scans helps to get the complete fringe skeleton. However the noise is not eliminated at this stage. Since the noise is scan direction dependent, when the images in Fig. 4.9a and Fig. 4.9b are logically "AND"ed, one obtains the complete fringe skeleton free of noise. The fringe skeleton is superimposed on the original image for comparison (Fig. 4.10). It is clear that the algorithm is capable of identifying local (not global) minimum intensity positions correctly. This is due to the fact that the algorithm locates minimum intensity between the fringe edges.

Thus, it is seen that the present algorithm is straightforward, applicable to fringes of any orientation and guarantees fringe skeleton of one pixel width free of discontinuities and noise.

4.7.3 Applications of the Algorithm to Actual Experimental Conditions

Unlike the test images, one requires to do pre-processing of the experimentally recorded images. Fringe processing is greatly simplified if the captured image is free of random noise. In order to eliminate electrical noise, it is usually recommended to take up to 16 images of the fringe pattern under

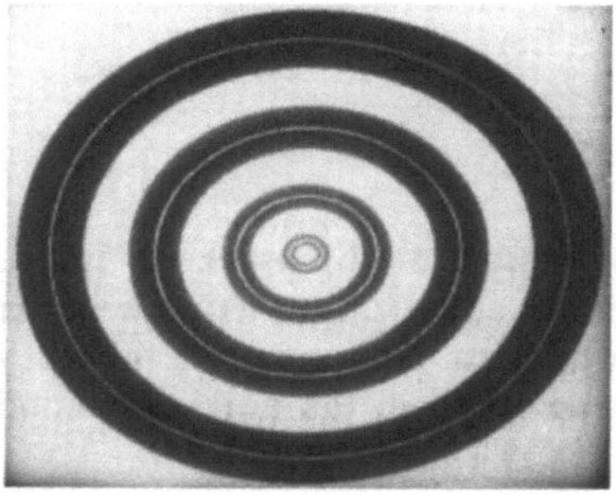

Fig. 4.10. Fringe skeleton obtained by logical AND operation of Figs. 4.9a and 4.9b. The fringe skeleton is superimposed on the original fringe pattern for comparison

consideration and perform time averaging of these images. Figure 4.11 shows the time-averaged image of crack-tip isochromatics. The pre-processing step includes noise suppression or fringe enhancement programs, which has to be properly employed according to the need. Repetition of spatial averaging followed by histogram equalisation three times is found to be sufficient for a variety of photoelastic images [23]. The following program has three macros to achieve this process. The macros have to be invoked sequentially.

Fig. 4.11. Time-averaged image of crack-tip isochromatics

Program 4.2. Spatial Averaging followed by Histogram Equalisation – Use of Macros

```
//clean1.pip (Macro 1)
average 0 0
histo 5200
//clean2.pip (Macro 2)
dq 1
setindex 0
cl 0 7
setindex 255
dhisto %1 10 962 400 255 255 5200
pause
dq 0
```

```
//clean3.pip (Macro 3)
modhist 0 %1 %2 0 5200 5200
dq 0
map 0 0 5200
```

Figure 4.12 shows the sequence of the histogram of the crack-tip image after each step of local averaging and histogram equalisation using a linear curve fitting. It is recommended that the optimum increment of the grey level values on either side of the histogram be taken in the neighbourhood of 10. The pre-processed image is shown in Fig. 4.13.

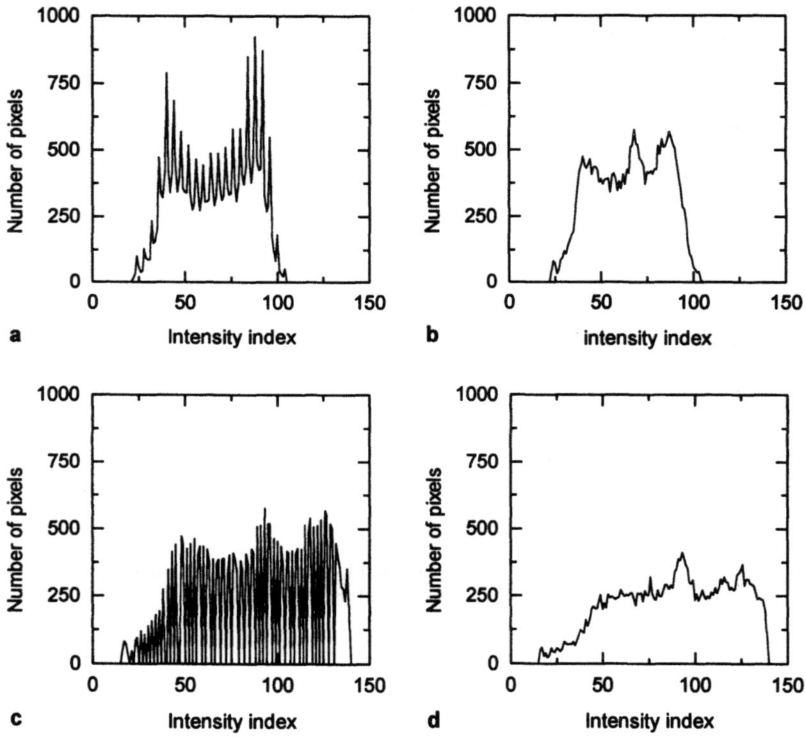

Fig. 4.12. Sequence of histograms after each stage of spatial averaging and histogram equalisation

Figure 4.14a shows a portion of the pre-processed image taken up for skeletonization and Fig. 4.14b shows the intensity variation along a typical vertical line close to the crack-tip. The intensity variation shows that a single global threshold cannot be used for retaining all the features of the image. Thus, it is desirable to use dynamic thresholding.

Fig. 4.13. Final pre-processed image of crack-tip isochromatics

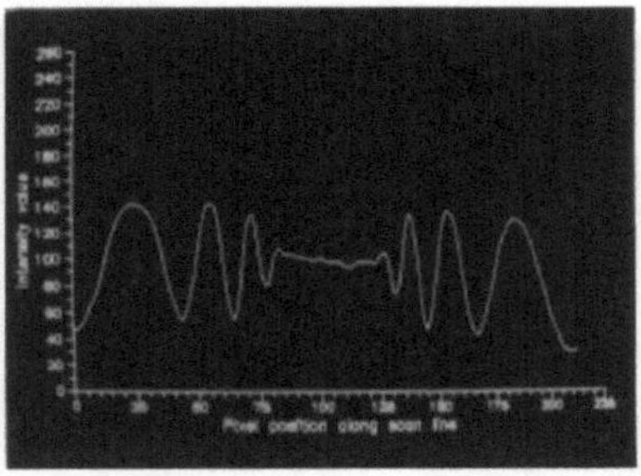

Fig. 4.14. **a** Portion of digitised image selected for further processing. **b** Intensity variation along the line shown in **a** as a function of pixel

Following this, semi-dynamic thresholding was employed on the image using the parameters $T_{max} = 140$, $T_{min} = 0$, and $\alpha = 3.5$. Figure 4.15 shows the fringe skeleton identified for 0-deg, 45-deg, 90-deg and 135-deg scans. Figures 4.16a and b show the 0-deg OR 90-deg, and 45-deg OR 135-deg image respectively. Figure 4.17a shows the final fringe-skeleton extracted. The intensity variation (Fig. 4.17b) along the line shown in Fig. 4.17a clearly indicates that the skeletonization algorithm is effective in extracting local minimum (not the global minimum) corresponding to the fringe point.

4.8 Further Improvements on the Global Thinning Algorithm

The procedure of identification of the minimum intensity point in the global thinning algorithm is developed with the premise that within a fringe, only

Fig. 4.15. Fringe skeletons that are obtained by 0-, 45-, 90-, and 135-deg scans

Fig. 4.16. Fringe skeleton obtained by **a** 0-deg OR 90-deg scans and **b** 45-deg OR 135-deg scans

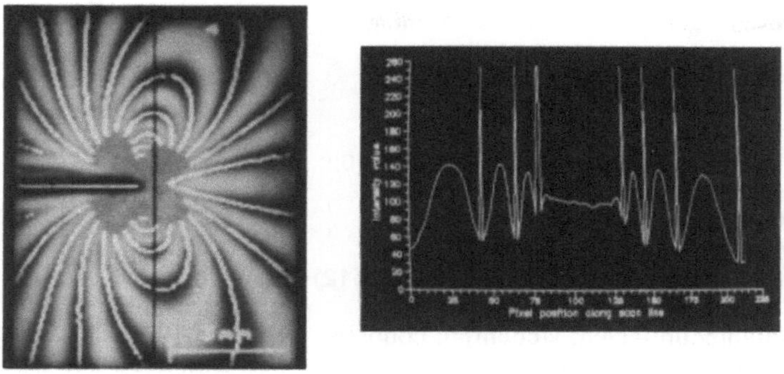

Fig. 4.17. **a** Fringe skeleton obtained by logical AND operation of Figs. 4.16a and 4.16b. The fringe skeleton is superimposed on the original image for comparison. **b** The intensity variation along the line shown in **a** is plotted as a function of the pixel location

one minimum intensity band can occur. However, situations occur particularly in the case of a *saddle point*, there can be more than one minimum intensity band along a scan direction within the fringe. This happens along line 2 in Fig. 4.18. Along line 1 the two branches are identified as two separate fringes and fringe minima would be identified correctly by the algorithm of Sect. 4.7.2.2. However, along line 2 it would only identify the second band as forming the fringe skeleton. This would result in fringe discontinuity. Even within the fringe band, there may be situations wherein more than one pixel may have the same minimum intensity forming a band. If the band is of odd number of pixels, the centre pixel is to be identified as the skeleton point and if it is of even number of pixels, then two pixels at the centre are to be identified forming the fringe skeleton. The improved version of this algorithm for row-wise scanning is as follows:

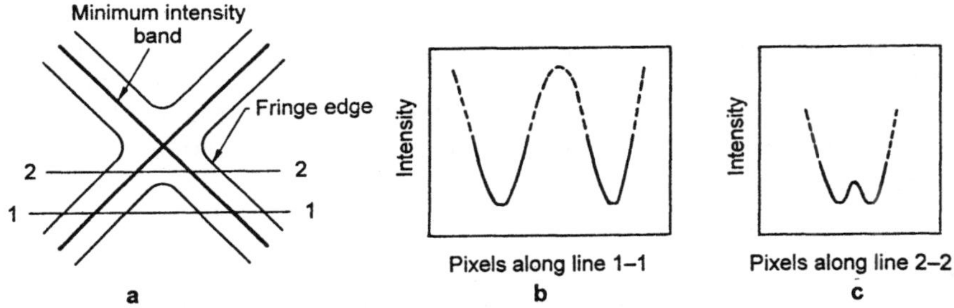

Fig. 4.18. Figure showing the saddle point and intensity distribution along lines 1 and 2. In **b** and **c**, the solid lines represent the intensity distribution within the fringe band and the dashed lines represent the intensity distribution outside the fringe band

Program 4.3. Improved algorithm for row-wise thinning

```
void rowthin(r,s,w,t)
int r,s,t,w;
{
int i,j,k,c,cext,count,min_ind,getmin,xcord=0,count2=0;
 for (j=qd2.y+winxy[1]+1;j<=qd2.y+winxy[3]-1;++j)
   { count = 0;  getmin = 0;  count2=0;
     for (i=qd2.x+winxy[0]+1;i<=qd2.x+winxy[2]-1;++i)
       {  c = fg_pixr(i,j);  cext = fg_pixr(((r*512)+i),((s*512)+j));
      if ((c >= 250) && (count > 0)) getmin = 1;
      if (c < 250) { inten[count]=cext; x[count]=i; count++;
             if (i == qd2.x+winxy[2]-1)  getmin = 1; }
      if (getmin == 1) { min_ind=inten[0]; count2=0;
         for (k = 1; k < count ; ++k)
         { if ( min_ind > inten[k] )  { min_ind=inten[k]; } }
         // finds initially the min_ind
         for (k = 0; k < count ; ++k)
          { if ((inten[k] == min_ind) && (count2 == 0))
         { xcord=x[k]; } if (inten[k] == min_ind) { count2++; }
        if (((inten[k] > min_ind) && (count2 > 0))
               || ((count2 >0 ) && (k == (count-1))))
         { fg_pixw(((w*512)+(xcord + count2/2)),((t*512)+j),250);
/*actual plotting is done here. k == (count-1) accounts for the case when
the last point on the fringe has the value min_ind. inten[k] > min_ind and
count2 > 0 accounts for the case if there are multiple minima segments of
value min_ind within the fringe*/
           if ((2*(count2/2) - count2) == 0 )
           fg_pixw(((w*512)+(xcord + count2/2-1)),((t*512)+j),250);
           count2 = 0; } } //if count2 is even prints one more pixel
        getmin = 0; count = 0;
              } //getmin == 1
       } // pixels in a row
   } // next row
}
```

On similar lines one can modify the header file *orthopip.h* for other scan directions. The improved version of this algorithm will be referred to as global thinning algorithm in the text hereafter. Thus, the evolutionary aspect of developing a DIP algorithm for a specific application is brought out here.

4.9 Performance Evaluation of Various Fringe Thinning Algorithms

It is desirable to make a comparative study of the various thinning algorithms discussed. The best way to assess the performance is that all the algorithms must be made available on one image processing system and be used to extract the fringe skeletons for the same set of images. A program code to implement this is given in the accompanying CD (*compare.c*).

The software *compare.c* returns the total processing time for each of the algorithms including the time taken for interactive operations (if any). While using the binary based algorithm, after ten iterations, the user has to interactively permit the algorithm to continue for another two iterations if need be.

4.9.1
Comparison of the Skeleton Extraction

4.9.1.1
Computer Generated Test Images

The test images should be such as to check the influence of the following aspects on the identification of the fringe skeleton point by various fringe skeletonization algorithms.

1. Overall fringe thickness.
2. Curvature of the fringe or the applicability of the algorithm to fringes of any orientation.
3. Performance of the algorithms near saddle points.
4. Influence of minimum intensity band rather than a minimum intensity point within the fringe.

To check the items (1) and (2), the test image ANNUL used earlier in Sect. 4.7.2.2 is good. Figure 4.19 shows the results of the skeletonization by the various algorithms. At higher fringe thickness, the binary-based algorithm introduces an extra branch, other than the actual skeleton, at 0° and 90°, whose length is a function of the fringe thickness. The overall performance of the mask-based algorithms (Figs. 4.19a and 4.19b) are good. However, they are sensitive to the curvature of the fringes and at certain curvatures, the skeleton is of more than one pixel thick. This feature is observable on close scrutiny. In Fig. 4.19b, in the neighbourhood of $\pm 45^{\circ}$, fringe skeleton is not of one pixel width. Only the global thinning algorithm is independent of fringe thickness and curvature in identifying the actual fringe skeleton.

To check the items 3 and 4, let us take two more images namely, PLUS1 and X3. The number indicates the width of the minimum intensity band and the name signifies the shape of the fringe pattern. The intensity variation across the fringe width for these images is simulated to be sinusoidal and the program codes to generate these are named respectively as *plus1.c* and *x3.c* in the accompanying CD.

Figure 4.20 show the results of PLUS1 by various algorithms. Only the global thinning algorithm (Fig. 4.20d) gives the correct representation of the fringe skeleton. Binary-based algorithm (Fig. 4.20c) also gives a reasonable representation of the fringe skeleton. But, there is loss of information at the beginning and at the end of the skeletons. This can be

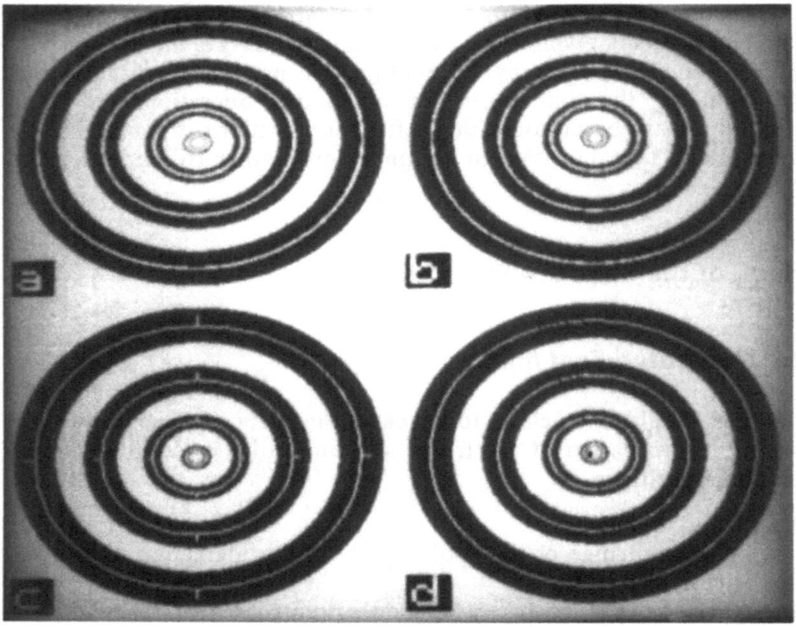

Fig. 4.19. Skeleton superimposed image for the test image ANNUL obtained by **a** mask-based algorithm-I **b** mask-based algorithm-II **c** binary-based algorithm and **d** global fringe-thinning algorithm

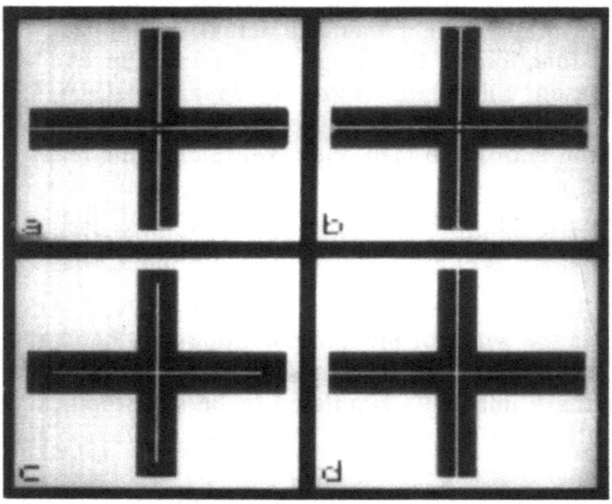

Fig. 4.20. Skeleton superimposed image for the test image PLUS1 obtained by **a** mask-based algorithm-I **b** mask-based algorithm-II **c** binary-based algorithm and **d** global fringe-thinning algorithm

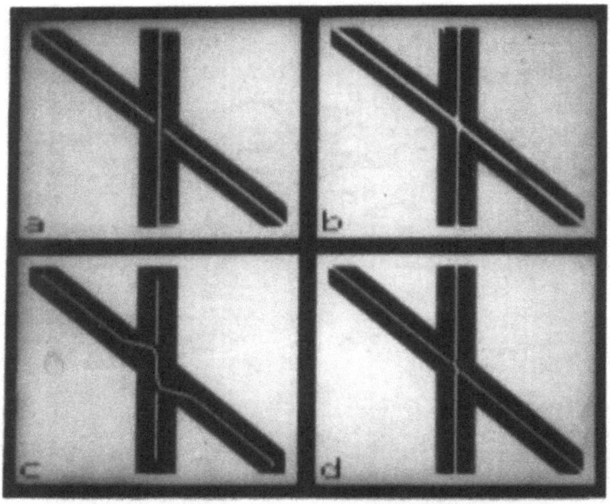

Fig. 4.21. Skeleton superimposed image for the test image X3 obtained by **a** mask-based algorithm-I **b** mask-based algorithm-II **c** binary-based algorithm and **d** global fringe-thinning algorithm

reasoned as follows. Consider the top portion of the vertical fringe. The topside elimination step continues to remove all the pixels until a stage occurs when left side and right side elimination thins the fringe to one pixel. Thus, the loss of fringe skeleton length is equal to half the width of the fringe. The performances of the mask-based algorithms are particularly not satisfactory at the centre and at the beginning and at the end of the fringe skeletons.

Figure 4.21 shows the results corresponding to the image X3. The results bring out the poor performance of the binary-based algorithm. The performance of the mask-based algorithms is again not satisfactory at the centre and also at the beginning and at the end of the skeletons. Further, on the inclined branches of the fringe, the skeleton is not of one pixel width. Here again, the advantages of the global thinning algorithm are seen. However, there is a small aberration at the centre of the fringe pattern X3 that is minimal in comparison to other algorithms.

In all the test images mentioned above, the minimum intensity point is also forming the centre of the fringe. The coincidence of the fringe skeleton extracted by the binary-based algorithm with other algorithms is due to this. If however, the centre point were not the minimum intensity point, then errors would be introduced in the exact location.

4.9.1.2 Image Recorded from Actual Experimental Situations

In order to test the performance of the above algorithms for actual experimental situations; let us consider the image ring under diametral compress-

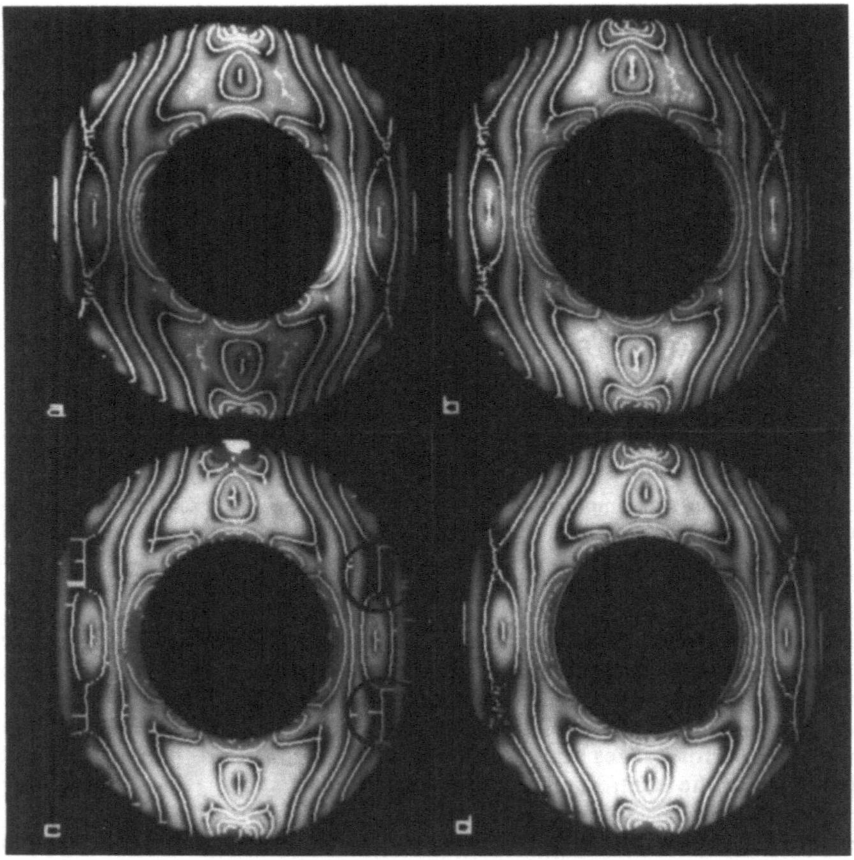

Fig. 4.22. Skeleton superimposed image for the experimentally recorded image RING obtained by **a** mask-based algorithm-I **b** mask-based algorithm-II **c** binary-based algorithm and **d** global fringe-thinning algorithm

sion (RING). The image is pre-processed with time averaging followed by contrast enhancement using histogram equalisation.

Figure 4.22 shows the results for the RING. The results of the binary-based algorithm have characteristic noise in the form of extra branches of the fringe. This is due to the variable thickness of the fringe patterns. It has been pointed out earlier that the results of the binary-based algorithm would be in error if the central point were not the minimum intensity point. This is clearly seen at the four saddle points (circles marked in Fig. 4.22c) of the RING. In comparison to other algorithms, the skeleton points are shifted inwards in the case of the binary-based algorithm. The mask based algorithms are sensitive to variations in the intensity of pixel in the white regions of the fringe patterns and have identified skeleton points in this region, which is, however, not desirable. Further in general, the fringe skeleton is not of one pixel width in the results obtained by these algorithms. In this context, the overall performance of the global thinning algorithm stands out and is definitely the best.

4.9.2 Comparison of the Computational Effort

Analysing the way in which each algorithm proceeds to identify the skeleton points, one can make certain observations on the required computational effort. For each scan direction, in the case of the global thinning algorithm, the computations necessary for identifying the pixel having minimum intensity is done only once between the fringe edges. This approach reduces the number of computations by an order of magnitude in comparison to other algorithms. For example, if a fringe has 10-pixel width, along scan direction, say, 1 (Figs. 4.3 and 4.4), the mask-based algorithms have to do the computations at least ten times for this scan direction alone to identify the minimum intensity point. Further, these algorithms do not recognise fringe areas initially and all the operations have to be done for the entire image irrespective of fringe density. Since the computations in global thinning are done only within the fringe edges, the computational effort is proportional to fringe density. Thus, enormous amount of time saving is achieved. Further the procedure for removing noise and discontinuities involves logical operations globally and computational effort is minimal for this. Thus, it is clear that the global thinning algorithm is more superior computationally. Obviously, the binary-based algorithm being iterative in nature would take more time than the other algorithms. In this, the total number of iterations depends on the maximum fringe thickness.

The above observations are reconfirmed by the results shown in Table 4.1. It is clear that the computational time (time per pixel) for the global thinning algorithm is less by an order of magnitude in comparison to other algorithms and the time taken by the binary based algorithm is the maximum. Further, the time required for processing is not only a function of the size but also depends on the nature of the image. It is to be noted that the results shown in Table 4.1 correspond to a PC-XT based image processing system and the actual processing time will be less if faster machines are used. Nevertheless, the relative comparison will still hold true.

4.10 Use of Tiling to Improve Information in Stress Concentration Zones

Information in stress concentration zones is usually lost due to the digitisation process. Further loss of information is possible, on the basis of the thinning scheme being employed. From the previous sections, it is clear

Table 4.1. Comparison of the time taken by various fringe thinning algorithms to process images on a PC-XT based system

Image	Size (pixels)	Mask-based algorithm-I		Mask-based algorithm-II		Binary-based algorithm		Global thinning algorithm	
		a (s)	b (s)	a (s)	b (s)	a (s)	b (s)	a (s)	b (s)
ANNUL	441 × 441	2501	0.013	2174	0.011	10 460	0.054	781	0.004
PLUS1	160 × 160	323	0.013	202	0.008	597	0.023	78	0.003
X3	160 × 160	323	0.013	195	0.008	539	0.021	110	0.004
RING	320 × 460	1855	0.013	2427	0.016	4833	0.033	620	0.004

a, Total time; *b*, Time per pixel.

that the global thinning approach is quite effective to identify fringe skeletons. In global thinning, additional information is lost due to the use of a single global threshold for edge detection. Because the edges are not resolved in the stress concentration zones, the fringe skeletons cannot be extracted in this region. Providing appropriate optical magnification, which is discussed in greater detail in Chap. 6, can reduce the loss of information in digitisation. Here, the advantage of tiling to improve the detection of edges better and thus extract more information is discussed.

The basic principle of tiling is that one can divide the image into several convenient windows and for each of them one can select the appropriate process parameters. In the context of fringe thinning, the process parameter is the threshold value for fringe identification and also the selection of appropriate scanning scheme. What is important in tiling is the book keeping of the information. This is greatly facilitated by the use of frame buffers as the intermediate results can be stored and accessed easily. A software code for fringe skeleton extraction by tiling is enclosed in the accompanying CD (*tilepip.c*)

In general, the high fringe density zone is brighter. Hence, to extract fringe edges one should use a higher threshold value. Figure 4.23a shows the image of crack-tip isochromatics for a radial crack, emanating from the outer boundary of a pressurised cylinder, processed using dynamic thresholding. The information very close to the crack-tip is not retained even with dynamic thresholding. Using the method of tiles this portion of original image is reprocessed with a new threshold value. The improvement achieved is shown in Fig. 4.23b. Further local discontinuities in the fringe pattern of Fig. 4.23a due to optical noise is also corrected with the method of tiles. Figure 4.23c shows the various tiles used for this purpose. The method of tiling offers a lot more flexibility to the user. Within a tile, one can choose any of the individual scans (0-deg, 45-deg, 90-deg or 135-deg)

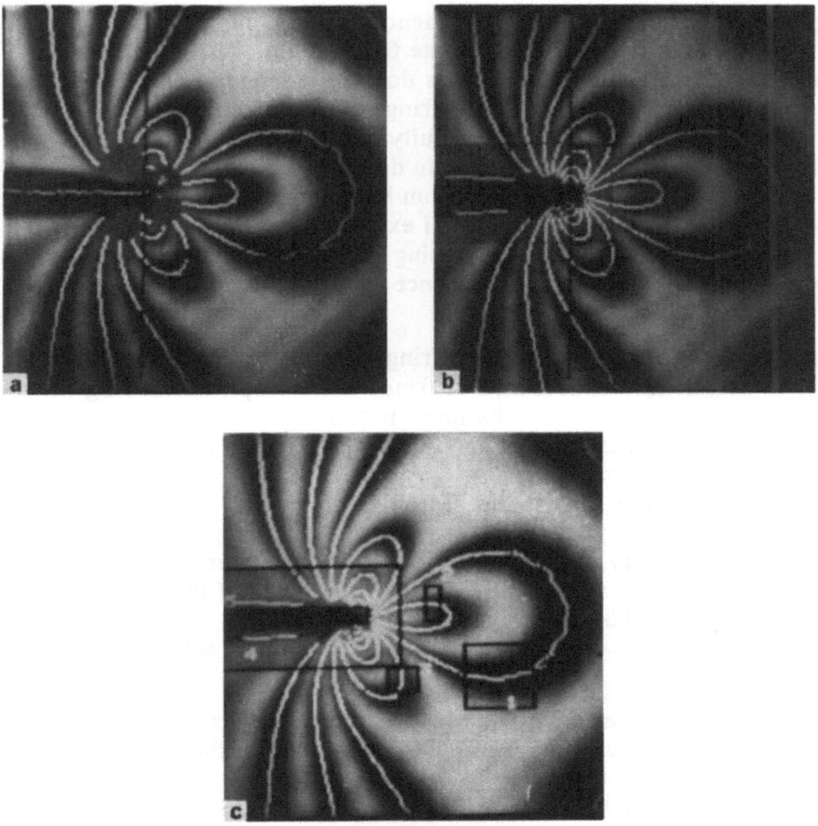

Fig. 4.23. Fringe thinning obtained by global thinning algorithm: **a** without tiling **b** skeletons improved by tiling and **c** figure showing the tiles used

to give better results corresponding to the orientation of the fringe. In Fig. 4.23c the tile no. 1 is processed with 90-deg. scan to remove the discontinuity.

4.11 Fringe Tracing Algorithm

To facilitate data collection, algorithms for fringe tracing and fringe clustering are needed. The fringe-tracing algorithm has to perform the task of identifying pixels forming a particular fringe so that data collection can be automated. The basic methodology is, if a fringe point is selected, the algorithm must be in a position to identify all the contiguous points forming a fringe. In fringe thinning algorithms, the fringe skeleton points are made white (250) and superimposed on the digitised image. Thus, in fringe tracing, the algorithm should simply identify contiguous points having an intensity of 250 and assign an appropriate fringe order.

This is actually a simple task, if it is done by a human being, as he can easily grasp the overall picture of the fringe skeleton and can also account for fringe thickness variation, discontinuity etc. However, to implement this digitally, one has to be very systematic in devising the algorithm.

Let us evolve a fringe tracing algorithm based on a mask of size 3 × 3. In general, more than one fringe pixel will exist in the mask. This is so if the fringe has several branches. The tracing algorithm must be capable of tracing all these branches. The sequence of steps involved in tracing a fringe contour is as follows:

1. Starting from a point $g(x, y)$ on a fringe (fringe order N), the algorithm has to search for the presence of fringe skeleton pixels (having intensity value equal to 250) in its immediate neighbourhood. Every pixel in the 3 × 3 mask is scanned and the fringe skeleton pixels determined are stored in an array.
2. Among the stored pixels, the search is continued in an ordered sequence. The x and y co-ordinates are set to the new values (x', y').
3. The pixel intensity of the point $g(x', y')$ is then changed from 250 to the intensity value numerically equal to the order of the fringe, which can be recognised by the program.
4. The steps (1), (2) and (3) are repeated for the new pixel location $g(x', y')$.

These steps are repeated until the end of the fringe is reached. This is indicated by the absence of a fringe point among any of the neighbourhood pixels within the mask.

When the end point of the fringe is encountered, the last point on the array is read, and the mask is set at this point and that part of the fringe is now traced. On encountering any additional branch points, they are stored to the array and the procedure is repeated until all those parts of the fringe are traced out. The count of the node points is decremented after the search is over. The mask is set at the previous branch point and the tracing continues. Thus, the branches are traced out in a reverse order i.e., the first node point being the last to be traced and so on. The program code is enclosed in the accompanying CD under the name *tracepip.c*.

It is to be noted that the algorithm is expected to fail at fringe discontinuities as no special care is taken in devising the algorithm. Figure 4.24a shows the application of the tracing algorithm for the problem of a ring under diametral compression. For each fringe only one point is interactively selected and assigned the fringe order. The tracing algorithm traces out the fringe, replacing the white skeleton points with grey level of zero. The algorithm has failed at a few points. This is due to the fact that the fringe skeleton is discontinuous which is visible at higher magnifications of the picture (Figs. 4.24b and 4.24c). The use of a dilation filter followed by an erosion filter will bridge such gaps and is to be applied to the skeleton superimposed image before fringe tracing.

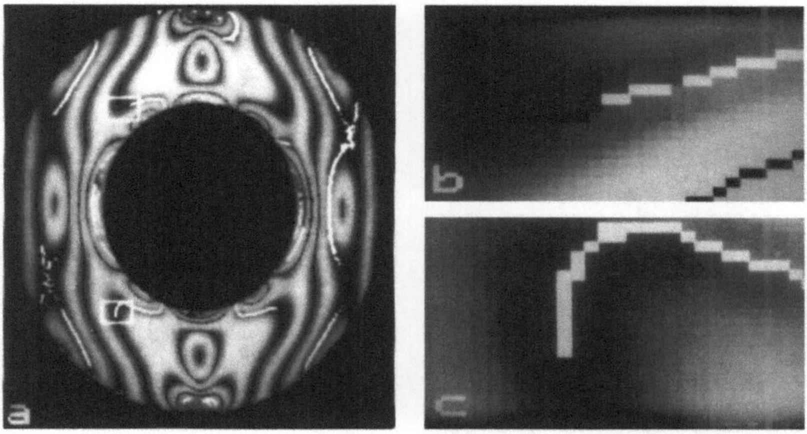

Fig. 4.24. **a** Figure showing the result of the fringe skeletons after employing the fringe tracing algorithm. The traced portions are made dark. For each fringe only one seed point is taken. The fringe tracing algorithm fails when there is fringe discontinuity as shown for typical cases in **b** and **c**

4.12 Ordering of Fringes

Fringe thinning is only an intermediate step in fringe data collection. For quantitative data analysis, as noted before, one needs to get fringe orders and the corresponding positional co-ordinates. This section will be devoted to a discussion on some of the early methods for fringe ordering used in conjunction with fringe thinning. The conversion of pixel co-ordinates to actual model co-ordinates will be discussed in Chapter 8.

The simplest approach to fringe ordering is the interactive assignment of fringe orders by the user. An approach of this nature demands knowledge of fringe ordering by the user. With an effective fringe-tracing algorithm, the interactive ordering needs to be done at as many points as the number of fringes in the field. In several problem situations (Chapter 8), all the points forming the fringe may not be required for data processing but only a few representative points may be needed. In such cases, even if a fringe tracing algorithm is not available, one can assign a fringe counter in the data collection software and interactively collect points on a particular fringe and then increment the fringe counter and collect points from the next fringe and so on.

In interactive fringe ordering only an experienced experimentalist can make use of the DIP system to quicken the data collection process. DIP approach to automating the data collection would be meaningful only if the user interaction is made minimal or at least the system should provide some guidance in data collection. This is possible if the fringe gradient directions are known.

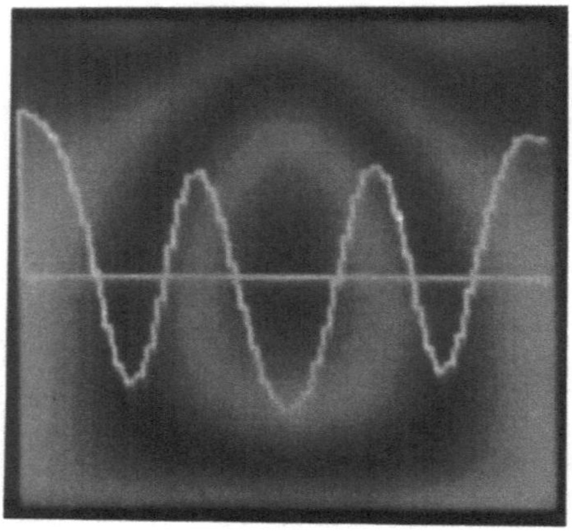

Fig. 4.25. Intensity distribution across a *sink*. Note that the minimum intensity is lower in comparison to the adjacent fringes

Ramesh and Kelkar [28, 29] used intensity information to identify fringe gradient directions even for fringe patterns recorded in Black & White. It is seen in Chapter 1 that identification of isochromatic fringe field characteristics such as *saddle point*, *source*, and *sink* could be effectively used for determining the fringe gradient directions. In Sects. 3.13.3 and 4.7.3, it is seen that the intensity of light transmitted increases in the direction of increasing fringe orders. A discussion on what causes this phenomenon is discussed in Chapter 6. Thus, using these features, sources in the fringe field could be identified. Even the presence of a *sink* in the fringe field could be identified on the basis of the intensity variation. It is shown in ref. [28] that the intensity of light transmitted decreases at the *sink* (Fig. 4.25). Thus, interactive fringe ordering can be done more easily if one has a utility to plot intensity variation of the digitised image on a chosen line.

Use of multiple wavelengths has also been used for identifying fringe gradient direction [22]. Figure 4.26 shows the simulated fringe patterns by PHOTOSOFT_H for a ring under diametral compression for two different wavelengths. Except the zeroth fringe order, the other fringes shift. The relative shift provides the information to identify the fringe gradient direction.

4.13 Closure

Fringe multiplication techniques help to get a quick look at the overall fringe filed variation over the model domain. For quantitative analysis, the fringe skeletons need to be identified by using one of the fringe thinning algorithms discussed. The fringe multiplication technique proposed by

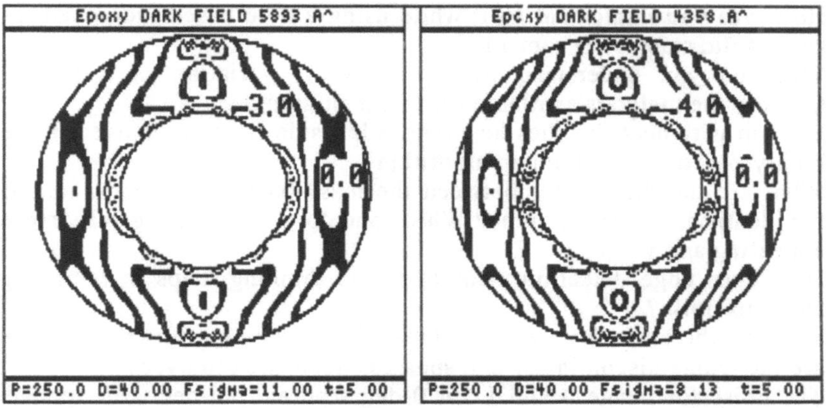

Fig. 4.26. Dark-field isochromatics for the problem of a ring under diametral compression using two different light sources. (Courtesy: PHOTOSOFT_H)

Chen et al. [9] requires only one photograph to be processed and hence, can be extended to analyse fringe fields recorded in dynamic studies, where usually only a bright-field image is recorded. On similar lines, all the fringe-thinning algorithms also require only one photograph to be analysed and hence can be used to automate data collection in dynamic studies. Among the fringe thinning algorithms, the intensity-based algorithms are generally faster and among these, the global thinning algorithm stands out both in quality of fringe skeleton extraction and processing time. Nevertheless, the binary-based algorithms do have a role to play and comes in handy for processing poorly recorded photographs.

In principle, the fringe thinning algorithms discussed are applicable for processing isoclinic fringe contours. Isoclinic fringe contours are much broader and also have a wider minimum intensity band within it. Attempts have been made to take these aspects into account for extracting isoclinic fringe skeleton [30].

Restricting data collection on fringe skeletons alone does not provide sufficient data for studies that finally require the evaluation of the stress tensor over the field. Modified form of HFP can provide data over the field; however, the technique is not elegant.

In this chapter, the digital techniques that have attempted to mimic the manual procedures used hitherto are discussed. In the subsequent chapters, newer and simpler approaches to data extraction that effectively use the DIP hardware will be discussed.

Exercises

4.1 Compare the image subtraction and image division approaches for digital fringe multiplication.

4.2 How will you implement HFP while using a tube based video camera and a solid state CCD camera?
4.3 What do you understand by the thickness variation in a fringe? Classify the methods available for fringe thinning.
4.4 Explain with neat sketches how does a left side edge is removed in the binary-based fringe-thinning algorithm.
4.5 What are the differences between the mask-based algorithm-I and II for identifying fringe minima? Which one do you feel is logically more appealing and why?
4.6 How does edge detection prior to fringe thinning helps in removing spurious noise?
4.7 Mention the steps involved in global fringe thinning algorithm. How are the fringe discontinuity and spurious noise are removed?
4.8 The noise for image ANNUL for 0-deg scan and 90-deg scan appears as lines whereas it appears like a band for 45-deg and 135-deg scans (Fig. 4.8). Why is it so? (Hint: Discuss based on the performance of simple scanning algorithm near saddle points).
4.9 Modify the header file orthopip.h included in the CD to incorporate the feature of improved row-wise thinning (Program 4.3) for all the scan directions needed for global thinning.
4.10 On the lines of the program *tilepip.c* (CD), develop a software such that one can choose any one of the various thinning methods discussed for processing a file.
4.11 Improve the *tracepip.c* (CD) to show colours for different fringe orders with a colour code for easy interpretation. Also include a module to tolerate up to 2 pixel gaps in the fringe contour.
4.12 Develop a data collection software using fringe tracing and interactive ordering of fringes. Build your software module such that the fringe gradient direction is identified using the 2-D intensity plot.

References

1. Post D (1955) Isochromatic fringe sharpening and fringe multiplication in photoelasticity. Proc SESA XII(2):143–156
2. Cloud GL (1968) Improvement in use of photometric methods for measurement of birefringence. Exp Mech 8:1–4
3. Brown AFC, Hickson VM (1950) Improvements in photoelastic technique obtained by the use of a photometric method. Brit J Appl Phys 1(2):39–44
4. Frocht MM, Pih H, Landsberg D (1954) The use of photometric devices in the solution of the general three-dimensional photoelastic problem. Proc SESA 12(1):181–190
5. Talukdar NK, Ghosh P (1975) On fringe multiplication by superposition of negatives. Exp Mech 15(6):143–156
6. Post D (1970) Photoelastic fringe multiplication: for tenfold increase in sensitivity. Exp Mech 10(8):305–312
7. Toh SL, Tang SH, Hovanesian JD (1990) Computerised photoelastic fringe multiplication. Exp Tech 14(4):21–23
8. Liu X , Yu Q (1993) Some improvement on digital fringe-multiplication methods. Exp Tech 17(1):26–29

9. Chen TY (1994) Digital fringe multiplication of photoelastic images – a new approach. Exp Tech 18(2):15–18
10. Chen TY (1995) Digital fringe multiplication in three-dimensional photoelasticity. J Strain Analysis for Engng Des 30(1):1–8
11. Voloshin AS, Burger CP (1983) Half fringe photoelasticity – a new approach to whole field stress analysis. Exp Mech 23(9):304–314
12. Wang WC, Chen TY (1989) Half-fringe photoelastic determination of opening mode stress intensity factor for edge cracked strips. Engng Fracture Mech 32(1):111–122
13. Miskioglu I, Mehdi-Soozani A, Burger CP, Voloshin AS (1987) Stress intensity factors for near edge cracks by digital image analysis. Engng Fracture Mech 27(3):329–343
14. Voloshin AS, Burger CP (1984) Half fringe photoelasticity for orthotropic materials. Fibre Science and Technology 21(4):341–351
15. Burger CP (1988) New approach to optical methods in experimental solid mechanics through digital image processing. Proc VI Int Conference on Experimental Mechanics:655–660
16. Müller RK, Saackel LR (1979) Complete automatic analysis of photoelastic fringes. Exp Mech 19(7):245–252
17. Seguchi Y, Tomita T , Watanabe M (1979) Computer aided fringe pattern analyser – a case of photoelastic fringes. Exp Mech 19(10):362–370
18. Chen TY, Taylor CE (1989) Computerised fringe analysis in photomechanics. Exp Mech 29(3):323–329
19. Smith RW (1987) Computer processing of line images: a survey. Pattern Recognition 20:7–15
20. Yatagai T, Nakadate S, Idesawa M, Saito H (1982) Automatic fringe analysis using digital image processing techniques. Opt Engng 21(3):432–435
21. Gillies AC (1988) Image processing approach to fringe patterns. Opt Engng 27(10):861–866
22. Umezaki E, Tamakai T, Takahashi S (1989) Automatic stress analysis of photoelastic experiment by use of image processing. Exp Tech 13(6):22–27
23. Ramesh K, Pramod BR (1992) Digital image processing of fringe patterns in photomechanics. Opt Engng 31(7):1487–1498
24. Ramesh K, Singh RK (1995) Comparative performance evaluation of various fringe thinning algorithms in photomechanics. J Electronic Imaging 4(1):71–83
25. Wei-chung Wang (1990) A digital imaging algorithm for extracting stress intensity factor from the photoelastic fringe pattern. Engng Fracture Mech 36(5):683–696
26. Ramesh K, Ganesan VR, Mullick SK (1991) Digital image processing of photoelastic fringes – a new approach. Exp Tech 15(5):41–46
27. Ramesh K, Pramod BR (1992) A new fringe thinning algorithm in photomechanics. Proc of VII Int Conference on Experimental Mechanics, Las Vegas, USA :1517–1523
28. Ramesh K, Kelkar AA (1995) Automatic ordering of isochromatic fringes – a new methodology. Strain 31(3):95–99
29. Ramesh K, Kelkar AA (1995) Automatic fringe ordering of photoelastic fringes – a new methodology. Proc of Int conference on Mechanics of Solids and Materials Engineering (MSME 95) Singapore:895–900
30. Yao JY (1990) Digital image processing and isoclinics. Exp Mech 30(3):264–269

5 Phase Shifting, Polarization Stepping and Fourier Transform Methods

5.1 Introduction

In the previous chapter, data acquisition by fringe skeleton identification was discussed. The skeleton identification became much simpler and effective if intensity variations over the fringe field were also taken into account. In view of skeleton identification, the data being collected is restricted to these zones. In the early stages of automatic acquisition of photoelastic data, several point-by-point methods were proposed which also utilised intensity information for automation. In these techniques, either the analyzer/polarizer or the compensator is rotated continuously to produce a modulated intensity signal at the point of interest. Data is recorded based on either the intensity signal is monitored for its minimum value or the phase of the modulated signal is compared with that of a reference signal [1–11]. Thus, the use of intensity information in some form has always attracted researchers to improve the methodology of data acquisition in photoelasticity.

Direct processing of intensity data for quantitative extraction of isochromatic and isoclinic parameter was not thought of earlier in view of the difficulties involved in recording the intensity data for the entire field. However, this problem is overcome with the advent of digital image processing hardware, which can record intensity data at video rates. This opened up the possibility of determining isoclinic and isochromatic parameter at every point (pixel) in the domain. The new methodology requires modifications in the experimental procedure and the way the experimental data needs to be recorded. In general, one may have to record a few images corresponding to different optical arrangements for a given experimental situation. These methodologies have come to be known as phase-shifting techniques – in view of the specific phase shifts introduced by the optical elements between the recorded images.

The phase shifting concept for photoelasticity was first introduced by Hecker and Morche [12] in 1986. Extending the work of Hecker and Morche, Patterson and Wang [13] reported in 1991 a fully automated procedure for photoelastic analysis. They basically used a circular polariscope. Several researchers have also proposed techniques based on a circular polariscope [14–20]. Special attempts have also been reported at improving isochromatic evaluation taking into account the low-modulation areas [17], influence of isoclinic interaction [18, 19], and influence of the quarter-wave plates [20]. Determination of continuous isoclinic contours has also received attention in the literature [21–23]. In 1992, Sarma et al. [24] reported a methodology using a plane polariscope. Later in 1993, Asundi [25] reported a new methodology, which extended the Tardy method of compensation from a point-by-point approach to evaluate the fractional fringe order on all points lying on the isoclinic line. Approaches have also been reported wherein, in a plane polariscope, the exit light from the model is analysed by a quarter-wave plate and an analyzer [26, 27].

Brown and Sullivan [28] introduced polarization stepping for photoelastic analysis in 1990. Polarization stepping is a simplified version of phase-shifting approach in which one records isoclinics in known steps in a dark-field plane polariscope. In view of this, no modifications need to be done for the conventional polariscope. Use of Fourier Transform for photoelastic analysis was introduced by Quan et al. [29]. Morimoto et al. [30] used polarization stepped images to separate isoclinics and isochromatics by Fourier transform. Ng [31] proposed the use of Fourier transforms in conjunction with load stepping.

In this chapter, a detailed discussion on several of these techniques is presented. The individual optical arrangements and the respective intensity equations used by them are tabulated for a quick comparison of the various methodologies and also to provide directions for new methodologies. For the sake of completeness, a brief review of the point-by-point methods for automation is also discussed.

5.2 Early Attempts for Automated Polariscopes

The first attempt to design a polariscope to determine automatically the isoclinic parameter and the isochromatic fringe order at a point was made by Zandman [1] in 1959. Zandman developed his system based on the measurement of the intensity of light from the polariscope. Absolute intensity was not used but the light intensity was monitored to detect a minimum intensity at the point of interest. To account for the background illumination, he used a 2-channel system, one for observing a stressed plastic and the other for an unstressed dummy. A servomotor was used to drive a Babinet compensator until the intensity of light received from the two channels was the same. The system was restricted to $\delta_N \leq 0.5$.

In 1961 Sapaly [2] introduced the use of a spinning analyzer in the polariscope. This modulated the light intensity emerging from a polariscope into a sinusoidal wave. The method introduced by Sapaly assumes the knowledge of the direction of principal stresses and is limited to measure fractional fringe orders. Nevertheless, the introduction of a spinning analyzer opened a fresh and fruitful approach. The method was further developed by Robert [3] who measured the phase and the amplitude of the sinusoidal wave emerging from the photoelastic model. From these readings, he was able to determine the azimuth and the ellipticity of the light from the polariscope. The technique is limited to retardations of one wavelength. A manual recognition of the orientation of the ellipse is required. It is important to note that though the intensity of light was monitored, its actual value was not used in the calculation but its wave characteristics were used for the measurement purpose.

Rather than the use of an external compensator, the use of Tardy method of compensation was proposed in conjunction with the identification of minimum intensity by Pant [4] and Young [5]. The analyzer was rotated with servomotors until the light intensity was a minimum. In order to extend the range of measurement beyond 0.5 fringe orders, they suggested the use of two different wavelengths to measure fractional fringe order. It is measured without sign and manual measurement is needed to determine it. Penkovsky [6] developed an automatic polariscope, which used a Faraday cell to modulate the light in the polariscope. The unit contained a multitude of gears and differentials with a logic circuit receiving the modulated signal and setting in motion all the motors and solenoids required to perform a point measurement.

In the 1970s newer systems were developed to overcome the limitations of the previously mentioned automatic polariscopes. Allison and Nurse [7] and Redner [8] have developed similar systems for complete automatic acquisition of both isochromatic and isoclinic parameter at a point. The systems were identical for measuring isoclinic data but the method of finding the isochromatic fringe orders were slightly different. In these systems, absolute intensity measurement at a point was not made but the wave characteristics viz., the phase shift of the modulated intensity signal is monitored with respect to a reference wave.

To determine the isoclinic parameter, a 0.1 mm diameter beam of white light (polychromatic light) was passed through the photoelastic model and the spinning crossed polarizer and analyzer. A definite relationship exists between the rotation frequency of the crossed polarizer and analyzer and the modulation of the intensity signal. It was found to be four times the rotation frequency of the crossed polarizer and analyzer. A silicon photocell converted this signal into an electrical one. A digital phase computer was used to determine the angular phase difference between this signal and that from a sine wave generator coupled to the polarizer drive motor. They showed that the phase difference between these two waves was directly proportional to the isoclinic angle. This provided one of the principal stress directions and an ambiguity exists on whether it corresponds to σ_1 or σ_2. A pre-calibration of the system is required to overcome this ambiguity. The use of white light for isoclinic determination was an intelligent step used by these researchers which accounted for the interaction of isochromatic fringes on isoclinic determination. This aspect will become clear in the later part of this chapter.

Allison and Nurse used a continuously rotating circular compensator while the polarizers were set at the previously determined isoclinic angle to measure the isochromatic fringe order. They sent a light beam, the colour of which was alternated at high frequency from red to green. The rotation of the compensator for minimum light intensity corresponded to the fringe order. Redner used two beams of light, polarised along the directions of the principal stresses and out of phase by $\pi/2$. The stress in the model produced a relative retardation between the two beams and the emergent light was elliptically polarised. The difference in phase between the two beams after they have passed through a spinning analyzer and a beam splitter was measured by phase detectors. The magnitude of the phase difference was proportional to the stress difference in the material. The phase difference is measured in the range $-\pi$ to π. This posed a restriction on the range of isochromatic fringe order measurement. This range is a function of the wavelengths of the light used. Redner [8] used monochrome filters corresponding to 577 nm and 632 nm. For these wavelengths, the maximum fringe order measurable was 5.4. He pointed out that if monochrome filters corresponding to 585 nm and 600 nm were used then the range could be extended to 20 fringe orders.

The polariscope designed by Redner reached commercial stage and till recently this polariscope was widely used. Later Allison and Nurse [9, 10] introduced interactive control to their polariscope by using two mini computers. One computer – the 'lower' level computer, operated in simple machine code and controlled the movement of the model and the sequence of the series of operations used in measuring isoclinic and isochromatic parameters. The second or the 'upper' computer performed the data processing functions.

Marston [11] has developed an automatic micro polariscope in which he fitted servomotors to drive the optical elements and the microscope stage. The user had to interact with the system to identify whole order fringes and to place the isoclinic angles in the correct quadrant.

All the above methods required cumbersome electronic equipment. Further these techniques can be used only for individual points in the fringe pattern and the main advantage of photoelasticity as a whole-field method is lost. This is where the most recent method of applying DIP scores over those old time efforts.

5.3 Phase Shifting in Photoelasticity

In optical techniques such as photoelasticity, holography and speckle used in experimental mechanics, the experimental data is recorded in the form of fringe patterns. The fringe patterns are nothing but the record of phase difference between light travelling in two different optical paths as intensity variation [25]. By varying the phase difference between the beams involved, in known steps, it is possible to generate sufficient number of equations to solve for the parameters involved. Phase-shifting approach has been applied to many classical interferometric techniques [32]. In most classical interferometers, the path length of the two interfering beams is distinct and separate. In general, phase differences can be introduced by altering the optical path length of any one of the light beams. Usually the phase of the reference light beam is altered in known steps.

Photoelasticity falls into a special category, in that the two light beams cannot be treated separately but always go together. Use of an external mirror as used in other interferometric techniques to introduce a phase shift cannot be applied here. In practice, a change in phase between the beams involved is achieved by appropriately rotating the optical elements of the polariscope. The simplest case to visualise this is the use of analyzer as a compensator in Tardy method of compensation, where the angular rotation of the analyzer can be related to the fractional retardation. It is important to point out that the retardation thus introduced is a function of the isoclinic angle. As the isoclinic angle is different for different points the modulation in phase information achieved by rotating the optical elements (here, the analyzer) will not be uniform as in other interferometric techniques but will be different from point to point. Thus, the challenge in employing phase shifting technique to photoelasticity arises in arriving at appropriate combination of optical elements to provide as high a modulation as possible over the domain. Further, calculation schemes have to be developed to appropriately account for the low modulation areas in the domain.

A distinction has to be made between the nature of phase information recorded in photoelasticity and other interferometric techniques. In most techniques, the phase information is related to a single physical information. For example, in holography one gets out-of-plane displacements and in Moiré one records either u or v displacement from a single experiment. On the other hand, in photoelasticity, the phase information is affected by both the difference in principal stresses and the orientation of the principal stress direction. This is referred to as isochromatic-isoclinic interaction and this affects the evaluation of isochromatics and isoclinics differently.

In conventional photoelasticity, one is concerned with very simple optical arrangements that provide a dark-field in a plane polariscope and both bright and dark-fields in the case of a circular polariscope. However, a great degree of flexibility in utilising the intensity information for proposing newer methodologies for data reduction is possible if the optical elements are kept at arbitrary positions. In the following discussions, the polariscope arrangement is classified based on the polarization of the incident light. Use of techniques based on plane and circular polariscopes have been reported in the literature. Usually, the optical elements after the model are kept at arbitrary positions for recording intensity data. In circular polariscope arrangement, the second quarter-wave plate and the analyzer are kept at arbitrary positions. In some techniques, the first quarter-wave plate is appropriately oriented to get either a right-handed or left-handed circularly polarised light. The most general arrangement in a plane polariscope is to keep both the polarizer and the analyzer at arbitrary positions. Techniques have also been attempted where the exit light from the model in a plane polariscope is analysed with the help of quarter-wave plate and an analyzer. A detailed study on the intensity of light transmitted for arbitrary orientations of the optical elements is necessary to appreciate the methodology of phase shifting in photoelasticity.

5.4 Intensity of Light Transmitted for a Generic Arrangement of a Plane Polariscope

Each optical element in a polariscope basically introduces a rotation and retardation. In Jones calculus (Sect. 1.11), these operations are handled in terms of matrices. It is easier to visualise the role of each optical element if its net effect is expressed in terms of separate matrices. In the subsequent discussions, δ is used to represent the retardation (in radians) introduced by the model, θ is the orientation of one of the principal stress directions with respect to the horizontal, $ke^{i\omega t}$ is the incident light vector, E_β and $E_{\beta+\pi/2}$ are the components of light vector along the analyzer axis and perpendicular to the analyzer axis respectively. The angular orientations of the various optical elements are referred with respect to the x-axis.

Figure 5.1 shows a photoelastic specimen kept in a plane polariscope with the polarizer and the analyzer kept at arbitrary angles of α and β respectively. Using Jones calculus, the components of light vector along the analyzer axis and perpendicular to the analyzer axis are obtained as

$$\begin{Bmatrix} E_\beta \\ E_{\beta+\pi/2} \end{Bmatrix} = \begin{bmatrix} \cos\beta & \sin\beta \\ -\sin\beta & \cos\beta \end{bmatrix} \begin{bmatrix} \cos\frac{\delta}{2} - i\sin\frac{\delta}{2}\cos 2\theta & -i\sin\frac{\delta}{2}\sin 2\theta \\ -i\sin\frac{\delta}{2}\sin 2\theta & \cos\frac{\delta}{2} + i\sin\frac{\delta}{2}\cos 2\theta \end{bmatrix} \times \begin{Bmatrix} \cos\alpha \\ \sin\alpha \end{Bmatrix} ke^{i\omega t} \tag{5.1}$$

Following the methodology of Sect. 1.13, the intensity of light transmitted I_p (the subscript 'p' denotes that the incident light is plane polarised) is obtained as

$$I_p = I_a\left[\cos^2\frac{\delta}{2}\cos^2(\beta-\alpha) + \sin^2\frac{\delta}{2}\cos^2(\beta+\alpha-2\theta)\right] \tag{5.2}$$

where I_a accounts for the amplitude of incident light vector and the proportionality constant. It is interesting to note that Eq. (5.2) is not altered when the model slow axis orientation is changed from θ to $\theta + \pi/2$.

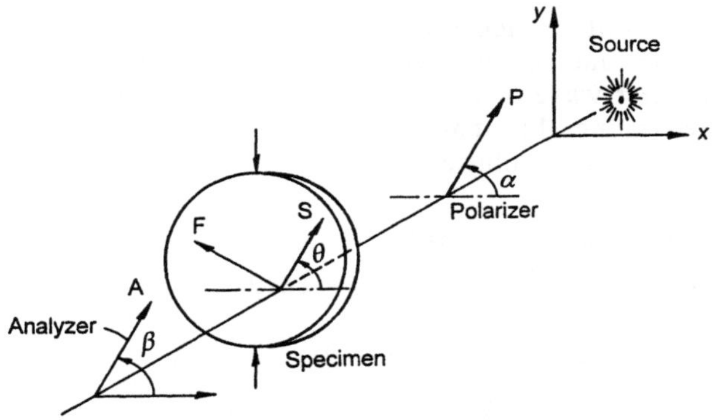

Fig. 5.1. Generic arrangement of plane polariscope to employ phase shifting methodology

If $\alpha = \beta + \pi/2$, Eq. (5.2) reduces to

$$I_p = I_a\left[\sin^2\frac{\delta}{2}\sin^2 2(\theta - \beta)\right] \tag{5.3}$$

Equation (5.3) represents the intensity of light transmitted for crossed positions of polarizer-analyzer combination.

Substituting $\alpha = \beta$ in Eq. (5.2), one gets the intensity of light transmitted in a bright-field plane polariscope as

$$I_p = I_a\left[\cos^2\frac{\delta}{2} + \sin^2\frac{\delta}{2}\cos^2 2(\beta - \theta)\right] \tag{5.4}$$

Upon simplification one gets

$$I_p = I_a\left[1 - \sin^2\frac{\delta}{2}\sin^2 2(\beta - \theta)\right] \tag{5.5}$$

If the polarizer is kept at 90-deg and the analyzer is kept at an arbitrary angle β, the intensity of light transmitted is

$$I_p = I_a\left[\sin^2\beta\cos^2\frac{\delta}{2} + \sin^2\frac{\delta}{2}\sin^2(2\theta - \beta)\right] \tag{5.6}$$

For dark-field arrangement, β is zero and Eq. (5.6) reduces to

$$I_p = I_a\left[\sin^2\frac{\delta}{2}\sin^2 2\theta\right] \tag{5.7}$$

which is same as that of Eq. (1.34) obtained in Chap. 1.

In certain class of phase-shifting techniques, a quarter-wave plate is introduced between the model and the analyzer. For the case when model slow axis is at θ and the quarter-wave plate slow axis is at η, the component of light vector transmitted along the analyzer and perpendicular to the analyzer can be obtained by Jones calculus as

$$\begin{Bmatrix} E_\beta \\ E_{\beta+\pi/2} \end{Bmatrix} = \frac{1}{\sqrt{2}} \begin{bmatrix} \cos\beta & \sin\beta \\ -\sin\beta & \cos\beta \end{bmatrix} \begin{bmatrix} 1-\mathrm{i}\cos 2\eta & -\mathrm{i}\sin 2\eta \\ -\mathrm{i}\sin 2\eta & 1+\mathrm{i}\cos 2\eta \end{bmatrix}$$
$$\times \begin{bmatrix} \cos\frac{\delta}{2} - \mathrm{i}\sin\frac{\delta}{2}\cos 2\theta & -\mathrm{i}\sin\frac{\delta}{2}\sin 2\theta \\ -\mathrm{i}\sin\frac{\delta}{2}\sin 2\theta & \cos\frac{\delta}{2} + \mathrm{i}\sin\frac{\delta}{2}\cos 2\theta \end{bmatrix} \begin{Bmatrix} \cos\alpha \\ \sin\alpha \end{Bmatrix} k e^{\mathrm{i}\omega t} \tag{5.8}$$

It is important to note that the light vector would be different if the model fast axis is at θ. This is left as an exercise to the reader. The intensity of light transmitted including the effect of model slow or fast axis orientation for this case is

$$I = \frac{I_a}{2}\Big\{\cos^2\frac{\delta}{2}[\cos^2(\alpha-\beta) + \cos^2(\alpha+\beta-2\eta)] + \sin^2\frac{\delta}{2}[\cos^2(\beta-\alpha-2\eta+2\theta)$$
$$+\cos^2(\alpha+\beta-2\theta)] + \sin\delta[\mp\cos(\alpha-\beta)\cos(\beta-\alpha-2\eta+2\theta) \tag{5.9}$$
$$\pm\cos(\alpha+\beta-2\theta)\cos(\alpha+\beta-2\eta)\Big\}$$

where the upper sign corresponds to the slow axis of the model and the lower sign corresponds to the fast axis of the model.

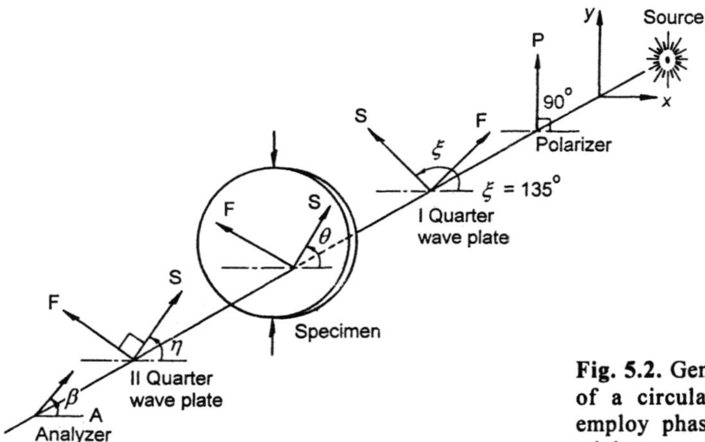

Fig. 5.2. Generic arrangement of a circular polariscope to employ phase shifting methodology

5.5 Intensity of Light Transmitted for a Generic Arrangement of a Circular Polariscope

In most data acquisition algorithms that use a circular polariscope, the second quarter-wave plate and the analyzer are kept at arbitrary positions (Fig. 5.2). The first quarter-wave plate is kept at either $\xi = 135$-deg or 45-deg. Using Jones calculus, the components of light vector along the analyzer axis and perpendicular to the analyzer axis (for $\xi = 135$-deg) are obtained as

$$\begin{Bmatrix} E_\beta \\ E_{\beta+\pi/2} \end{Bmatrix} = \frac{1}{2}\begin{bmatrix} \cos\beta & \sin\beta \\ -\sin\beta & \cos\beta \end{bmatrix}\begin{bmatrix} 1-\mathrm{i}\cos 2\eta & -\mathrm{i}\sin 2\eta \\ -\mathrm{i}\sin 2\eta & 1+\mathrm{i}\cos 2\eta \end{bmatrix}$$
$$\times\begin{bmatrix} \cos\frac{\delta}{2}-\mathrm{i}\sin\frac{\delta}{2}\cos 2\theta & -\mathrm{i}\sin\frac{\delta}{2}\sin 2\theta \\ -\mathrm{i}\sin\frac{\delta}{2}\sin 2\theta & \cos\frac{\delta}{2}+\mathrm{i}\sin\frac{\delta}{2}\cos 2\theta \end{bmatrix}\begin{bmatrix} 1 & \mathrm{i} \\ \mathrm{i} & 1 \end{bmatrix}\begin{Bmatrix} 0 \\ 1 \end{Bmatrix} k e^{\mathrm{i}\omega t} \tag{5.10}$$

The combination of the quarter-wave plate and the polarizer for $\xi = 135$-deg produces a left circularly polarised light. Denoting the intensity of light transmitted as I_ℓ, the intensity expression is obtained following the methodology of Sect. 1.13 as

$$I_\ell = \frac{I_a}{2} + \frac{I_a}{2}[\sin 2(\beta-\eta)\cos\delta - \sin 2(\theta-\eta)\cos 2(\beta-\eta)\sin\delta] \tag{5.11}$$

where I_a accounts for the amplitude of light vector and the proportionality constant. If in Fig. 5.2, the model fast axis is kept at θ (i.e., slow axis at $\theta + \pi/2$), then the intensity of light transmitted is

$$I_\ell = \frac{I_a}{2} + \frac{I_a}{2}[\sin 2(\beta-\eta)\cos\delta + \sin 2(\theta-\eta)\cos 2(\beta-\eta)\sin\delta] \tag{5.12}$$

The Eqs. (5.11) and (5.12) can be combined and represented as a single equation of the form

$$I_\ell = \frac{I_a}{2} + \frac{I_a}{2}[\sin 2(\beta-\eta)\cos\delta \mp \sin 2(\theta-\eta)\cos 2(\beta-\eta)\sin\delta] \tag{5.13}$$

where the upper sign corresponds to the case for which the model slow axis is at θ and the lower sign corresponds to the case for which the model fast axis is at θ.

When the polariscope is arranged for Tardy compensation arrangement (θ = 0-deg and η = 45-deg) with the analyzer at an arbitrary angle β, the intensity of light transmitted is

$$I_\ell = \frac{I_a}{2} - \frac{I_a}{2}\cos(2\beta \pm \delta) = I_a \sin^2\left(\beta \pm \frac{\delta}{2}\right) \tag{5.14}$$

where the upper and the lower signs correspond to the cases mentioned before. For the dark-field arrangement (η = 45-deg and β = 0-deg), Eq. (5.13) reduces to

$$I_\ell = \frac{I_a}{2} - \frac{I_a}{2}\cos\delta = I_a \sin^2\frac{\delta}{2} \tag{5.15}$$

and for the bright-field arrangement (η = 45-deg and β = 90-deg), Eq. (5.13) reduces to

$$I_\ell = \frac{I_a}{2} + \frac{I_a}{2}\cos\delta = I_a \cos^2\frac{\delta}{2} \tag{5.16}$$

Equations (5.15) and (5.16) are same as those of Eqs. (1.38) and (1.39) of Chap. 1. Inspection of Eqs. (5.11)–(5.16) show that the intensity of light transmitted depends on the orientation of the model fast or slow axis for the cases when the optical elements after the model are kept at arbitrary orientations. As the model fast or slow axes are not known *a priori*, this leads to an ambiguity in the interpretation of intensity information. Hence, this aspect needs special attention in automated techniques based on intensity measurement.

In Fig. 5.2, if the I-quarter-wave plate is rotated such that the slow axis is kept at 45-deg, then one gets a right circularly polarised light being incident on the model. Following the development of Eq. (5.13) the intensity of light transmitted I_r (the subscript 'r' denotes that the incident light is right circularly polarised), including the effect of model slow or fast axis orientation for this case

$$I_r = \frac{I_a}{2} - \frac{I_a}{2}[\sin 2(\beta - \eta)\cos\delta \mp \sin 2(\theta - \eta)\cos 2(\beta - \eta)\sin\delta] \tag{5.17}$$

where the upper sign corresponds to the case for which the model slow axis is at θ and the lower sign corresponds to the case for which the model fast axis is at θ. The intensity of light transmitted for Tardy method of compensation, bright and dark-fields could be obtained by appropriately substituting the values of θ, η and β in Eq. (5.17). While doing so, one would observe that for η = 45-deg and β = 90-deg one gets a dark-field arrangement instead of a bright-field arrangement. Here again it is important to note that the conventional dark and bright-fields are not influenced by the model fast or slow axis.

From the foregoing discussions, it is clear that the conventional dark and bright-field arrangements are insensitive to the orientation of the model fast or slow axis. This makes the interpretation of data simple in conventional photoelasticity. The only case where the model axis influences the interpretation is while applying Tardy Method of compensation. The observer usually takes care of the ambiguity by noting which fringe order moves to the point of interest when the analyzer is rotated clockwise/anti-clockwise.

In conventional photoelasticity, the optical arrangement is usually set based on the background field. The relative orientations of the optical elements are more important (Table 1.1, Chap. 1) and their absolute orientations are less significant in most studies. However, in phase shifting methodology, the absolute orientations of the optical elements do play a significant role. Thus, every effort must be made to correctly align the polariscope such that one to one correspondences exists between the arrangement used for theoretical development of intensity equations and the arrangement used for experimental recording of intensities.

The intensity equations developed in this section and the previous sections show that when the extinction conditions are satisfied, the intensity of light transmitted should be zero. However, this never happens in an experiment (refer to the 2-D intensity plots in Chap. 3). In order to account for this behaviour, many investigators have explicitly added a term I_b to account for the stray light/background illumination. In view of the optical inhomogenity and non-uniform characteristics of the light source and the diffuser, the quantities I_a and I_b are functions of the spatial co-ordinates viz., $I_a(x, y)$ and $I_b(x, y)$. With this explicit understanding these will be represented as I_a and I_b in further text.

While discussing the basic optical arrangement used by various investigators, the angles α and β refer to the polarizer and analyzer (axis of polarization) axes with the horizontal. The angles ξ and η refer to the slow axis of the I and II quarter-wave plates respectively. The angle θ refers to the slow axis of the model unless otherwise stated. Though, the model slow or fast axis orientation plays an important role on the intensity equations, in subsequent discussions this aspect is not highlighted for every technique discussed. The equations based on slow axis are used for describing the techniques. In subsequent discussions, the intensity of light transmitted will be denoted as I_i for the ith optical arrangement. The orientation of the optical element that needs to be rotated will also appear with a subscript i. In all the techniques, the parameters of interest are determined by inverse trigonometric functions. Solution of these are multi-valued and in the subsequent discussions, the principal value of the parameter evaluated will be indicated with a subscript 'c'.

5.6 Evaluation of Fractional Fringe Order along an Isoclinic Contour

As noted before, phase shifts between the beams involved in photoelasticity could be achieved by rotating the analyzer when the polariscope is set at the Tardy compensation arrangement. In conventional photoelasticity, Tardy method of compensation is used for point-by-point analysis. However, by using intensity data, it could be extended to all points lying in the respective isoclinic. For this a few images have to be recorded for data processing.

The intensity of light transmitted for arbitrary positions of the analyzer in Tardy compensation arrangement is given by Eq. (5.14). If a light intensity of I_b is added to account for background/stray light illumination, then Eq. (5.14) becomes

$$I_i = I_b + I_a \sin^2\left(\beta_i + \frac{\delta}{2}\right) \tag{5.18}$$

Table 5.1 summarises the intensity equations for various values of β.

Table 5.1. Optical arrangement and the intensity equations to find fractional retardation along an isoclinic contour

S. No.	β	Intensity equation
1	0	$I_1 = I_b + \frac{I_a}{2}(1 - \cos\delta)$
2	$\pi/4$	$I_2 = I_b + \frac{I_a}{2}(1 + \sin\delta)$
3	$\pi/2$	$I_3 = I_b + \frac{I_a}{2}(1 + \cos\delta)$
4	$3\pi/4$	$I_4 = I_b + \frac{I_a}{2}(1 - \sin\delta)$

Simplifying the four equations shown in Table 5.1, one gets the fractional retardation δ as

$$\delta_c = \tan^{-1}\left(\frac{I_2 - I_4}{I_3 - I_1}\right) = \tan^{-1}\left(\frac{\sin\delta}{\cos\delta}\right) \tag{5.19}$$

where, δ_c denotes the principal value of the inverse trigonometric function.

The following are the steps involved in this algorithm

1. Record a particular isoclinic fringe using a plane polariscope.
2. Change the set-up to a circular polariscope such that the polarizer axis is aligned parallel to one of the principal stress directions.
3. Keep the analyzer in four different positions as per Table 5.1 and record the intensity data. Care must be taken while selecting the direction of rotation (clockwise or anti-clockwise) of analyzer such that the lower fringe order moves toward the higher order [10]. This accounts for the ambiguity of the orientation of fast or slow axis of the model.

5.7 Whole Field Evaluation of Photoelastic Data by Using a Plane Polariscope

The phase-shifting methods reported by Kihara [26], Dupré et al. [27], Sarma et al. [24] and Nurse [33] use a plane polariscope for data acquisition. Of these methods, Kihara and Dupré et al. have analysed the exit light from the model by using a quarter-wave plate and analyzer combination. The various optical arrangements and the respective intensity equations for these positions for the first three methods are summarised in Table 5.2.

The expressions for the determination of isoclinic and the fractional retardation based on the arrangement used by Sarma et al. are

$$\theta_c = \frac{1}{2}\tan^{-1}\left(\frac{2I_2}{I_1 + I_2 - 2I_3}\right) \quad \text{for } \sin^2\delta/2 \neq 0 \tag{5.20}$$

where θ_c is the principal value of the inverse trigonometric function.

$$\delta_c = \cos^{-1}(C) \tag{5.21}$$

where

$$C = \frac{A - A^2 - B^2}{1 - A}; \quad \text{and} \quad A = \frac{I_1 - I_2}{I_1 + I_2}; \quad B = \frac{I_1 + I_2 - 2I_3}{I_1 + I_2} \tag{5.22}$$

Sarma et al. have not explicitly taken into account the effect of background /stray light intensity by introducing a parameter I_b in the intensity equation. Instead, they proposed that one has to initially view the unstressed model in the polariscope and capture an image when polarizer and analyzer are kept in crossed position. The intensity values of this image are to be subtracted from all other images before any processing is done.

Table 5.2. Optical arrangements and intensity equations for various methods based on plane polariscope

α	η	β	Intensity equation
Sarma et al.			
$\pi/2$	-	$\pi/2$	$I_1 = I_a\left[\cos^2\frac{\delta}{2} + \sin^2\frac{\delta}{2}\cos^2 2\theta\right]$
$\pi/2$	-	0	$I_2 = I_a\left[\sin^2\frac{\delta}{2}\sin^2 2\theta\right]$
$\pi/2$	-	$\pi/4$	$I_3 = \frac{I_a}{2}\left[1 - \sin^2\frac{\delta}{2}\sin 4\theta\right]$
Kihara			
0	$\pi/2$	$-\pi/4$	$I_1 = \frac{I_a}{2}[1 + \sin\delta\sin 2\theta]$
0	$\pi/2$	$\pi/4$	$I_2 = \frac{I_a}{2}[1 - \sin\delta\sin 2\theta]$
$-\pi/4$	$\pi/2$	$-\pi/4$	$I_3 = \frac{I_a}{2}[1 + \sin\delta\cos 2\theta]$
$-\pi/4$	$\pi/2$	$\pi/4$	$I_4 = \frac{I_a}{2}[1 - \sin\delta\cos 2\theta]$
0	$\pi/2$	0	$I_5 = I_a\left[\cos^2\frac{\delta}{2} + \sin^2\frac{\delta}{2}\cos^2 2\theta\right]$
0	$\pi/2$	$\pi/2$	$I_6 = I_a\left[\sin^2\frac{\delta}{2}\sin^2 2\theta\right]$
$-\pi/4$	$\pi/4$	$-\pi/4$	$I_7 = I_a\left[\cos^2\frac{\delta}{2} + \sin^2\frac{\delta}{2}\sin^2 2\theta\right]$
$-\pi/4$	$\pi/4$	$\pi/4$	$I_8 = I_a\left[\sin^2\frac{\delta}{2}\cos^2 2\theta\right]$
Dupré et al.			
0	-	0	$I_1 = I_a$ Model not loaded
0	-	0	$I_2 = I_a\left[\cos^2\frac{\delta}{2} + \sin^2\frac{\delta}{2}\cos^2 2\theta\right]$
0	-	$\pi/4$	$I_3 = \frac{I_a}{2}\left[1 + \sin^2\frac{\delta}{2}\sin 4\theta\right]$
$\pi/4$	-	$\pi/4$	$I_4 = I_a\left[\cos^2\frac{\delta}{2} + \sin^2\frac{\delta}{2}\sin^2 2\theta\right]$
0	$3\pi/4$	0	$I_5 = \frac{I_a}{2}[1 + \sin 2\theta\sin\delta]$

For the optical arrangement used by Kihara, the expressions for determining isoclinic and fractional retardation are

$$\theta_c = \frac{1}{2}\tan^{-1}\left(\frac{I_1 - I_2}{I_3 - I_4}\right) = \frac{1}{2}\tan^{-1}\left(\frac{\sin\delta\sin 2\theta}{\sin\delta\cos 2\theta}\right) \quad \text{for } \sin\delta \neq 0 \tag{5.23}$$

$$\delta_c = 2\cos^{-1}\left[\sqrt{\frac{1}{2}\left(\frac{I_5 - I_6}{I_5 + I_6} + \frac{I_7 - I_8}{I_5 + I_6}\right)}\right] \tag{5.24}$$

Since, in Eqs. (5.21) and (5.24) δ_c is evaluated using an inverse cosine function, ambiguity exists regarding the sign of δ_c. This ambiguity cannot be resolved from images recorded using one wavelength alone. Sarma et al. suggested the use of two different wavelengths to identify the full fringe order as proposed in ref. [34]. Another approach to resolve this ambiguity by using three wavelengths with a specific inter-relationship between them has been proposed by Buckberry and Towers [35] (see Sect. 5.10.2).

For the arrangement used by Dupré et al., the expressions are respectively

$$\theta_c = \frac{1}{4}\tan^{-1}\left(\frac{2I_3 - I_1}{I_2 - I_4}\right) = \frac{1}{4}\tan^{-1}\left(\frac{\sin^2\frac{\delta}{2}\sin 4\theta}{\sin^2\frac{\delta}{2}\cos 4\theta}\right) \quad \text{for } \sin^2\frac{\delta}{2} \neq 0 \tag{5.25}$$

$$\delta_c = \sin^{-1}\left(\frac{2I_5 - I_1}{I_1\sin 2\theta}\right) \tag{5.26}$$

5.8 Whole Field Evaluation of Photoelastic Data by Using a Circular Polariscope

5.8.1 The Generic Procedure

The phase shifting concept for photoelastic analysis was first introduced by Hecker and Morche [12]. For fractional retardation determination, they used a circular polariscope. The basic optical arrangement used by them is shown in Fig. 5.2. The governing intensity equation is Eq. (5.11) and by adding a term I_b to account for background illumination/stray light one gets

$$I_i = I_\mathrm{b} + \frac{I_\mathrm{a}}{2} + \frac{I_\mathrm{a}}{2}[\sin 2(\beta_i - \eta_i)\cos\delta - \sin 2(\theta - \eta_i)\cos 2(\beta_i - \eta_i)\sin\delta] \quad (5.27)$$

The specific optical arrangements used by them and the appropriate intensity equations are summarised in Table 5.3. From the equations listed in Table 5.3, the fractional retardation is obtained as

$$\delta_\mathrm{c} = \tan^{-1}\left(\frac{2\sqrt{2}\sqrt{(I_1 - I_2)^2 + (I_2 - I_3)^2 + (I_3 - I_1)^2}}{3(I_4 - I_5)}\right) \quad (5.28)$$

Table 5.3. Polariscope arrangement and intensity equations for various phase shifting methodologies based on circular polariscope

ξ	η	β	Intensity equation
Hecker and Morche			
$3\pi/4$	0	0	$I_1 = I_\mathrm{b} + \frac{I_\mathrm{a}}{2}(1 - \sin 2\theta \sin\delta)$
$3\pi/4$	$\pi/3$	$\pi/3$	$I_2 = I_\mathrm{b} + \frac{I_\mathrm{a}}{2} + \frac{I_\mathrm{a}}{4}(\sin 2\theta + \sqrt{3}\cos 2\theta)\sin\delta$
$3\pi/4$	$-\pi/3$	$-\pi/3$	$I_3 = I_\mathrm{b} + \frac{I_\mathrm{a}}{2} + \frac{I_\mathrm{a}}{4}(\sin 2\theta - \sqrt{3}\cos 2\theta)\sin\delta$
$3\pi/4$	$3\pi/4$	0	$I_4 = I_\mathrm{b} + \frac{I_\mathrm{a}}{2}(1 + \cos\delta)$
$3\pi/4$	$\pi/4$	0	$I_5 = I_\mathrm{b} + \frac{I_\mathrm{a}}{2}(1 - \cos\delta)$
Patterson and Wang			
$3\pi/4$	0	$\pi/4$	$I_1 = I_\mathrm{b} + \frac{I_\mathrm{a}}{2}(1 + \cos\delta)$
$3\pi/4$	0	$3\pi/4$	$I_2 = I_\mathrm{b} + \frac{I_\mathrm{a}}{2}(1 - \cos\delta)$
$3\pi/4$	0	0	$I_3 = I_\mathrm{b} + \frac{I_\mathrm{a}}{2}(1 - \sin 2\theta \sin\delta)$
$3\pi/4$	$\pi/4$	$\pi/4$	$I_4 = I_\mathrm{b} + \frac{I_\mathrm{a}}{2}(1 + \cos 2\theta \sin\delta)$
$3\pi/4$	$\pi/2$	$\pi/2$	$I_5 = I_\mathrm{b} + \frac{I_\mathrm{a}}{2}(1 + \sin 2\theta \sin\delta)$
$3\pi/4$	$3\pi/4$	$3\pi/4$	$I_6 = I_\mathrm{b} + \frac{I_\mathrm{a}}{2}(1 - \cos 2\theta \sin\delta)$

Patterson and Wang [13] extended the work of Hecker and Morche and used different steps of optical rotation of the elements to produce a higher modulation over the field. The specific optical arrangement used by them and the respective intensity equations are summarised in Table 5.3.

Using the equations listed in Table 5.3, the isoclinic parameter is obtained as

$$\theta_c = \frac{1}{2}\tan^{-1}\left(\frac{I_5 - I_3}{I_4 - I_6}\right) = \frac{1}{2}\tan^{-1}\left(\frac{I_a \sin\delta \sin 2\theta}{I_a \sin\delta \cos 2\theta}\right) \quad \text{for } \sin\delta \neq 0 \qquad (5.29)$$

From the equations shown in Table 5.3, the fractional retardation is obtained as

$$\delta_c = \tan^{-1}\left(\frac{I_4 - I_6}{(I_1 - I_2)\cos 2\theta_c}\right) = \tan^{-1}\left(\frac{I_a \cos 2\theta_c \sin\delta}{I_a \cos 2\theta_c \cos\delta}\right) \quad \text{for } \cos 2\theta_c \neq 0 \qquad (5.30)$$

$$\delta_c = \tan^{-1}\left(\frac{I_5 - I_3}{(I_1 - I_2)\sin 2\theta_c}\right) = \tan^{-1}\left(\frac{I_a \sin 2\theta_c \sin\delta}{I_a \sin 2\theta_c \cos\delta}\right) \quad \text{for } \sin 2\theta_c \neq 0 \qquad (5.31)$$

The exact calculation of isoclinic angle and fractional retardation is not trivial from the above equations. In the earlier sections, the respective equations to evaluate these were given but no effort has been made to critically analyse the accuracy of these equations in the calculation of parameters. This is because, in the early studies, the research efforts were purely exploratory and arriving at different optical arrangements was the prime concern of research. Only later, issues related to accuracy found a place. In this sense, the most researched schemes in phase shifting is the set of equations proposed by Patterson and Wang. Though the intensity equations obtained by various researchers were the same, the specific optical arrangements have been different [17, 20]. For the same incident light used by Patterson and Wang, if the analyzer and the second quarter-wave plate are kept in steps of $\pi/4$, the various optical combinations of these providing the intensity equations I_1 to I_6 are summarised in Table 5.4.

Equation (5.29) shows that the isoclinic value is indeterminate at those points where $\delta_c = 0, \pi, 2\pi, 3\pi \ldots$. The implication is that the isoclinic contour will not be continuous over the domain. When δ_c is not exactly equal to 0 or π or 2π etc., but very close to these values, θ_c does not become indeterminate but the value of θ_c determined will be unreliable and appears as noise in isoclinic fringe when plotted. This is known as isochromatic-isoclinic interaction in the calculation of the isoclinic parameter. Several ways to overcome this have been proposed. These schemes basically involve shifting the location where $\delta_c = 0, \pi, 2\pi, 3\pi \ldots$ occur by using either multiple loads or multiple wavelengths and combining the results from these multiple experiments to get a continuous isoclinic contour.

Table 5.4. Table showing the multiple optical arrangements to obtain intensity equations I_1 to I_6 of Patterson and Wang's algorithm

η \ β	0	$\pi/4$	$\pi/2$	$3\pi/4$	π
0	I_3	I_1	I_5	I_2	I_3
$\pi/4$	I_2	I_4	I_1	I_6	I_2
$\pi/2$	I_3	I_2	I_5	I_1	I_3
$3\pi/4$	I_1	I_4	I_2	I_6	I_1
π	I_3	I_1	I_5	I_2	I_3

Considering that θ_c is calculated correctly by Eq. (5.29), the calculation of δ_c by Eqs. (5.30) or (5.31) should result in the same value of δ_c if continuous variables are used in the arctangent calculations. In practical applications, this is not the case due to intensity digitisation and quantization and only discrete values are used in the arctangent calculations. This can lead to error in the calculations. Further, in Eqs. (5.30) and (5.31), δ_c has low-modulation regions when $\cos 2\theta_c$ or $\sin 2\theta_c$ are small. In these regions, the intensity differences $(I_4 - I_6)$ or $(I_5 - I_3)$ will also be small. The evaluation of δ_c will involve the ratio of two small discrete numbers and will be in considerable error in such regions. This is termed as isoclinic-isochromatic interaction in the calculation of the fractional retardation. Apart from this, the principal value of θ_c evaluated by Eq. (5.29) lies in the range $-\pi/4$ to $\pi/4$ and an ambiguity exists on whether θ_c corresponds to σ_1 or σ_2 direction over the domain. This problem is a little more complex to handle and is discussed later.

It is instructive to note that regions of low values of $\cos 2\theta_c$ coincide with high values of $\sin 2\theta_c$ and vice versa. Therefore, for every point in the region of interest, one has two estimates of δ_c with different degrees of error depending on the region. Thus, the results must be combined appropriately to get the best possible values of δ_c. For high values of $\cos 2\theta_c$ or $\sin 2\theta_c$ the selection is obvious. However, the transition between these regions has to be handled with care for correct estimation. A new expression for δ_c is obtained for the whole domain [17] as

$$\delta_c = \tan^{-1}\left\{\frac{(I_5 - I_3)\sin 2\theta_c + (I_4 - I_6)\cos 2\theta_c}{(I_1 - I_2)}\right\} = \tan^{-1}\left\{\frac{I_a \sin\delta}{I_a \cos\delta}\right\} \quad (5.32)$$

Unlike Eqs. (5.30) or (5.31), Eq. (5.32) always guarantees a high modulation over the field. By combining Eqs. (5.29)–(5.31), δ_c can also be found as

$$\delta_c = \tan^{-1}\left\{\frac{\pm\sqrt{(I_5 - I_3)^2 + (I_4 - I_6)^2}}{(I_1 - I_2)}\right\} = \tan^{-1}\left\{\frac{\pm I_a \sin\delta}{I_a \cos\delta}\right\} \tag{5.33}$$

Equation (5.33) also provides a high modulation, equivalent to Eq. (5.32), over the field. Since the numerator involves an expression under a square root, both +ve and –ve roots are admissible for the calculation. In further text, these will be referred to as +ve square root result and –ve square root result. Since Eqs. (5.29) through (5.33) are obtained from six different optical arrangements, it can be referred to as a six-step phase shifting method.

Efforts have also been made to reduce the number of images needed for the calculation of the parameters. Barone et al. [16] and Patterson and Wang [36] have proposed use of a four-step phase shifting method. Barone et al. used the first four arrangements of the six-step phase shifting method. The isoclinic and isochromatic parameters are obtained as

$$\theta_c = \frac{1}{2}\tan^{-1}\left(\frac{I_1 + I_2 - 2I_3}{2I_4 - (I_1 + I_2)}\right) = \frac{1}{2}\tan^{-1}\left(\frac{I_a \sin\delta \sin 2\theta}{I_a \sin\delta \cos 2\theta}\right) \quad \text{for } \sin\delta \neq 0 \tag{5.34}$$

$$\begin{aligned}\delta_c &= \tan^{-1}\left(\frac{(I_1 + I_2 - 2I_3)\sin 2\theta_c + [2I_4 - (I_1 + I_2)]\cos 2\theta_c}{(I_1 - I_2)}\right) \\ &= \tan^{-1}\left\{\frac{I_a \sin\delta}{I_a \cos\delta}\right\}\end{aligned} \tag{5.35}$$

Patterson and Wang [36] used the arrangements corresponding to the first, third, fourth and fifth of the six-step phase shifting method to find the isoclinic and isochromatic parameters. The derivation of the respective equations is left as an exercise to the reader. Patterson and Wang [36] devised a polariscope to simultaneously record the four intensity patterns with the help of beam splitters and four CCD cameras. The use of beam splitter leads to absorption of the polarised beam and this has been accounted for in the calculation of the parameters. For analysis of such optical systems, the use of Muller matrices is appropriate [37].

5.8.2
Calculation and Representation of Whole Field Data

The optical arrangements represented in Table 5.3 are unconventional. It has been noted before that in intensity based calculations, the optical arrangement used for theoretical development and the experimental arrangement should be identical. In conventional photoelasticity, the background field usually verifies the optical arrangement. Such a simplistic approach is not possible here. Hence, it is desirable to have a reference set of images for aligning the polariscope. Figure 5.3 shows the sequence of six images recorded for a circular disc under diametral compression (dia 60.14 mm, thickness 6.3 mm, load 669.4 N and F_σ = 10.29 N/mm/fr). From the intensity recordings, the values of isoclinic parameter and the isochromatic parameter have to be evaluated using the Eqs. (5.29)–(5.33).

It is to be noted that Eqs. (5.29)–(5.33) involve inverse trigonometric operations and thus one has multi-valued solutions and this poses a problem in correctly evaluating δ and θ over the domain. The use of atan function to evaluate θ_c and use of atan2 function to evaluate δ_c by Eq. (5.30)–(5.32) has been found to give good results for most of the region in the model [38]. In the subsequent discussions δ_c is calculated by Eq. (5.32) due to its advantages pointed out earlier. Use of atan function for evaluating θ_c by Eq. (5.29) gives θ_c in the range of $-\pi/4$ to $\pi/4$. The use of atan2 function (Microsoft-C or FORTRAN Library), which account for the correct quadrant, for δ_c in Eq. (5.29)–(5.32) gives it in the range of $-\pi$ to π.

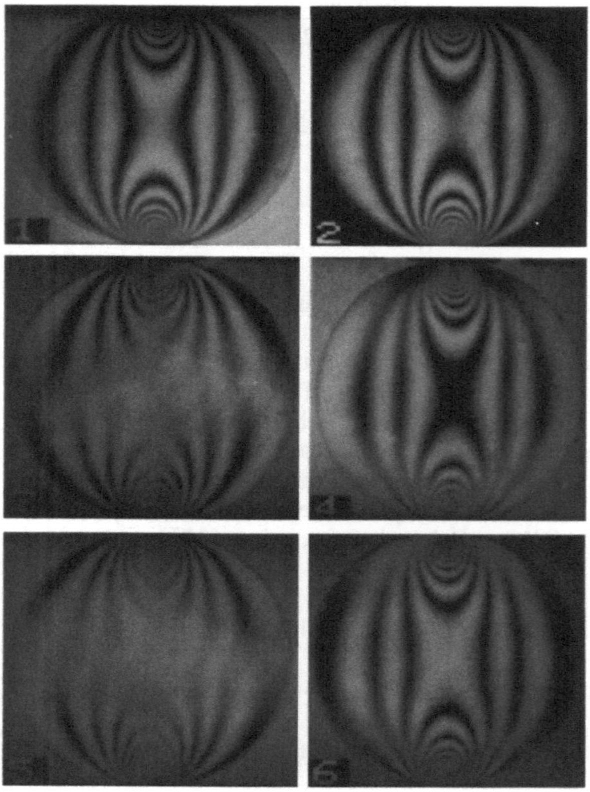

Fig. 5.3. The sequence of six images recorded for employing phase shifting. Note that the first image corresponds to bright-field fringe patterns and the second image corresponds to dark-field fringe patterns. No whole field interpretation can be given for other images

The accuracy of evaluation of various parameters could be qualitatively assessed by obtaining a whole field plot of the interested parameter. For the whole domain, the fractional retardation, δ_c as obtained by Eqs. (5.30) to (5.32), can be represented either in the form of dark-field or bright-field isochromatics or a phase map. Phase map is a convenient form of representation of fractional retardation, which could be eventually unwrapped to provide total retardation/fringe order. The dark and bright-field representations are usually used to counter check the quality of the intensity data recorded. No unwrapping procedure can be employed to find the total retardation by them. Dark-field isochromatics can be obtained as

$$g(x,y) = \frac{255}{\pi}\,\text{abs}(\delta_c) \tag{5.36}$$

where $g(x, y)$ is the grey level of the pixel at the location (x, y). To obtain phase map, fractional retardation has to be first expressed in the range of 0 to 2π. This is then converted into grey levels between 0 to 255 for graphical plotting. The zero fractional retardation corresponds to pitch black and 2π fractional retardation corresponds to pure white i.e., 255. The phase map is plotted using the following relations

$$\begin{aligned} \delta_p &= \begin{cases} \delta_c & \text{for } \delta_c > 0 \\ 2\pi - \delta_c & \text{for } \delta_c \le 0 \end{cases} \\ g(x,y) &= \frac{255}{2\pi}\delta_p \end{aligned} \tag{5.37}$$

The bright-field isochromatics can be obtained as

$$\begin{aligned} \delta_b &= \begin{cases} \delta_c & \text{for } \delta_c > 0 \\ 2\pi - \delta_c & \text{for } \delta_c \le 0 \end{cases} \\ g(x,y) &= \frac{255}{\pi}\,\text{abs}(\delta_b - \pi) \end{aligned} \tag{5.38}$$

Figure 5.4 shows the simulated dark-field isochromatics, bright-field isochromatics and phase map. The dark and bright-field isochromatics are quite good even in stress concentration zones, whereas, the phase map is not good in these regions. This needs special attention and is discussed in Sect. 5.8.3.

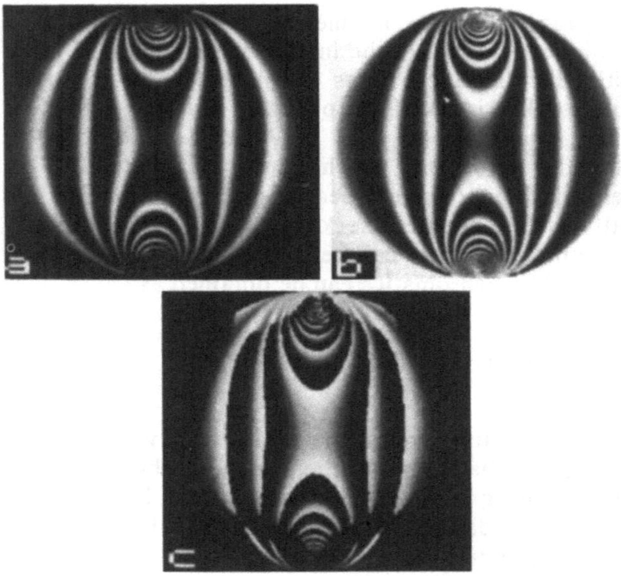

Fig. 5.4. Whole field representation of fractional retardation: **a** simulated dark-field **b** simulated bright-field **c** phase map. Note the characteristic intensity variations in the phase map

5.8.3 Parameters Affecting the Generation of Phase Map and its Solution

5.8.3.1 Influence of Local Oscillations of Isoclinic Parameter on Fractional Retardation Calculation

Use of Eqs. (5.29)–(5.32) involve the explicit use of θ_c in the calculations. Thus, in principle any error in evaluating θ_c would indirectly affect the calculation of δ_c. The whole field data plotted in Fig. 5.4 is reasonably good over the domain. Though, θ_c influences the calculation of δ_c, its influence is negligible. This can be well understood by plotting the variation of θ_c and δ_c calculated for a specific line. Figure 5.5 shows the comparison between theoretical and experimentally obtained value of θ for the diametral line. The evaluation of θ_c by phase shifting has sharp hikes and is well documented [20, 38, 39]. Despite such hikes, the estimation of retardation is quite good for this line (Fig. 5.6). Hence, one can conclude that the error in θ is not propagating to the calculation of retardation significantly (Fig. 5.6).

In Fig. 5.6 one can observe that the fractional retardation increases in the direction of the increase in the total fringe order. In Fig. 5.6, fractional retardation has been calculated using Eq. (5.32). The same will be true for Eqs. (5.30) and (5.31) if judiciously combined. This is a remarkable achievement of equations (5.30)–(5.32) and makes the unwrapping of the phase map much simpler.

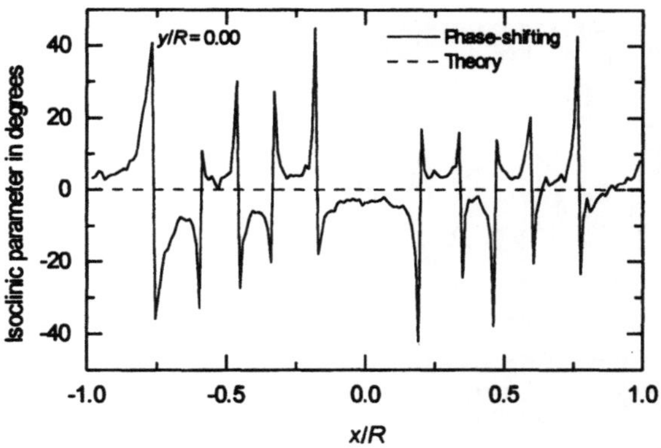

Fig. 5.5. Comparison between theoretical and experimentally obtained value of θ for the diametral line of the disc

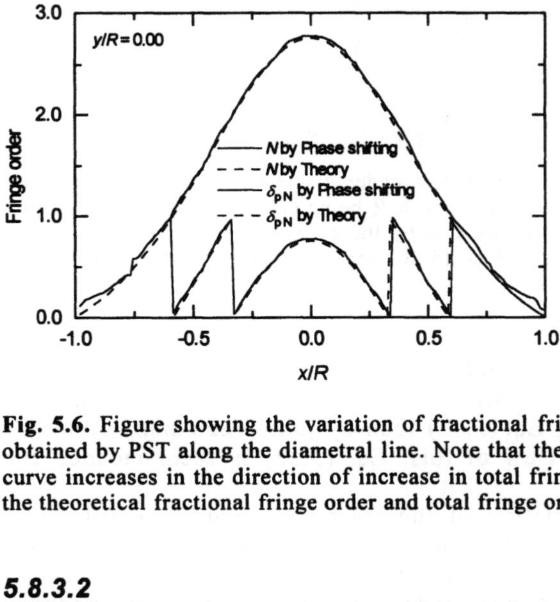

Fig. 5.6. Figure showing the variation of fractional fringe order and total fringe order obtained by PST along the diametral line. Note that the slope of fractional fringe order curve increases in the direction of increase in total fringe order. The figure also shows the theoretical fractional fringe order and total fringe order variations

5.8.3.2 Importance of Isoclinic Parameter Representing Either σ_1 or σ_2 Direction Over the Domain

It is worthwhile to analyse how the Eqs. (5.30)–(5.32) perform for theoretically obtained θ_c^t and experimentally obtained θ_c. Figure 5.7 shows the phase map obtained by Eq. (5.32) using experimentally evaluated θ_c (Fig. 5.7a) and theoretical θ_c^t corresponding to σ_1 direction (Fig. 5.7b). Figures 5.7 c and 5.7d shows these plots by judicious use of Eqs. (5.30) and (5.31). The figure shows that for experimentally determined θ_c by Eq. (5.29) the use of either Eq. (5.32) or the judicious combination of Eqs. (5.30) and (5.31) provide the same results. However, for theoretical θ_c^t

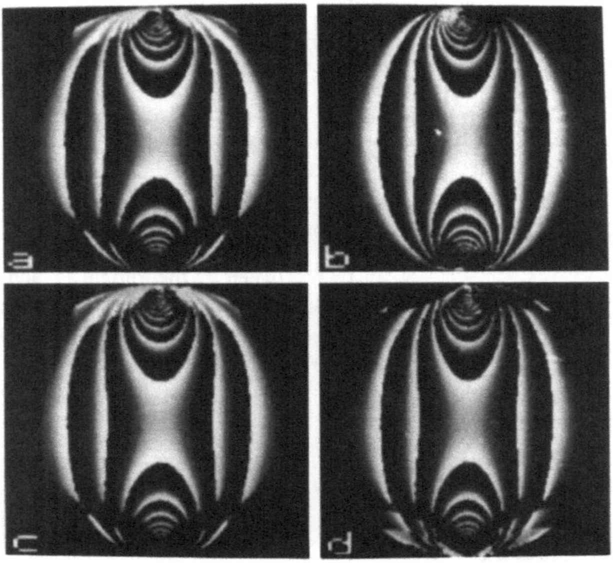

Fig. 5.7. Comparison of phase map of a disc under diametral compression obtained using Eqs. (5.30) and (5.31) and Eq. (5.32) with θ_c being calculated by various means: **a** θ_c by Eq. (5.29) and δ_c by Eq. (5.32) **b** θ_c by theory and δ_c by Eq. (5.32) **c** θ_c by Eq. (5.29) and δ_c by judicious use of Eqs. (5.30) and (5.31) **d** θ_c by theory and δ_c by judicious use of Eqs. (5.30) and (5.31)

use of Eq. (5.32) shows a remarkable improvement in the stress concentration zone. This indicates that the transition zone of low-modulation areas is better combined while using Eq. (5.32) than by judicious combinations of Eqs. (5.30) and (5.31). In view of this, Eq. (5.32) is recommended for general use.

Figure 5.8 shows the variation of the fractional retardation (obtained using experimental θ_c). In this, leaving the central zone (about ±5.0 mm), there is no correspondence between the variation of the fractional retardation and the total fringe order and hence phase-unwrapping algorithms would fail. Even when a higher optical magnification is used for stress concentration zone, the phase map will not improve [40]. Thus, this error is not just due to the difficulties in the measurement of intensities in steep stress concentration zones.

Since use of θ_c^t in Eq. (5.32) has provided good phase map in stress concentration zones, one has to now compare the difference between θ_c by Eq. (5.29) and θ_c^t. It has been noted earlier that θ_c by Eq. (5.29) lies in the range $-\pi/4$ to $\pi/4$. This means that it can represent either σ_1 or σ_2 direction over the domain. On the other hand, θ_c^t is so calculated to represent only σ_1 direction over the domain.

The problem is well illustrated by obtaining the phase map of the problem of a ring under diametral compression. The model of ring is ideal

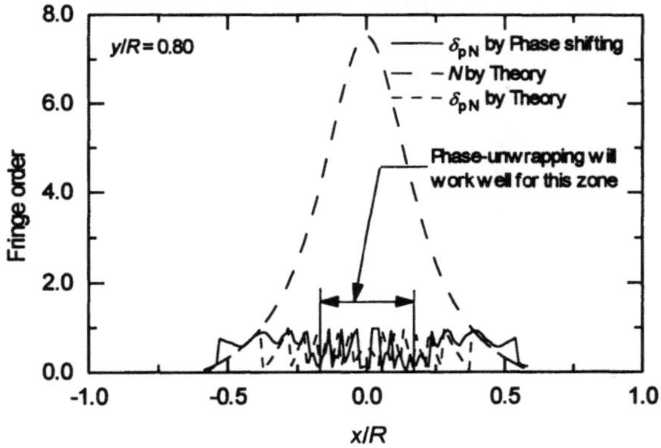

Fig. 5.8. Figure showing the relationship between the slope of partial fringe order obtained by PST with total fringe order variation by theory along a line passing through the stress concentration zone ($y/R = 0.80$)

to choose since it has all the fringe features such as *source*, *sink*, *saddle point*, *isotropic point* and *singular point* in its domain [41]. In view of this, fringe ordering is generally difficult for this problem and the fringe gradient has all possible combinations. A ring (outer diameter 80.00 mm, inner diameter 40.0 mm, thickness 6.26 mm) subjected to a compressive load of 523.51 N having a material stress fringe value of 13.5 N/mm/fringe is taken as another specimen. Six images corresponding to six positions shown in Table 5.3 are captured. Using these intensity patterns, for one quarter of the ring dark-field isochromatics, bright-field isochromatics and phase map are obtained by Eqs. (5.29) and (5.32) which are shown in Fig. 5.9. The bright and dark-field images are good over the complete domain but the phase map is good only in some zones. Close scrutiny of Fig. 5.9 reveals that in some zones, phase map has reversal of intensities.

Figure 5.10 shows the phase map obtained by theoretically evaluated θ_c^t by various means. For theoretical estimation of θ_c^t, one normally uses the following equation

$$\tan 2\theta_c^t = \frac{2\tau_{xy}}{(\sigma_x - \sigma_y)} \tag{5.39}$$

It is to be noted that the Eq. (5.39), used most popularly for evaluating principal stress directions, is ambiguous in providing the orientation of either σ_1 or σ_2 always. Use of atan function for evaluating Eq. (5.39) will give θ_c^t in the range $-\pi/4 < \theta_c^t < \pi/4$ and the corresponding phase map is shown in Fig. 5.10a. The phase map has distinct zones. The boundary of these zones is at 45° isoclinic. The different zones could be attributed to the

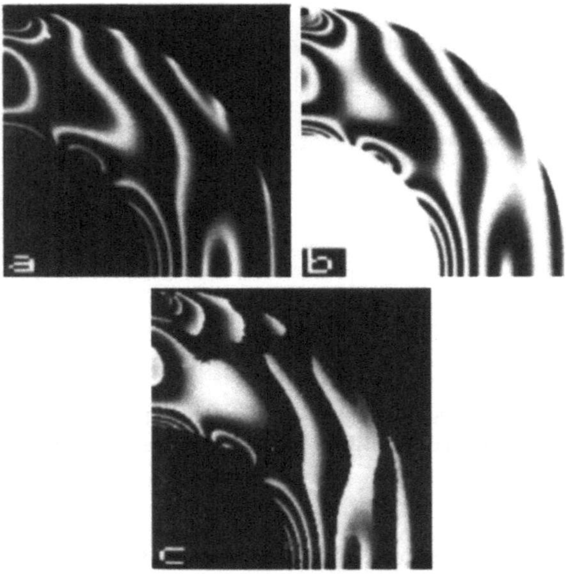

Fig. 5.9. Whole field representation of fractional fringe order for the problem of a ring under diametral compression: **a** simulated dark-field **b** simulated bright-field **c** phase map

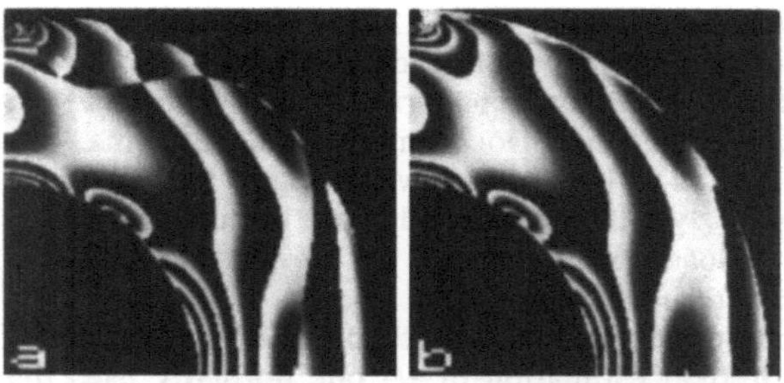

Fig. 5.10. Phase map obtained by PST for a ring under diametral compression using Eq. (5.32) with theoretically evaluated value of by θ_c^t various means: **a** θ_c^t obtained by atan function of Eq. (5.39) **b** $\theta_c^t = \theta_1$ obtained by eigen value approach **c** $\theta_c^t = \theta_2$ obtained by eigen value approach **d** θ_c^t by atan function but replaced by $\theta_c^t + \pi/2$

fact that in some zones θ_c^t corresponds to σ_1 direction and in other zones it corresponds to σ_2 direction.

This ambiguity can be mathematically overcome in theoretical evaluation by representing the evaluation of principal stress and their orientation as an eigen value problem (*eigen.c*, CD). By eigen value approach, if only θ_1 corresponding to σ_1 is used, then the phase map is as shown in Fig. 5.10b. Figure 5.10c shows if only θ_2 is used for the whole domain. Figure 5.10d shows the phase map where θ_c^t is replaced as $\theta_c^t + \pi/2$ in Eq. (5.39). The use of $\theta_c^t + \pi/2$ reverses the intensity variations in the zones showed in Fig. 5.10a.

Figure 5.10 clearly shows that one should use θ corresponding to either σ_1 or σ_2 direction to get the correct phase map. It is worthwhile to note that in Fig. 5.10b the fractional retardation increases in the direction of increasing total fringe order but in Fig. 5.10c, it decreases in the direction of increasing total fringe order. This fact has to be used in devising the phase unwrapping algorithm to construct total fringe order variation from the phase map.

5.8.3.3
Ambiguity in Experimentally Evaluating the Isoclinic Parameter

Normally, in photoelastic analysis, only the relative retardation plays a crucial role than the way retardation is introduced in the model. However, it was pointed out in Sect. 5.5 that for arbitrary orientations of optical elements after the model, the orientation of model slow or fast axis has an influence on the intensity of light transmitted. Let us now consider that the model fast axis is at angle θ. The governing equation is Eq. (5.12) and as before the term I_b can be added to account for stray light/background illumination. For the six optical arrangements indicated before, one obtains a new set of equations, which are summarised in Table 5.5. Comparing Table 5.5 with Table 5.3, one can observe that the expressions for intensity for the first two optical arrangements are the same for both the cases. However, the intensity equations for other optical arrangements are different. For example I_3' is actually I_5 and I_4' is I_6 and vice versa.

If Eq. (5.29) is used to find θ_c from equations listed in Table 5.5, since both the numerator and denominator signs are reversed, the net result does not change if atan function is used. However, while using Eq. (5.32), only the sign of the numerator changes and this leads to a different estimation of δ_c. Expressing this problem in other words, Eq. (5.29) is insensitive to the orientation of the fast or slow axis of the model, whereas, Eq. (5.32) is sensitive to this. In a model, *a priori*, one does not know whether the fast axis or the slow axis lie in the range $-\pi/4 < \theta < \pi/4$. This ambiguity leads to the problem of unsuccessful estimation of δ_c in certain zones of the model.

Comparing Fig. 5.10 with Fig. 5.9c, whenever θ_c^t is in the range $-\pi/4 < \theta_c^t < \pi/4$, the phase maps are identical. Beyond this zone, in some regions the phase map is reversed and in some others it is not good. The key point to note is that θ_c, despite its local oscillations should represent either σ_1 or

σ_2 direction uniformly over the domain to get a correct phase map for the whole domain.

Table 5.5. Intensity equation for 6-step phase shifting technique if the model fast axis is at an angle θ

ξ	η	β	Intensity equation
$3\pi/4$	0	$\pi/4$	$I'_1 = I_b + \frac{I_a}{2}(1+\cos\delta)$
$3\pi/4$	0	$3\pi/4$	$I'_2 = I_b + \frac{I_a}{2}(1-\cos\delta)$
$3\pi/4$	0	0	$I'_3 = I_b + \frac{I_a}{2}(1+\sin 2\theta \sin\delta)$
$3\pi/4$	$\pi/4$	$\pi/4$	$I'_4 = I_b + \frac{I_a}{2}(1-\cos 2\theta \sin\delta)$
$3\pi/4$	$\pi/2$	$\pi/2$	$I'_5 = I_b + \frac{I_a}{2}(1-\sin 2\theta \sin\delta)$
$3\pi/4$	$3\pi/4$	$3\pi/4$	$I'_6 = I_b + \frac{I_a}{2}(1+\cos 2\theta \sin\delta)$

This problem is partly understood in the literature and it is reported that this happens because θ_c is in the range of $-\pi/4 < \theta_c < \pi/4$ and could be solved if unwrapped in the range $-\pi/2 < \theta_c < \pi/2$. Since the numerator of the Eq. (5.29) is proportional to $\sin 2\theta_c$ and the denominator is proportional to $\cos 2\theta_c$, one can, in principle, use atan2 function of the Microsoft-C Library which accounts for the correct quadrant and θ_c can be obtained in the range $-\pi/2 < \theta_c < \pi/2$ mathematically. This leads to a phase map as shown in Fig. 5.11a. Obviously Fig. 5.11a shows that unwrapping θ_c in this sense does not solve the problem. Use of fuzzy sets to solve this has been attempted by Patterson and Wang [18]. The fuzzy sets were not quite exhaustive to include all possibilities. Its usefulness is shown only for the problem of a circular disc under diametral compression, which is a trivial example.

Another approach to solve this is possibly by combining Figs. 5.10a and 5.10d. The Figs. 5.10a and 5.10d are complementary to each other. It is in principal possible to combine these to get a phase map similar to either Fig. 5.10b or Fig. 5.10c. The Figs. 5.10a and 5.10d are obtained by using θ_c^t and $\theta_c^t + \pi/2$ respectively. This brings out that if θ_c evaluation can be made accurate, then, in principal, one can get a phase map by combining phase maps corresponding to θ_c and $\theta_c + \pi/2$. However, in reality, θ_c is not as good as one wants it to be. If $\theta_c + \pi/2$ is used in Eq. (5.32) for the ring problem, the phase map is shown in Fig. 5.11b. Obviously by combining Figs. 5.9c and 5.11b, one cannot get a phase map of either Fig. 5.10b or

Fig. 5.10c. This could be attributed to the fact that apart from sign reversals, local oscillation of θ_c is also influencing the result in these zones. The use of fuzzy sets may also not solve this, as it is difficult to arrive at comprehensive fuzzy sets for encompassing all situations. Though the local oscillations in θ_c calculation affect the estimation of δ_c in some sense, the requirement that θ_c should always represent either σ_1 or σ_2 direction is more critical from phase-unwrapping point of view.

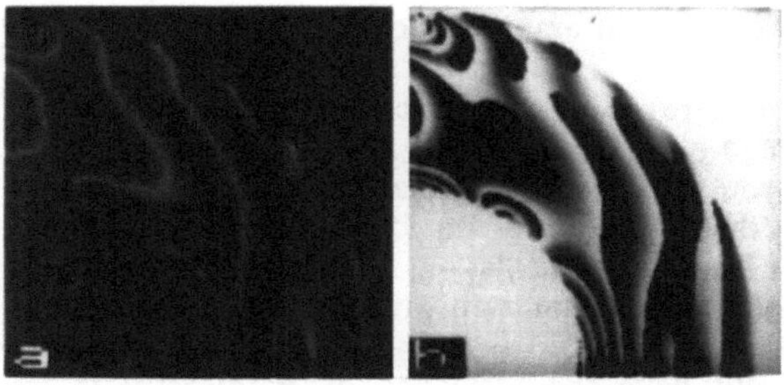

Fig. 5.11. Phase map of a ring under diametral compression obtained using Eq. (5.32) with θ_c obtained by Eq. (5.29) by various means: **a** θ_c obtained by atan2 function **b** θ_c by atan function but replaced by $\theta_c + \pi/2$

In theoretical estimation of θ_c^t, it is possible to evaluate only σ_1 or σ_2 direction. On the other hand, θ_c evaluation by Eq. (5.29) can never guarantee either σ_1 or σ_2 direction alone and this is the core of the problem.

5.8.3.4
Interactive Approach to Obtain a Good Phase Map

The previous sections have brought out the difficulties in the use of Eqs. (5.30)–(5.32) for finding the fractional retardation for the entire domain. For most of the regions of the problem, these equations can provide the correct phase map. Phase unwrapping is possible by just providing total fringe order for one of the points in the good zones and phase unwrapping is to be done within the good zone. This will work well for the problem of a plate with a hole subjected to uniaxial tension and for major portions for the problem of a circular disc under diametral compression but will be quite inaccurate for problems such as a ring under diametral compression [19]. An interactive approach to improve the phase map obtained by these equations is discussed in this section.

The problem in fractional retardation determination is attributed to the fact that Eq. (5.29) is insensitive to the fast or slow axis of the model but Eqs. (5.30)–(5.32) are sensitive to this. This could be overcome, if Eqs. (5.29) and (5.30)–(5.32) are recast such that they are either sensitive to the fast and slow axis of the model or totally insensitive uniformly. It is important to note that the conventional/unconventional optical arrangements for getting dark or bright-field are insensitive to the orientation of the fast or slow axis of the model. Hence, this aspect did not receive any attention in conventional photoelasticity.

In the way the Eq. (5.29) is constructed, it is not possible to make it sensitive to fast or slow axis of the model. Hence, it would be easier to recast the equation for evaluating the fractional retardation as insensitive to fast or slow axis of the model. If the evaluation of δ_c is to be made insensitive to the fast or slow axis of the model, then one has to use Eq. (5.33). However, its use introduces a different ambiguity on whether to use a +ve or –ve square-root result.

In Eq. (5.33), for +ve square root, the fractional retardation, δ_{pN} (fractional retardation in fringe orders), value will always lie between 0 to 0.5 and for the –ve square root, the fractional retardation, δ_{pN}, value will vary between 0.5–1.0. Using this limitation of Eq. (5.33), advantageously, a new concept of continuity of fractional retardation, considering the slope in the scan direction, is proposed [19], which will give fractional retardation varying from 0.0–1.0.

Though Eq. (5.33) does not have the influence of θ_c directly in its formulation, the advantage of automatically knowing fringe order gradient direction while using Eqs. (5.30)–(5.32) is lost. Thus, apart from providing total fringe order at a point of interest, the direction of fringe gradient is also need to be supplied by the user. This is a limitation of the technique. But by considering its advantage, for an experimentalist with some knowledge of photoelasticity, it is not difficult to give. The analysis becomes much simpler, if one uses phase shifting in colour domain [42] as colour code can be used to identify the fringe gradient direction. Alternatively, one can use 2-D intensity plots [43, 44] to identify the fringe gradient direction.

To provide flexibility in the analysis, the technique is implemented on a user-defined window called a tile (*2dtile.c*, CD). As different zones of the image will require different gradient direction to be supplied, one can choose as many tiles as one wants to solve the problem. Within a tile, one has to just state whether the total fringe order is increasing or decreasing along the scan direction. To accommodate fringes of all curvature, scanning is provided for left to right, right to left, top to bottom, bottom to top, diagonal up and diagonal down. The basic scheme of identifying fractional retardation within a tile for a scan direction is summarised in Table 5.6 where $d\delta_1$ represents the change in the fractional retardation of current pixel in the scan direction being evaluated by the +ve square root result and $d\delta_2$ represents the change for the –ve square root result [Eq. (5.33)].

Figure 5.12 shows the phase map obtained by various number of tiles for the problem of a ring under diametral compression. Figure 5.12a gives the phase map by using Eqs. (5.29) and (5.32) by first tile of Table 4. The phase map of Fig. 5.12a is improved by next three tiles in Fig. 5.12b, six tiles in Fig. 5.12c and by eight tiles in Fig. 5.12d. The figure shows that the phase map can be totally improved by interactive processing. The more the number of tiles, the better is the phase map. Figure 5.13 shows the location of the tiles on the originally obtained phase map by using experimental θ_c and δ_c calculated by Eqs. (5.29) and (5.32). Table 5.7 summarises the type of processing used to recover the phase map in each of these tiles.

Table 5.6. Table showing the selection procedure of +ve or –ve square root value for the case where the total fringe order is increasing in the scan direction

S No.	Conditions	Chosen value of δ_{pN} by
1	$d\delta_1/dx < d\delta_2/dx$ and $d\delta_1/dx > 0$ and $d\delta_2/dx > 0$	+ve sqrt result
2	$d\delta_1/dx < 0$ and $d\delta_2/dx > 0$	–ve sqrt result
3	$d\delta_1/dx < 0$ and $d\delta_2/dx < 0$	+ve sqrt result

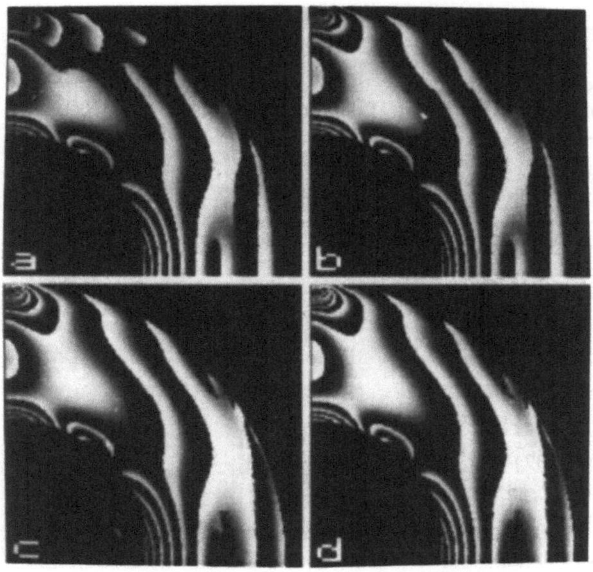

Fig. 5.12. Figure showing the improvement in the phase map for the problem of a ring under diametral compression obtained by the interactive approach for various number of tiles listed in Table 5.7: **a** Original phase map obtained by Eqs (5.29) and (5.32), Tile-1 **b** Tiles 1–4 **c** Tiles 1–7 **d** Tiles 1–9

Here, the new methodology is basically used to improve the phase map obtained by Eqs. (5.29) and (5.32). Alternatively, one can directly use Eq. (5.33) to obtain the phase map if fringe gradient direction is known for the problem domain.

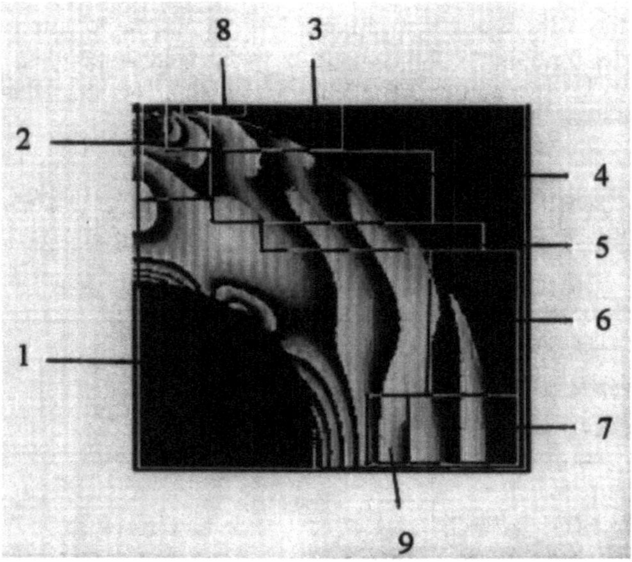

se map
. (5.29)
he tiles
obtain
map of

Table 5.7. Table showing the type of processing used to recover the phase map for various tiles for the problem of ring under diametral compression

Tile No.	Choice of scanning	Left and right corners of tile (pixels)	Fringe order gradient	Initial value chosen
1	Eq. (5.32)	(8,1)–(212,253)	-	-
2	Top to bottom	(8,1)–(72,66)	Decreasing	+ve sqrt result
3	Right to left	(52,1)–(132,31)	Increasing	"
4	"	(73,32)–(173,82)	Increasing	"
5	"	(96,83)–(196,101)	Increasing	"
6	"	(172,101)–(212,201)	Decreasing	"
7	"	(163,202)–(213,253)	Decreasing	"
8	Diagonal down	(59,1)–(88,8)	Increasing	"
9	Left to right	(142,202)–(163,253)	Decreasing	"

5.9
Error Sources and Methods to Minimise their Influence

There are numerous sources of error that affect the accuracy of phase measurements in phase shifting algorithms. Error sources generally fall into three categories: (1) those associated with the data acquisition process, (2) those associated with defects in optical and mechanical design and fabrication (3) environmental effects. The data acquisition process includes errors due to non-linearities in the detection system, amplitude and frequency stability of the source, quantization errors introduced in the analog-to-digital conversion process and errors in the phase shift process.

The choice of the hardware particularly the camera and the digitiser used have an influence on the accuracy of the results obtained. Sarma et al. used RCA ultracon tube black & white camera and Patterson and Wang used CCTV (Toshiba CC1200) camera. Haake et al. [45, 46] while evaluating the methodology proposed by Patterson and Wang, suggested that the noise due to electronics of the camera can be alleviated by switching from a tube camera to a solid state camera such as a CCD (charge coupled device) camera. Use of CCD cameras for image acquisition is becoming very common nowadays. Asundi used a CCD camera and Ramesh et al. also used CCD cameras for data acquisition.

Modern CCD sensors operate linearly over a dynamic range of nearly 100:1 at room temperature. The range can be easily extended by an order of magnitude by appropriate cooling arrangement (Sect. 3.18). Though normally one does expect a CCD sensor to behave linearly, at high levels of illumination due to saturation of the sensing elements, non-linearities could occur. Thus, in experimentation every effort must be made to avoid high illumination.

Stability of the light source is important both in frequency as well as amplitude. Random intensity fluctuations of the source can be mitigated by resorting to time averaging of the images as discussed in Chap. 3.

Digitisation and quantization error of the video signal is an important aspect to be considered. It is noted in Chap. 3 that a cosmetically pleasing black & white TV picture needs 128 grey levels (7-bit) with 512 × 512 pixels but in digital photoelasticity, use of 256 grey levels (8-bit) for quantization and 512 × 512 pixels for spatial digitisation are quite common. Though this level of digitisation and quantization may be quite sufficient for low fringe gradient areas, in zones of stress concentration, the resolution will be quite poor. One alternative to overcome this problem is to switch to a 10-bit system or higher. In a 10-bit system, the spatial resolution will be 1024 × 1024 and the grey level quantization will be 1024. This approach will lead not only to increase in initial cost but also in the running costs as the speed and cost of handling a 10-bit image is quite cumbersome. Another cost-effective alternative is to go for optically enhanced tiling (OET) [40], which is discussed in Chap. 6.

Errors in phase-shift process essentially presents itself as the angular alignment of the optical elements and imperfections in the optical elements. Error in angular alignment could be reduced to a great extent by resorting to a computer assisted stepper motor control of the rotation of optical elements [47, 48]. Imperfections in the optical elements can be of two categories. One is, inhomogeneity of the optical elements which is common to all optical elements and the other is, specific to quarter-wave plates. In view of the use of large field commercial quality optical elements in photoelasticity, their properties may not be uniform over the complete field. Error due to this could be minimised, say, by grabbing images corresponding to a particular intensity equation by multiple optical arrangements shown in Table 5.4 and then taking their average as the intensity values for further calculations.

The quarter-wave plates may not exactly match with the wavelength of light used as a source. This may be due to fabrication errors in the quarter-wave plates or the algorithm demands image grabbing at different wavelengths but only one set of quarter-wave plates is used.

The foregoing discussions have broadly brought out the role of various parameters affecting phase shifting calculations. It may be worthwhile to obtain quantitatively the influence of various parameters on phase-shifting calculations. The errors due to detector non-linearity, source instability, digital-to-analog conversion and errors due to digitisation can all be modelled as error in measuring intensities. The influence of quarter-wave plate mismatch needs special treatment. It was pointed out that among the various algorithms for phase shifting the 6-step phase shifting method has stood the test of time. In the following error modelling for this method is given.

5.9.1 Influence of Error in Measuring Intensities

It is noted previously that several factors contribute to the correct measurement of intensities. In view of practical difficulties the intensity values used in Eqs. (5.29)–(5.32) may not be exact. The purpose of this section is to show how the error in measuring intensities influences the final estimation of the parameters. The error in estimation of isoclinic parameter as a function of error in intensity measurement can be expressed as

$$D\theta_{\mathrm{c}} = \sum_{j=3}^{6} \frac{\partial \theta_{\mathrm{c}}}{\partial I_j} DI_j \tag{5.40}$$

Where $D\theta_{\mathrm{c}}$ gives the error in θ_{c} and DI_j represent the error in intensity value. Substituting the required partial derivatives of Eq. (5.29) in Eq. (5.40), one gets

$$D\theta_{\mathrm{c}} = \frac{1}{2}\left[\frac{(I_4 - I_6)(DI_5 - DI_3) + (I_5 - I_3)(DI_6 - DI_4)}{(I_4 - I_6)^2 + (I_5 - I_3)^2}\right] \tag{5.41}$$

The intensity values vary widely over the problem domain. It may be worthwhile to express the error in intensity as a percentage of the intensity amplitude, I_{a}. The magnitude of I_{a} is obtained as

$$I_{\mathrm{a}} = \sqrt{(I_4 - I_6)^2 + (I_5 - I_3)^2 + (I_1 - I_2)^2} \tag{5.42}$$

For maximum error in the desired parameter, the error terms $(DI_5 - DI_3)$

and $(DI_6 - DI_4)$ may be considered equal to εI_a where ε is the percentage in intensity measurement. Since, I_a is a variable and its intensity value will also vary from pixel to pixel in the domain, the above assumption is realistic. Putting the above assumption in Eq. (5.41), the absolute maximum error in isoclinic can be written as

$$D\theta_{c|max} = \frac{\varepsilon I_a}{2}\left\{\frac{(I_4 - I_6) + (I_5 - I_3)}{(I_4 - I_6)^2 + (I_5 - I_3)^2}\right\} \tag{5.43}$$

Similarly, the error in fractional retardation of Eq. (5.32) can be expressed as the addition of error due to intensity measurement and that due to error in θ_c evaluation as

$$D\delta_c = \sum_{j=1}^{6}\frac{\partial\delta_c}{\partial I_j}DI_j + \frac{\partial\delta_c}{\partial\theta_c}D\theta_{c|max} \tag{5.44}$$

Substituting the required partial derivatives of Eq. (5.32) in Eq. (5.44), one gets

$$\begin{aligned} D\delta_c = {} & \frac{(I_5 - I_3)\sin 2\theta + (I_4 - I_6)\cos 2\theta}{(I_1 - I_2)^2 + \{(I_5 - I_3)\sin 2\theta + (I_4 - I_6)\cos 2\theta\}^2}(DI_2 - DI_1) \\ & + \frac{(I_1 - I_2)\sin 2\theta(DI_5 - DI_3) + (I_1 - I_2)\cos 2\theta(DI_4 - DI_6)}{(I_1 - I_2)^2 + \{(I_5 - I_3)\sin 2\theta + (I_4 - I_6)\cos 2\theta\}^2} \\ & + \frac{2(I_1 - I_2)\{(I_5 - I_3)\cos 2\theta - (I_4 - I_6)\sin 2\theta\}}{(I_1 - I_2)^2 + \{(I_5 - I_3)\sin 2\theta + (I_4 - I_6)\cos 2\theta\}^2}D\theta_{c|max} \end{aligned} \tag{5.45}$$

The error in intensity difference are replaced by εI_a, Eq. (5.45) reduces to

$$D\delta_{|max} = \frac{\begin{aligned}&\varepsilon I_a\{(I_1 - I_2 + I_5 - I_3)\sin 2\theta + (I_1 - I_2 + I_4 - I_6)\cos 2\theta\} \\ &+ 2D\theta_{c|max}(I_1 - I_2)\{(I_5 - I_3)\cos 2\theta - (I_4 - I_6)\sin 2\theta\}\end{aligned}}{(I_1 - I_2)^2 + \{(I_5 - I_3)\sin 2\theta + (I_4 - I_6)\cos 2\theta\}^2} \tag{5.46}$$

The modelling of the error expression obtained by Eqs. (5.43) and (5.44) have to be verified first. For the present analysis ε is taken as 10 percent. For verifying error modelling, the diametral line of a disc under diametral compression is taken. For this line the theoretical isoclinics, experimental isoclinics and the error value in isoclinic as obtained by Eq. (5.43) are shown in Fig. 5.14. In Fig. 5.14, hikes in experimental isoclinics are occurring at the half and full fringe orders. The value of error in isoclinic plot using Eq. (5.43) is also giving the same trend and thus the modelling of error is realistic. The two curves don't exactly match since in theoretical

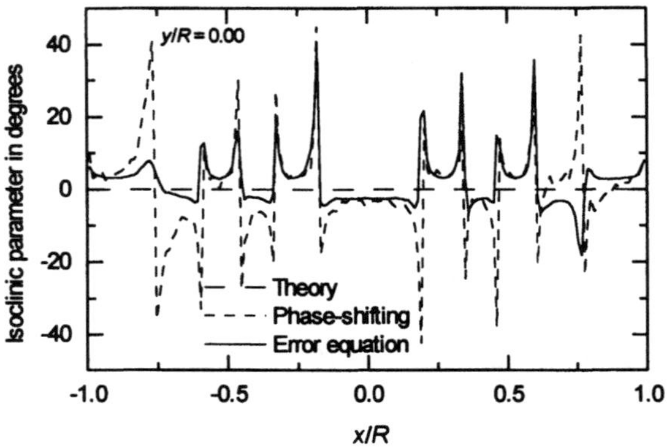

Fig. 5.14. Comparison of theoretical isoclinics, experimental isoclinics and the error in isoclinic as obtained by Eq. (5.43) for the diametral line of the disc under diametral compression

modelling, a uniform error of 10% is assumed. However, the error in experiment varies continuously and is quite high at $\delta_c = \pi, 2\pi, 3\pi, \ldots$ etc.

The error in retardation does not remain a constant but varies from point to point. Error predicted by Eq. (5.46) is plotted in Fig. 5.15. While plotting, the experimental values of intensity are used, and $D\theta_c$ is evaluated by Eq. (5.43) and the value of θ in Eq. (5.46) is taken from theory. The graph clearly shows that the error in retardation varies from point to point and varies between –0.037 to +0.037 fringe order. Instead of using $D\theta_c$ evaluated by Eq. (5.43), if the error in θ_c is obtained by taking the difference of theoretical and experimentally obtained θ, then error in retardation variation is shown in Fig. 5.16 which varies from –0.050 to +0.050 fringe order except for an occasional spike. Comparison of Figs. 5.15 and 5.16 shows that the error modelling by Eq. (5.46) is reasonably good. One can also observe from Figs. 5.15 and 5.16, the contribution to the total error of fringe order from error in isoclinic angle is not very significant.

Extending this idea for evaluating error in retardation by the use of Eq. (5.33), one gets the error as

$$D\delta|_{\text{square}} = \frac{\varepsilon\{(I_1 - I_2)(I_4 - I_6) + (I_1 - I_2)(I_5 - I_3) + (I_4 - I_6)^2 + (I_5 - I_3)^2\}}{I_a\sqrt{(I_4 - I_6)^2 + (I_5 - I_3)^2}} \quad (5.47)$$

The plot of the error is shown in Fig. 5.17. The error in retardation varies from –0.025 to +0.025 fringe order. Figure 5.17 clearly brings out that the overall error by using Eq. (5.47) is lower than that by Eq. (5.46). In view of this, the use of Eq. (5.33) to calculate fractional retardation is expected to give better results over the domain.

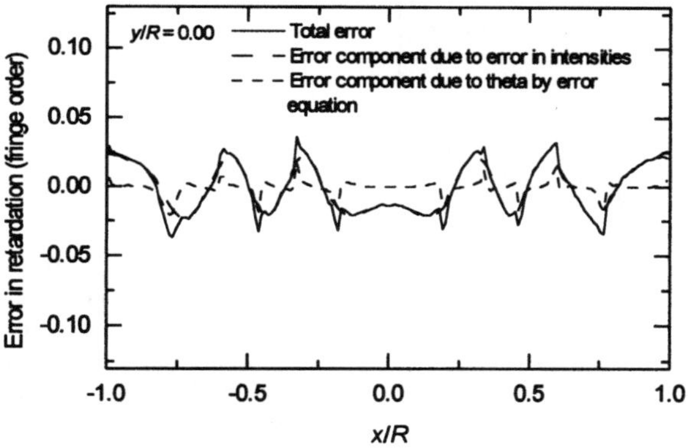

Fig. 5.15. Error predicted by Eq. (5.46) for evaluating fringe order along the diametral line of the disc under diametral compression. The figure also shows the total error, error component due to error in measuring intensities and error in finding θ_c. Here $D\theta_c$ predicted by Eq. (5.43) is used

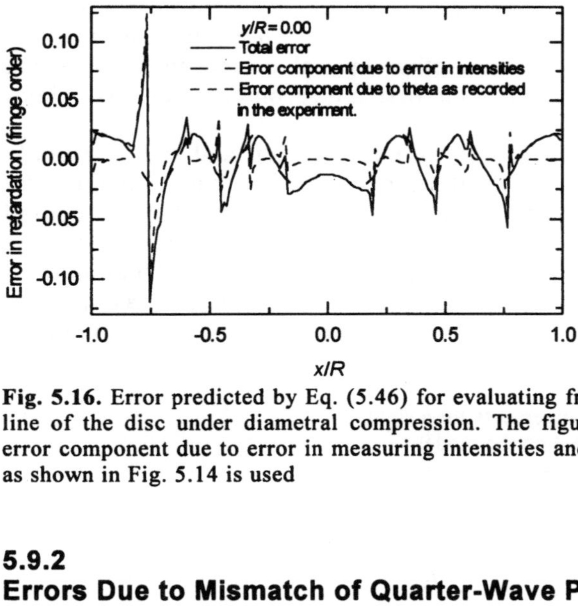

Fig. 5.16. Error predicted by Eq. (5.46) for evaluating fringe order along the diametral line of the disc under diametral compression. The figure also shows the total error, error component due to error in measuring intensities and error in finding θ_c. Here $D\theta_c$ as shown in Fig. 5.14 is used

5.9.2 Errors Due to Mismatch of Quarter-Wave Plates

Mismatch of quarter-wave plates play a significant role in influencing the phase-shifting calculations. In conventional photoelasticity, to account for fabrication problems associated with the quarter-wave plates, it is usually recommended to go for only those optical arrangements where the quarter-wave plates are crossed. Referring to Table 5.3, in the optical arrangements proposed by Patterson and Wang, the first and second quarter-wave plates

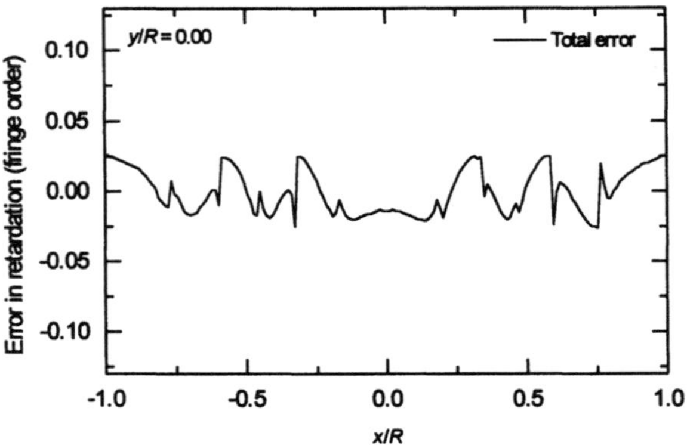

Fig. 5.17. Error in fractional retardation predicted by Eq. (5.47) for the diametral line of a circular disc under diametral compression

are not crossed. However, referring to Table 5.4, leaving the intensity equations I_3 & I_5, for all other optical arrangements, one can identify optical arrangements where the first and second quarter-wave plates are crossed. If the incident light is changed from left circularly polarised to right circularly polarised light then one can construct a table similar to Table 5.4 and the respective arrangements are shown in Table 5.8. Here again leaving the intensity equations I_3 & I_5, for all other optical arrangements, one can identify optical arrangements where the first and second quarter-wave plates are crossed.

Ajovalasit et al. [20] proposed the use of both left and right circularly polarised light to arrive at an optical arrangement that minimises the error due to mismatch of quarter-wave plates. The optical arrangements proposed by them and the intensity equations in the absence of quarter-wave plate error are shown in Table 5.9. Let the quarter-wave plate introduces an error of ε. In Eq. (5.10) if the retardation introduced by the quarter-wave plate is taken as $\pi/2 + \varepsilon$ one can get the intensity of light transmitted incorporating

Table 5.8. Multiple optical arrangements for a right circularly polarised light incident on the model

η \ β	0	$\pi/4$	$\pi/2$	$3\pi/4$	π
0	I_5	I_2	I_3	I_1	I_5
$\pi/4$	I_1	I_6	I_2	I_4	I_1
$\pi/2$	I_5	I_1	I_3	I_2	I_5
$3\pi/4$	I_2	I_6	I_1	I_4	I_2
π	I_5	I_2	I_3	I_1	I_5

Table 5.9. Polariscope arrangement to minimise the effect of quarter-wave plate mismatch in phase shifting calculation

ξ	η	β	Intensity equation
Ajovalasit et al.			
$3\pi/4$	$\pi/4$	$\pi/2$	$I_1 = I_b + \frac{I_a}{2}(1+\cos\delta)$
$3\pi/4$	$\pi/4$	0	$I_2 = I_b + \frac{I_a}{2}(1-\cos\delta)$
$3\pi/4$	0	0	$I_3 = I_b + \frac{I_a}{2}(1-\sin 2\theta \sin\delta)$
$3\pi/4$	$\pi/4$	$\pi/4$	$I_4 = I_b + \frac{I_a}{2}(1+\cos 2\theta \sin\delta)$
$\pi/4$	0	0	$I_5 = I_b + \frac{I_a}{2}(1+\sin 2\theta \sin\delta)$
$\pi/4$	$3\pi/4$	$\pi/4$	$I_6 = I_b + \frac{I_a}{2}(1-\cos 2\theta \sin\delta)$

Table 5.10. Intensity equations including quarter-wave plate error for the optical arrangement listed in Table 5.9

Image No.	Intensity equation
1	$I_1 = I_b + I_a\left[(1-\cos^2 2\theta \sin^2\varepsilon)\cos^2\frac{\delta}{2} + \cos^2 2\theta \sin^2\varepsilon\right]$
2	$I_2 = I_b + I_a\left[(1-\cos^2 2\theta \sin^2\varepsilon)\sin^2\frac{\delta}{2}\right]$
3	$I_3 = I_b + \frac{I_a}{2}\left[1+(\cos^2 2\theta + \sin^2 2\theta \cos\delta)\sin\varepsilon - \sin 2\theta \cos\varepsilon \sin\delta\right]$
4	$I_4 = I_b + \frac{I_a}{2}\left[1+(1-\cos\delta)\sin 2\theta \cos 2\theta \sin\varepsilon + \cos 2\theta \cos\varepsilon \sin\delta\right]$
5	$I_5 = I_b + \frac{I_a}{2}\left[1+(\cos^2 2\theta + \sin^2 2\theta \cos\delta)\sin\varepsilon + \sin 2\theta \cos\varepsilon \sin\delta\right]$
6	$I_6 = I_b + \frac{I_a}{2}\left[1+(1-\cos\delta)\sin 2\theta \cos 2\theta \sin\varepsilon - \cos 2\theta \cos\varepsilon \sin\delta\right]$

the effect of quarter-wave plate mismatch. Similarly one can repeat the exercise for a right circularly polarised light. These equations are very lengthy for a generic arrangement. If appropriate values of the orientation of second quarter-wave plate and the analyzer are substituted then the respective equations are much simpler (Table. 5.10).

From the equations listed in Table 5.10, the isoclinic parameter is obtained as

$$\theta_c = \frac{1}{2}\tan^{-1}\left(\frac{I_5 - I_3}{I_4 - I_6}\right) = \frac{1}{2}\tan^{-1}\left(\frac{I_a \sin\delta \sin 2\theta \cos\varepsilon}{I_a \sin\delta \cos 2\theta \cos\varepsilon}\right) \quad \text{for } \sin\delta \neq 0 \quad (5.48)$$

Though the error in the quarter-wave plates modifies the intensities, it is clear from Eq. (5.48) the evaluation of θ_c is not affected by it. It is shown in ref. [20], that if the intensity equations based on only left circularly polarised light is used, the value of θ_c is significantly affected.

The fractional retardation is obtained as

$$\begin{aligned}\delta_c &= \tan^{-1}\left\{\frac{(I_5 - I_3)\sin 2\theta_c + (I_4 - I_6)\cos 2\theta_c}{(I_1 - I_2)}\right\} \\ &= \tan^{-1}\left\{\frac{I_a \sin\delta \cos\varepsilon}{I_a[\cos\delta(1-\cos^2 2\theta_c \sin^2\varepsilon) + \cos^2 2\theta_c \sin^2\varepsilon]}\right\}\end{aligned} \quad (5.49)$$

The inspection of Eq. (5.49) shows that the mismatch in quarter-wave plates has a direct influence in the evaluation of δ_c. The nature of this error is that the maximum value of the phase is spatially shifted and its value is slightly less than predicted by theory.

For the common tolerances of ± 10 nm of the retardance of the quarter-wave plates, the error ε for sodium vapour is within ± 0.1 radians ($\pm 6^\circ$). It is shown in ref. [20], that for this typical error in quarter-wave plates, the common error in retardation is within ± 0.017 fringe orders for a 4-step or 6-step phase shifting method based on left circularly polarised light. Though the error in retardation is small, the error in isoclinics is quite significant for these arrangements.

The fact that the mismatch of quarter-wave plates significantly affects the isoclinic calculations and less significantly the retardation calculations, has also been noted in ref. [39]. Ji and Patterson [39] studied the use of monochrome filters in conjunction with white light using only one set of quarter-wave plates. In view of using different wavelengths, there will be significant mismatch of the quarter-wave plate for specific wavelengths.

The error in isoclinic calculations due to quarter-wave plate mismatch can be totally eliminated if in 6-step phase shifting algorithm both left and right circularly polarised lights are used. Further, for the common tolerances of the quarter-wave plates, the error in retardation calculation is within ± 0.0018 fringe orders only, which is an order of less than the previously mentioned methods.

Thus, the use of left and right circularly polarised light for data acquisition has advantages and is recommended for general use.

5.10 Evaluation of Isoclinic Value by Phase Shifting Technique

Phase-shifting techniques discussed so far provide isochromatic fringe order with high accuracy but isoclinic determination is not satisfactory. One generally gets discontinuous isoclinic fringes due to isochromatic-isoclinic interaction. To bridge the gap, use of interpolation approach [49], multiple loads [21] or multiple wavelengths [33, 50] is reported.

5.10.1 Use of Two Loads to Get Continuous Isoclinic Contours

In this, the polarizer and analyzer are kept at arbitrary positions. The intensity of light transmitted is given by Eq. (5.2). If the term I_b is added to account for stray light/background illumination, one gets

$$I_i = I_b + I_a\left[\cos^2\frac{\delta}{2}\cos^2(\beta_i-\alpha_i)+\sin^2\frac{\delta}{2}\cos^2(\beta_i+\alpha_i-2\theta)\right] \qquad (5.50)$$

If one takes all possible combinations of α and β in steps of $\pi/4$, then only six intensity equations turn out to be unique (Table 5.11) and the rest repeat any one of the six intensity equations (Table 5.12).

Table 5.11. Table showing that in a plane polariscope, only six intensity equations turn out to be unique for all possible combinations of α and β in steps of $\pi/4$

α \ β	0	$\pi/4$	$\pi/2$	$3\pi/4$	π
0	I_6	I_3	I_1	I_2	I_6
$\pi/4$	I_3	I_5	I_2	I_4	I_3
$\pi/2$	I_1	I_2	I_6	I_3	I_1
$3\pi/4$	I_2	I_4	I_3	I_5	I_2
π	I_6	I_3	I_1	I_2	I_6

Table 5.12. Table showing the intensity equations for different orientations of the polarizer and analyzer

α	β	Intensity equation
0	$\pi/2$	$I_1 = I_b + I_a \sin^2\frac{\delta}{2}\sin^2 2\theta$
$\pi/4$	$\pi/2$	$I_2 = I_b + \frac{I_a}{2}\left[1 - \sin^2\frac{\delta}{2}\sin 4\theta\right]$
0	$\pi/4$	$I_3 = I_b + \frac{I_a}{2}\left[1 + \sin^2\frac{\delta}{2}\sin 4\theta\right]$
$\pi/4$	$3\pi/4$	$I_4 = I_b + I_a \sin^2\frac{\delta}{2}\cos^2 2\theta$
$\pi/4$	$\pi/4$	$I_5 = I_b + I_a\left[\cos^2\frac{\delta}{2} + \sin^2\frac{\delta}{2}\sin^2 2\theta\right]$
0	0	$I_6 = I_b + I_a\left[\cos^2\frac{\delta}{2} + \sin^2\frac{\delta}{2}\cos^2 2\theta\right]$

From Table 5.12, using only the first four intensity equations, the isoclinic parameter can be obtained as

$$\theta_c = \frac{1}{4}\tan^{-1}\left(\frac{I_3 - I_2}{I_4 - I_1}\right) = \frac{1}{4}\tan^{-1}\left(\frac{I_a \sin^2\frac{\delta}{2}\sin 4\theta}{I_a \sin^2\frac{\delta}{2}\cos 4\theta}\right) \quad \text{for } \sin^2\frac{\delta}{2} \neq 0 \quad (5.51)$$

Use of atan2 function is recommended for evaluating θ_c from Eq. (5.51) (*2dplane2.c*, CD). The isoclinic evaluated will have discontinuities when $\delta = 0, 2\pi, 4\pi, \ldots$. To bridge the gap, if the experiment is repeated for two different loads, the extinction location of fractional retardation can be spatially shifted. A load factor of 1.5 between the experiments is found to be good. Figure 5.18a shows the whole field plot of isoclinics drawn in steps of 10-deg for a circular disc (diameter 60 mm, thickness 5 mm) under a diametral load of 335.05 N. The discontinuities, which appear in Fig. 5.18a, are around full fringe orders. A tolerance of ±0.20 fringe order is used to plot them. Figure 5.18b shows a similar plot of isoclinics obtained for a diametral load of 502.5 N. Continuous isoclinics for the whole field is obtained by performing a logical 'OR' operation between the images shown in Figs. 5.18a and 5.18b (*oring.c*, CD). It is seen that a considerable shift in isochromatic fringes can be obtained by multiple loads. Though the methodology can be readily applied for analysing two-dimensional problems, the technique is not suitable for analysing stress frozen slices. Use of multiple wavelengths may be desirable in such cases.

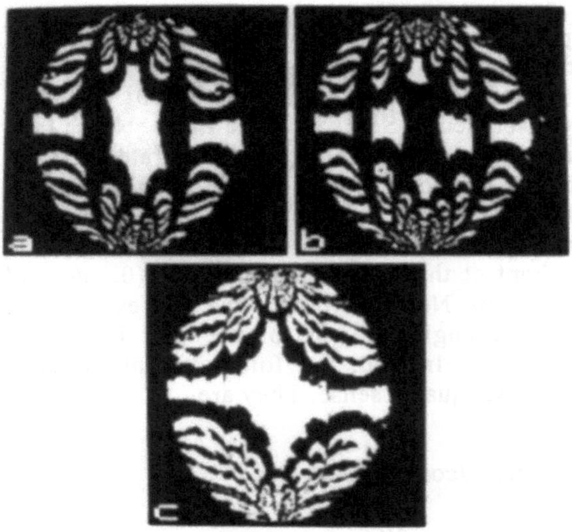

Fig. 5.18. Generation of continuous isoclinics (plotted in steps of 10-deg) by two-load method: **a** isoclinics for load 335.05 N **b** isoclinics for load 502.5 N **c** continuous isoclinic obtained by logical 'OR' operation of 'a' and 'b' (Courtesy of BSSM [21])

5.10.2 Use of Multiple Wavelengths to Get Continuous Isoclinic Contours

It is well known that the retardation introduced by the model is a function of the wavelength. This property is used to shift the isochromatic fringe orders in the multi-wavelength technique.

Two approaches have been reported on the generation of different wavelengths. Kihara [50] used a laser source for illumination. He used He-Ne laser and Ar-ion laser to obtain wavelengths of 632.8 nm, 514.5 nm and 457.9 nm. Nurse [33] obtained different wavelengths by using a white light source coupled with monochrome filters. He used three monochrome filters with the central wavelengths as 580 nm, 520 nm and 470 nm.

A plane polariscope arrangement is well suited for the determination of isoclinics. If one attempts to get fractional retardation, only an arccosine function is possible to evaluate the fractional retardation. This leads to an ambiguity in the sign of the fractional retardation. Following the work of Buckberry and Towers [35], Nurse [33] used the following relationship between the wavelengths:

$$\lambda_2 = \frac{2\lambda_1\lambda_3}{(\lambda_1 + \lambda_3)} \tag{5.52}$$

Equation (5.52) helps in phase unwrapping the fractional retardation information. The details are discussed in Chap. 6.

The multiple wavelength technique proposed by Kihara [50] is just a repetition of the methodology discussed in Sect. 5.7 three times. Nurse [33] proposed an over-deterministic method for evaluating the parameters. He used six optical arrangements and recorded in all eighteen images for data processing. The six optical arrangements are obtained by keeping the polarizer (α) at two different angles ($\pi/2$ and $3\pi/4$) and for each of these positions, the analyzer (β) is kept at three different positions (0, $\pi/4$ and $\pi/2$). No attempt has been made by Nurse to obtain the expressions for fractional retardation and isoclinic angle in terms of the light intensities recorded as in other methodologies. Instead, the following intermediate parameters are determined in a least squares sense. They are

$$\begin{aligned} V_1 &= I_a \sin 2\theta \cos 2\theta \sin^2\frac{\delta}{2} \\ V_2 &= I_a \sin^2 2\theta \sin^2\frac{\delta}{2} \\ V_3 &= I_a \cos^2 2\theta \sin^2\frac{\delta}{2} \\ V_4 &= I_a \end{aligned} \tag{5.53}$$

The isoclinic parameter is obtained by

$$\theta_c = \begin{cases} \dfrac{1}{2}\dfrac{\tan^{-1}(V_2/V_1)}{\left|\tan^{-1}(V_2/V_1)\right|}\sin^{-1}\left(\dfrac{V_2}{V_2+V_3}\right) & \text{for } \sin 2\theta < \cos 2\theta \\ \text{else} & \\ \dfrac{1}{2}\tan^{-1}\left(\dfrac{V_2}{V_1}\right) = \dfrac{1}{2}\tan^{-1}\left(\dfrac{\sin 2\theta \sin^2\frac{\delta}{2}}{\cos 2\theta \sin^2\frac{\delta}{2}}\right) & \end{cases} \tag{5.54}$$

The values of V_i in Eq. (5.54) are substituted for that wavelength for which the modulation is high. The fractional retardation is obtained as

$$\delta_c = \cos^{-1}(1 - 2(V_2 + V_3)/V_4) \tag{5.55}$$

In view of the use of arccosine function for δ, the sign of the fractional retardation is not known. This ambiguity is resolved by choosing the wavelengths following Eq. (5.52) the details of which are discussed in Chap. 6.

5.11 Polarization Stepping for Isoclinic Determination

Brown and Sullivan introduced use of polarization stepping for isoclinic determination in 1990. In polarization stepping, the isoclinic images are recorded in known steps to provide enough equations to solve for the determination of the isoclinic parameter. The advantage of such an approach is that the commercial polariscope that has been hitherto used for conventional photoelasticity could directly be used for polarization stepping without any modifications. The governing intensity equation is Eq. (5.3) and by adding a term I_b to account for stray light/background illumination one gets

$$I_i = I_b + I_a \sin^2\frac{\delta}{2}\sin^2 2(\theta - \beta_i) \tag{5.56}$$

The polarization steps used by Brown and Sullivan and the appropriate intensity equations are summarised in Table 5.13. The isoclinic parameter is obtained as

$$\theta_c = \frac{1}{4}\tan^{-1}\left(\frac{I_4 - I_2}{I_3 - I_1}\right) = \frac{1}{4}\tan^{-1}\left(\frac{I_a \sin^2\frac{\delta}{2}\sin 4\theta}{I_a \sin^2\frac{\delta}{2}\cos 4\theta}\right) \quad \text{for } \sin^2\frac{\delta}{2} \neq 0 \tag{5.57}$$

As in phase shifting, here again Eq. (5.57) shows that the isoclinic parameter is undefined when $\delta = 0, 2\pi, 4\pi, \ldots$. To overcome this, Brown and Sullivan suggested that in recording the polarization-stepped images, the loading should be adjusted such that the maximum isochromatic fringe order does not exceed 0.5. Buckberry and Towers suggested the use of white light as a source and recorded polarization stepped images by using two monochrome filters of 550 nm and 633 nm. In view of the dependence of the retardation on the wavelength of light used, the undefined zones will be different for different wavelengths. The results of isoclinic contours for two different wavelengths are appropriately combined to get continuous isoclinic image over the domain.

If the influence of background light is not considered, Buckberry and Towers observed that the fractional retardation could be obtained by grabbing one more image in bright-field plane polariscope. Inspection of Eqs. (5.3) and (5.5) that addition of these gives I_a. The fractional retardation is obtained as

$$\delta_c = \cos^{-1}\left(\frac{I_a - 2(I_1 + I_3)}{I_a}\right) \tag{5.58}$$

Table 5.13. Polariscope arrangements and intensity equations for polarization stepping schemes

β	Intensity equation
Brown and Sullivan, Buckberry and Towers (Polarizer & Analyzer crossed; $\alpha = \beta + \pi/2$)	
0	$I_1 = I_b + I_a \sin^2\frac{\delta}{2}\sin^2 2\theta$
$\pi/8$	$I_2 = I_b + \frac{I_a}{2}\sin^2\frac{\delta}{2}(1-\sin 4\theta)$
$\pi/4$	$I_3 = I_b + I_a \sin^2\frac{\delta}{2}\cos^2 2\theta$
$3\pi/8$	$I_4 = I_b + \frac{I_a}{2}\sin^2\frac{\delta}{2}(1+\sin 4\theta)$
Mawatari et al. (Polarizer & Analyzer crossed; $\alpha = \beta + \pi/2$)	
0	$I_1 = I_b + I_a \sin^2\frac{\delta}{2}\sin^2 2\theta$
$\pi/12$	$I_2 = I_b + \frac{I_a}{4}(\sqrt{3}\sin 2\theta - \cos 2\theta)^2 \sin^2\frac{\delta}{2}$
$\pi/3$	$I_3 = I_b + \frac{I_a}{4}(\sqrt{3}\cos 2\theta + \sin 2\theta)^2 \sin^2\frac{\delta}{2}$
$\pi/4$	$I_4 = I_b + I_a \sin^2\frac{\delta}{2}\cos^2 2\theta$
Chen and Lin (Polarizer & Analyzer parallel; $\alpha = \beta$)	
0	$I_1 = I_a\left[1-\sin^2 2\theta \sin^2\frac{\delta}{2}\right]$
$\pi/8$	$I_2 = I_a\left[1-\frac{1}{2}(\sin 2\theta - \cos 2\theta)^2 \sin^2\frac{\delta}{2}\right]$
$\pi/4$	$I_3 = I_a\left[1-\cos^2 2\theta \sin^2\frac{\delta}{2}\right]$
	Also record intensity data by keeping an unloaded model for each of these positions

To facilitate unwrapping of the fractional retardation, Buckberry and Towers [35] used wavelengths of 550 nm, 589 nm and 633 nm following the relationship of Eq. (5.52).

An interesting feature of the polarization stepping scheme is that an appropriate processing of the intensities recorded can help to generate whole-field isochromatic contours. This was first brought into focus by Mawatari et al. who showed that the addition of intensities of any two polarization stepped images separated by $\pi/4$ gives the isochromatic fringe field. They did not consider the effect of background light/stray light for this calculation. However, they considered the effect of background illumination for the calculation of isoclinic parameter. Leaving the influence of term I_b, the sum of intensities I_1 and I_3 of Table 5.13 gives

$$I_1 + I_3 = I_a \sin^2\frac{\delta}{2} \tag{5.59}$$

The intensity equation thus obtained represents the isochromatic fringe field. Petrucci [51] proposed a different equation by considering the effect of background illumination and the isochromatic fringe field is obtained by

$$I_a \sin^2\frac{\delta}{2} = \sqrt{(I_3 - I_1)^2 + (I_4 - I_2)^2} \tag{5.60}$$

The intensities referred in Eqs. (5.59) and (5.60) correspond to the polarization steps used by Brown and Sullivan.

Mawatari et al. [52] proposed a scheme to evaluate the isoclinic angle from four polarization stepped images recorded at angles of $\beta = 0$, $\pi/12$, $\pi/3$ and $\pi/4$. These angles are chosen such that two sets of polarization stepped images separated by $\pi/4$ are obtained. The intensity equations thus obtained were not sufficient to get the isoclinic angle as an arctangent function. They, however, proposed an elaborate technique to process the intensity information and finally got the expression for evaluating θ as an arcsine function. The discontinuity of isoclinic fringe orders due to isochromatic interaction has been taken into account by using appropriate interpolation functions.

Chen and Lin [53] proposed another polarization-stepped scheme. In this, initially three bright-field images in plane polariscope arrangement were obtained. From these bright-field images, normalised dark-field images are constructed using trigonometric identity and the concept of image division. For performing image division, the intensity corresponding to unloaded images are also to be recorded. The orientation of β and the appropriate intensity equations are summarised in Table 5.13. The intensity equation corresponding to dark-field is obtained by

$$I_i' = 1 - \frac{I_i}{I_{ui}} \tag{5.61}$$

where, I_{ui} is the intensity recorded using an unloaded model. The isoclinic parameter is then obtained as

$$\theta_c = \frac{1}{4}\tan^{-1}\left(\frac{I_1' + I_3' - 2I_2'}{I_3' - I_1'}\right) \tag{5.62}$$

Use of atan2 function to evaluate θ_c from Eq. (5.62) is recommended.

5.12 Fourier Transform Methods for Photoelastic Data Acquisition

5.12.1 Use of Carrier Fringes

Use of carrier fringes for photoelastic analysis was first introduced by Rupeng [54]. The advantage of the approach is that by introducing a high frequency spatial carrier fringes, the fringe orders in the model are made to increase monotonically over the domain. This makes the fringe ordering simpler. A quartz wedge can be introduced after the model to superimpose spatial carrier fringes. Let the retardation introduced by the wedge (fast axis along the horizontal) for a particular point in the model be δ_1, then the intensity of light transmitted is obtained as

$$I = I_a\left[\sin^2\left(\frac{\delta-\delta_1}{2}\right)\sin^2\theta + \sin^2\left(\frac{\delta+\delta_1}{2}\right)\cos^2\theta\right] \tag{5.63}$$

The orientation of the carrier fringes has an influence on the way the model fringes are modulated. Figure 5.19 shows the theoretically simulated fringe patterns in a circular disc under diametral compression with carrier fringes

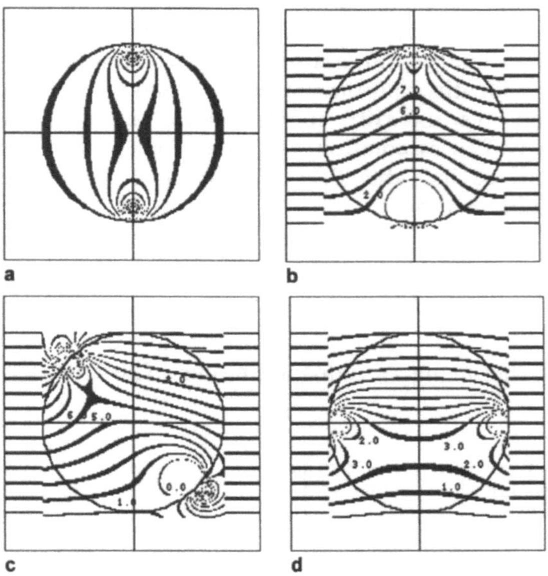

Fig. 5.19. Role of carrier fringes in modulating the photoelastic fringe patterns: **a** Dark-field isochromatics **b** with carrier fringes at 90-deg to the loading direction **c** at 45-deg and **d** at 0-deg. To facilitate theoretical plotting, the disc is rotated with respect to the carrier fringes

of different orientation. The carrier fringes increase in the vertical direction from 0 to 13 fringe orders. To facilitate plotting, instead of the carrier fringes, the loading in the model is rotated. The figure clearly brings out the interaction of carrier fringes with the model fringes. The model boundary is clearly indicated in the figure. Due to numerical difficulties, the zone immediately outside this boundary has spurious fringes. At a little distance away, one can clearly see the carrier fringes monotonically increasing from bottom to top. For the purpose of illustration, a very low frequency carrier fringes of 0.5 fr/mm is used for theoretical simulation. In practice, it is desirable to use as high a carrier frequency as possible.

Quan et al. [29] have shown that by using carrier fringes and operating in the frequency domain, it is possible to evaluate the fractional fringe orders using only one photograph. They introduced a quartz wedge after the model such that the carrier fringes increase in the horizontal direction. Let the spatial frequency of the carrier fringes be f_s and at position x (measured from the tip of the quartz edge) the retardation introduced (δ_1) is $2\pi f_s x$. For the purpose of Fourier fringe analysis, Eq. (5.63) is rewritten in the following form

$$I(x,y) = I_b(x,y) + C(x,y)e^{2\pi f_s x} + C^*(x,y)e^{-2\pi f_s x} \tag{5.64}$$

where $C^*(x, y)$ is the complex conjugate of $C(x, y)$ and it contains the phase information δ. Fourier transform is obtained for every horizontal line in the image. The introduction of the carrier frequency permits the separation of the terms of the FT related to $I_b\,(x, y)$. By filtering the FT in the frequency domain and evaluating the inverse FT, the retardation is obtained as

$$\delta_c = \tan^{-1}\left(\frac{\mathrm{Im}(C(x,y))}{\mathrm{Re}(C(x,y))}\right) \tag{5.65}$$

In view of the use of arctangent function the phase information is similar to Eq. (5.32). Since, a single photograph is sufficient, the technique could be applied for dynamic studies. The technique requires further development for it to be suitable for practical applications.

5.12.2 Use of Multiple Polarization Stepped Images

Morimoto et al. [30] used the Fourier transform technique to separate isochromatics and isoclinics from ninety polarization stepped images recorded in steps of 1 degree from –45 to 44 degrees (Fig. 5.20). The Eq. (5.3) can be rewritten as

$$I(\beta) = \frac{I_a}{2}[1 - \cos 4(\theta - \beta)]\sin^2\frac{\delta}{2} \tag{5.66}$$

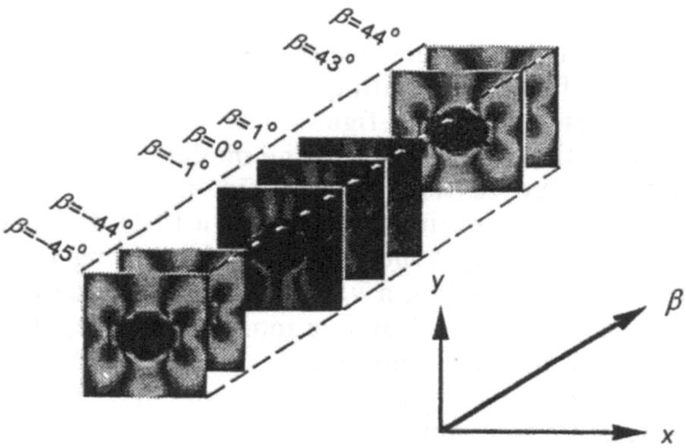

Fig. 5.20. The sequence of polarization-stepped images being used to generate three-dimensional data to perform Fourier transforms (Courtesy: ref. [30])

Equation (5.66) shows that at each point of the fringe field the intensity is a periodic function of β with a period $T = \pi/2$. Equation (5.66) is expressed in Fourier series with respect to β as

$$I(\beta) = \sum_{n=-\infty}^{\infty} c_n \exp(in\omega_\circ \beta) \tag{5.67}$$

where

$$c_n = \frac{1}{T} \int_{-T/2}^{T/2} I(\beta) \exp(-in\omega_\circ \beta) d\beta \tag{5.68}$$

and

$$\omega_\circ = \frac{2\pi}{T} = \frac{2\pi}{\pi/2} = 4 \tag{5.69}$$

Morimoto et al. showed that in the frequency domain, $\omega = 0$ corresponds to isochromatics (Fig. 5.21). The Fourier coefficient corresponding to $\omega = -\omega_\text{o}$ can be written as

$$c_{-1} = -I_\text{a} \sin^2 \frac{\delta}{2} e^{\text{i}4\theta} \tag{5.70}$$

The isoclinic parameter is obtained as

$$\theta_\text{c} = \frac{1}{4} \tan^{-1} \left(\frac{\text{Im}(c_{-1})}{\text{Re}(c_{-1})} \right) \tag{5.71}$$

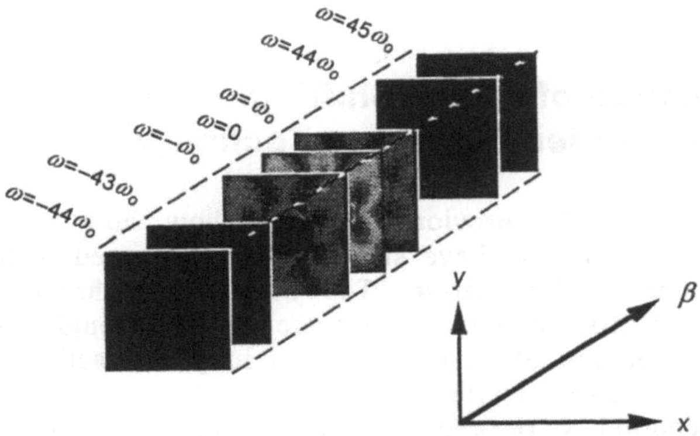

Fig. 5.21. The β directional frequency spectrum of image data of Fig. 5.20 (Courtesy: ref. [30])

5.12.3 Use of Load Stepping

Ng proposed the use of load stepping to determine the fractional retardation. In this approach, dark and bright-field images from a conventional circular polariscope are recorded for various load steps. By subtracting the dark-field intensity [Eq. (5.15)] from bright-field intensity [Eq. (5.16)] one gets

$$I(x,y) = I_a \cos\delta(x,y,k) \tag{5.72}$$

where k denotes the load stepping number. Equation (5.72) can be expressed as

$$I(x,y) = C(x,y,k) + C^*(x,y,k) \tag{5.73}$$

The fractional retardation is obtained as

$$\delta_c = \tan^{-1}\left(\frac{\mathrm{Im}[C(x,y,k)]}{\mathrm{Re}[C(x,y,k)]}\right) \tag{5.74}$$

Ng used a 32-step load procedure and in all grabbed 64 images to obtain the fractional retardation. The choice of the load step influences the accuracy of the method and he suggested that the load step should be so selected that it introduces a change of phase of π or less.

5.13 Comparative Evaluation of Phase Shifting, Polarization Stepping and Fourier Transform Techniques

The techniques presented in the previous sections show how data acquisition methods in photoelasticity have significantly progressed with the effective use of intensity information. The numerous techniques mentioned also indicate the different ways the researchers have sought to solve the problem of automation of data acquisition in photoelasticity. Although it is desirable to have a comprehensive quantitative comparison of these techniques, no reported work exists on evaluating the relative performance of phase shifting, polarization stepping and Fourier transform methods. Nevertheless, several observations on these techniques could be made based on the following aspects:

1. Accounting for background light
2. Suitability of the technique to determine both isoclinic and isochromatic parameters.
3. Ease of phase unwrapping of fractional retardation
4. Capability to generate continuous isoclinics and whether it can always provide orientation of one of the principal directions
5. Accuracy of the evaluated parameters
6. The total number of images needed.

Most of the phase shifting techniques based on a plane polariscope does not take into account the influence of background light in the computation of the parameters. This can in turn affect the accuracy of the evaluated parameters. Further, the fractional retardation is evaluated by using an arccosine function and unwrapping of this is little cumbersome. In general one may need to go for recording the data for multiple wavelengths to unwrap. For getting continuous isoclinics, the use of techniques based on plane polariscopes are ideal. The use of two-load method will be useful for 2-D problems and it also takes into account the effect of background illumination. For analysing 3-D problems using stress freezing and slicing, one may have to resort to the use of multiple wavelengths.

All the phase shifting techniques based on a circular polariscope (4-step, 5-step, 6-step and modified 6-step to account for quarter-wave plate mismatch) take into account explicitly the influence of the background light. Hence, the results are expected to be more accurate. Leaving the 5-step method (Hecker and Morche) all other techniques in this category provide both isoclinic and isochromatic data. In view of the use of arctangent function for evaluating the parameters, the phase unwrapping is simpler. There is no need to go for the use of multiple wavelengths to effect phase unwrapping. The methods based on circular polariscope are the ones very actively pursued by various investigators and issues related to accuracy [19, 55], accounting for low-modulation [17] and mismatch of quarter-wave plates has been addressed [20]. A quantitative comparative study of the phase shifting methods [56] based on both plane and circular polariscopes has brought out the elegance of the 6-step phase shifting method. In general it is recommended to use the modified 6-step phase shifting method as it accounts for the quarter-wave plates mismatch. In case, the objective is to record slowly varying problem situations, the use of 4-step method may be used at the cost of accuracy.

Polarization stepping methods are basically suited for evaluating the isoclinics. The background illumination is accounted for in most of the techniques. The number of images needed is comparable to the previously mentioned techniques.

Fourier transform methods are basically computationally intensive. They can provide either isoclinic or isochromatic parameters only. In view of the arctangent function for parameter evaluation, phase unwrapping is simple. The number of images is very high except for the technique proposed by Quan et al.

5.14 Closure

In this chapter the various phase shifting, polarization stepping and Fourier transform methods based on monochrome image processing systems have been presented. A comprehensive presentation of the development of the intensity equations for these techniques are given. Newer methodologies are continuously appearing in the literature [57, 58] and their methodology could be well understood based on the intensity equations presented in this chapter.

For getting continuous isoclinics and for unwrapping fractional retardation in some of the techniques, the use of multiple wavelengths has been indicated. The actual employment of these would become much simpler if colour image processing techniques are used. The methodology can be further simplified if a proper light source in conjunction with a colour image processing system is used. Aspects related to this are discussed in Chap. 7.

None of the techniques discussed in this chapter can provide principal stress direction corresponding to only one of principal stresses in the complete model domain. This has to be the focus of future research. Phase unwrapping of fractional retardation could be completely automated if this is achieved.

Exercises

5.1 How intensity information is used in early attempts for designing automatic polariscopes?

5.2 Describe in simple terms the methodology of phase shifting in photoelasticity. What are the differences between phase shifting applied to other interferometric techniques and photoelasticity?

5.3 Can you find fractional fringe orders based on dark and bright-field fringe patterns alone? Explain the potential of this to solve real problem situations.

5.4 Why there are so many techniques for determining fractional retardation and isoclinic angle? Compare the methodologies based on a plane polariscope.

5.5 Develop a software code to implement the various plane polariscope based algorithms.

5.6 How low-modulation areas are accounted for in the 6-step phase shifting method based on a circular polariscope?

5.7 Discuss the importance of using isoclinic angle corresponding to either σ_1 or σ_2 in phase shifting calculations. Are there methods available as of now to provide this over the whole field?

5.8 Mention the role of different error sources and methods to minimise them in practice.

5.9 What is the role of quarter-wave plate mismatch in phase-shifting technique? How can this be minimised?

5.10 Compare the various methodologies available to determine continuous isoclinic over the domain.

5.11 What is the interesting result that can be derived from intensity equation of polarization-stepping schemes?

5.12 Given a choice, which of the phase shifting techniques would you choose to solve a practical problem?

5.13 Mention the research goal of future developments in phase shifting technique applied to photoelasticity.

References

1. Zandman F (1959) Photostress analysis. Product Engng
2. Sapaly J (1961) Contribution a l'Etude de la Photoextensometric Statique et dynamique, Th Sc Phys Series A3761 (4612)
3. Robert AJ (1967) New methods in photoelasticity. Exp Mech 7:224–232
4. Pant B (1963) A method for determining integral fringe orders in photoelastic analysis. Exp Mech 3:173–174
5. Young WC (1969) An automated process for three-dimensional photoelastic analysis. 9:275–280
6. Penkovsky AI, Adagamova RT, Kulikov NN, Kapalov WP, Kuzmina RI (1971) Automatic apparatus for photoelastic investigations. Proc of VII all union Conference on Photoelasticity Tallin:106–110
7. Allison and Nurse P (1971) Optical data acquisition for an automatic polariscope. Proc of VII all union Conference on Photoelasticity Tallin:93–105
8. Redner S (1974) New automatic polariscope system. Exp Mech 14:486–491
9. Allison, Nurse P (1979) Interactive control for an automatic polariscope. Strain 15:90–94
10. Patterson EA (1988) Automated photoelastic analysis. Strain 24(1):15–20
11. Marston RE (1985) An automatic micropolariscope its design, development and use for tubular joint stress analysis. PhD thesis, University of Nottingham

12. Hecker FW, Morche B (1986) Computer-aided measurement of relative retardations in plane photoelasticity. In: Wieringa H (ed) Experimental stress analysis. Martinus Nijhoff, Dordrecht, The Netherlands, pp 535–542
13. Patterson EA, Wang ZF (1991) Towards full field automated photoelastic analysis of complex components. Strain 27(2):49–56
14. Ramesh K, Mangal SK (1998) Data acquisition techniques in digital photoelasticity: a review. Optics and Lasers in Engng 30(1):53–75 Errata (1999) 31:85
15. Ajovalasit A, Barone S, Petrucci G (1998) A review of automated methods for the collection and analysis of photoelastic data. J of Strain Analysis for Engng Des 33(2):75–91
16. Barone S, Burriesci G, Petrucci G (1997) Automated photoelasticity by phase stepping technique. Proc of XIV Imeko World Congress, Tampere IXA:57–62
17. Quiroga JA, González-Cano A (1997) Phase measuring algorithm for extraction of isochromatics of photoelastic fringe patterns. Appl Optics 36(32):8397–8402
18. Wang ZF, Patterson EA (1995) Use of phase-stepping with demodulation and fuzzy sets for birefringence measurement. Optics and Lasers in Engng 22:91–104
19. Ramesh K, Mangal SK (1999) Phase-shifting calculations in 2-D photoelasticity: revisited. Unpublished work
20. Ajovalasit A, Barone S, Petrucci G (1998) A method for reducing the influence of the quarter-wave plate error in phase-shifting photoelasticity. J Strain Analysis for Engng Des 33(3):207–216
21. Mangal SK, Ramesh K (1999) Use of multiple loads to extract continuous isoclinic fringes by phase-shifting. Strain 35(1):15–17 Errata (1999). 35(2):76
22. Ekman MJ, Nurse AD (1998) Absolute determination of the isochromatic parameter by load-stepping photoelasticity. Exp Mech 38(3):189–195
23. Ekman MJ, Nurse AD (1998) Completely automated determination of two-dimensional photoelastic parameters using load stepping. Opt Engng 37(6): 1845–1851
24. Sarma VSSSR, Pillai SA, Subramanian G, Varadan TK (1992) Computerized image processing for whole-field determination of isoclinics and isochromatics. Exp Mech 32(1):24–29
25. Asundi A (1993) Phase shifting in photoelasticity. Exp Tech 17(1):19–23
26. Kihara T (1990) Automatic whole-field measuremet of photoelasticity using linear polarised incident light. Proc of IX Int Conference of Experimental Mechanics Copenhagen 2:821–827
27. Dupré JC, Bremand F, Lagarde A (1993) Photoelastic data processing through digital image processing: isostatics and isochromatics reconstruction. Presented at the Int Conference on Photoelasticity: New Instrumentation, Materials and Data Processing Techniques, London
28. Brown GM, Sullivan JL (1990) The computer-aided holophotoelastic method. Exp Mech 30(2):135–144
29. Quan C, Bryanston-Cross PJ, Judge TR (1993) Photoelasticity stress analysis using carrier fringe and FFT techniques. Optics and Lasers in Engng 18:79–108
30. Morimoto Y, Morimoto Jr Y, Hayashi T (1994) Separation of isochromatics and isoclinics using Fourier transform. Exp Tech 18(5):13–17
31. Ng TW (1997) Photoelastic stress analysis using an object step loading method. Exp Mech 37:137–141
32. Creath K (1988) Phase-measurement interferometry techniques. In: Wolf E (ed) Progress in optics. Elsevier, Amsterdam 26:349–393
33. Nurse AD (1997) Full-field automated photoelasticity using a three-wavelength approach to phase-shifting. Appl Optics 36:5781–5786
34. Srinath LS, Sarma VSSSR (1973) Determination of integral fringe orders in photoelasticity. Exp Mech 13:138–141

35. Buckberry C, Towers D (1996) New approaches to the full-field analysis of photoelastic stress patterns. Optics and Lasers in Engng 24:415–428
36. Patterson EA, Wang ZF (1998) Simultaneous observation of phase-stepped images for automated photoelasticity. J Strain analysis for Engng Des 33(1):1–15
37. Theocaris PS, Gdoutos EE (1979) Matrix theory of photoelasticity. Springer-Verlag, Berlin Heidelberg
38. Ramesh K, Ganapathy V (1996) Phase-shifting methodologies in photoelastic analysis – the application of Jones calculus. J Strain analysis for Engng Des 31(6):423–432
39. Ji W, Patterson E A (1998) Simulation of error in automated photoelasticity. Exp Mech 38(2):132–139
40. Ramesh K, Sreedhar D (1998) Optically enhanced tiling (OET) in digital fringe pattern analysis. Strain 34(4):127–131
41. Ramesh K (1997) PHOTOSOFT_H: a comprehensive photoelasticity simulation module to teach the technique of photoelasticity. Int J Mechanical Engng Education 25(4):306–324
42. Ramesh K, Deshmukh SS (1997) Automation of white light photoelasticity by phase shifting technique using colour image processing hardware. Optics and Lasers in Engng 28(1):47–60
43. Ramesh K, Kelkar AA (1995) Automatic ordering of isochromatic fringes – a new methodology. Strain 31(3):95–99
44. Ramesh K, Kelkar AA (1995) Automatic fringe ordering of photoelastic fringes – a new methodology. Proc of Int conference on Mechanics of Solids and Materials Engineering (MSME 95) Singapore:895–900
45. Haake SJ, Wang ZF, Patterson EA (1993) Evaluation of full field automated photoelastic analysis based on phase stepping. Exp Tech 17(6):19–25
46. Haake SJ, Patterson EA (1995) Photoelastic analysis using automated polariscopes. Proc of Int conference on Mechanics of Solids and Materials Engineering (MSME 95) Singapore:884–888
47. Carazo-Alvarez J, Haake SJ, Patterson EA (1994) Completely automated photoelastic fringe analysis. Optics and Lasers in Engng 21:133–149
48. Manish Jhawaar, Arun N, Ajay Singh (1990) Design and fabrication of an automatic polariscope. B. Tech project, Dept of Mech Engng, IIT Kanpur
49. Patterson EA, Ji W, Wang ZF (1997) On image analysis for birefringence measurement in photoelasticity. Optics and Lasers in Engng 28(1):17–36
50. Kihara T (1994) Automatic whole-field measurement of principal stress directions using three wavelengths. Proc of X Int Conference on Experimental Mechanics Lisbon 1:95–99
51. Petrucci G (1997) Full-field automatic evaluation of an isoclinic parameter in white light. Exp Mech 37(4):420–426
52. Mawatari S, Takashi M, Toyada Y, Kunio T (1990) A single valued representative function for determination of principal stress direction in photoelastic analysis. Proc of IX Int conference on Experimental Mechanics Copenhagen 5:2069–2078
53. Chen TY, Lin (1996) An improved method for whole field automatic measurement of principal stress direction. Abst proc VIII Int Conference on Experimental Mechanics:178–179
54. Rupeng W (1987) The image photo-carrier theory and its application to the determination of principal stress direction. Proc of SPIE on Photomechanics and Speckle Meterology 814:257–260
55. Ji W, Patterson EA (1998) A simulation of a polariscope for error analysis in automated photoelasticity. Proc of XI Int Conference on Experimental Mechanics, University of Oxford, Allison (ed) Balkema, Rotterdam:521–526

56. Ramesh K, Ganapathy V (1995) Comparative performance evaluation of various phase shifting algorithms in photoelasticity. Proc Of ICAME Held at IISc Bangalore, Narosa publishing House, New Delhi:1191–1204
57. Sarma AVSSSR, Pillai SA, Subramanian G, Varadan TK (1995) Digital image processing for the determination of isochromatics in photoelasticity. J Aero Soc India 47(3):127–130
58. Plouzennec N, Dupré JC, Lagade A (1999) Whole field determination of isoclinic and isochromatic parameters. Exp Tech 23:30–33

6 Phase Unwrapping and Optically Enhanced Tiling in Digital Photoelasticity

6.1 Introduction

In the previous chapter, various techniques to obtain fractional retardation over the complete model domain have been presented. In these techniques, the fractional retardation is represented as a phase map. In the phase map, the fractional fringe orders in the range of 0–1 are represented as an intensity map of 0–255. For practical utilisation of the data, one has to find the total fringe order over the domain. This is achieved by a process called phase unwrapping. For phase unwrapping to be effective, the phase map should be free of noise and discontinuities.

In Chapter 5, the reasons for discontinuities in the phase map and the method to overcome has been discussed in detail for the 6-step phase shifting algorithm (Sect. 5.8). In this chapter, a novel scheme for noise removal in phase maps is given. Since one is interested to collect data only within the model domain, the software structure needed for accounting for the boundary of the object is also discussed. For effecting phase unwrapping over the model domain, appropriate scanning schemes have to be developed. For simple model geometries, a row-wise scanning of the model domain is sufficient for phase unwrapping. However, for arbitrarily shaped models the use of tiling for phase unwrapping is needed. The details regarding this are also presented in this chapter.

One of the most common problems in digital fringe pattern analysis is that information is usually lost in high fringe density zones due to digitisation and quantization errors. This is mainly due to the fact that the image is identified as an assembly of pixels and for each pixel, the intensity is usually quantized to 256 grey level shades between pitch black and pure white. In commonly available cameras (PAL-system) the spatial resolution is only 512×512. By increasing both the spatial resolution and the quantization levels, one can enhance the picture. Of late, very high-resolution cameras of the order of 4096×4096 with quantization levels of 1024 grey levels are available. However, these are quite expensive for general use. It is also possible to enhance the picture in high fringe density zones by appropriate optical zooming. In such a case, the spatial resolution is substantially enhanced and it depends entirely on the range of lenses available with the experimentalist.

The methodology to enhance the information in high fringe density areas by the use of a normal resolution camera equipped with a zoom lens and a range extender set is also discussed in this chapter. The usefulness of the approach is illustrated with the improvement of total fringe order extraction in digital photoelasticity using phase shifting technique.

6.2 Boundary Detection

To limit the determination of the field parameter within the region of interest (within the specimen), the image has to be segmented. Segmentation of the image is done by detecting the boundary. In DIP literature, boundary of the image is detected using spatial filters like Roberts, Prewitt, Sobel etc. These filters detect the edges based on abrupt change in grey levels. The edge detection algorithms not only detect the boundary, but also detect the regions in the image where there is abrupt change in grey levels. The set of pixels obtained from the above filters seldom characterises a boundary completely. This is because, non-uniform illumination causes noise and breaks in the boundary. These edge detection algorithms are typically followed by other boundary detection procedures designed to assemble and link the edge pixels into meaningful boundaries. In DIP applications like astronomy and defence where image recognition is the only criterion and the accuracy of the boundary is not critical, such an approach may be meaningful. Unlike the above applications, in photoelasticity, the shape of the object is known and the accurate detection of the boundary is critical to apply algorithms for stress-separation (Chap. 9). Hence, instead of using the above general edge detection algorithms, the geometric properties of the model can be used to detect the boundary accurately.

In bright-field arrangement of the polariscope, the boundary of the image can be seen clearly. Using this image, it is suggested to detect the boundary based on geometric primitives such as line, circle, ellipse and 3-point arc etc. The boundary is to be drawn using the primitives developed, such that the outer side pixels of the line/curve drawn represents the boundary of the specimen. For ease of book keeping, it is better to assign a specific grey level to the detected boundary. A grey level value of 255 may be chosen for this purpose. The boundary detection process can be made faster if the actual dimensions of the specimen are available. For example to identify the boundary of a circular disc, one has to specify the centre and the radius. Based on the optical magnification and aspect ratio of the video monitor, the boundary can be drawn. For all computational purposes, only the outermost pixel must be recognised as the boundary.

Once the boundary is identified, to make further computations, the boundary data has to be expressed in a suitable representation (*boundary.c*, CD). The positions of the pixels, which have an intensity of 255, are written into two separate intermediate files, one for horizontal scanning (scanx.dat) and the other for vertical scanning (scany.dat). In view of spatial quantiza-

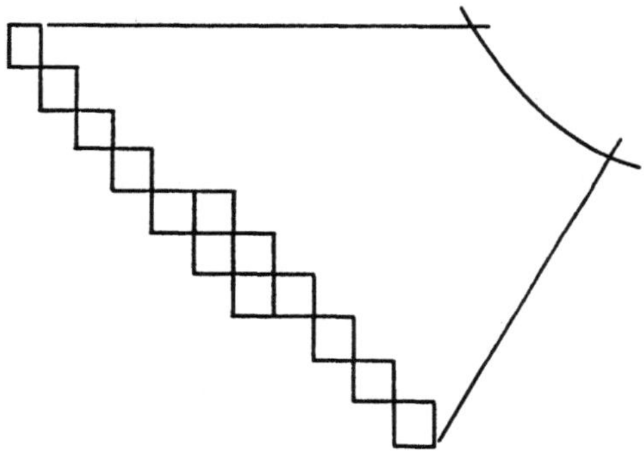

Fig. 6.1. Magnified pixel representation of a curved boundary

tion, to represent the segments of the boundary having a curvature, more than one pixel is needed (Fig. 6.1). Further processing of these files are done in which the first and last columns for every line from these files are extracted and stored into two separate files for horizontal scanning (∗.ybn) and vertical scanning (∗.xbn). In the examples presented, the analysis of the image is limited to the first quarter (256 × 256) of the quadrant.

6.3 Noise Removal in Phase Maps

The phase maps in general may have *salt and pepper noise* (binary noise) at a few points in the domain. To remove noise from images which have sharp edges, usually a median filter is suggested as other filters can destroy the sharp edges. The median filter is a non-linear spatial filter. In this, the grey level of each pixel is replaced with the median of the grey levels in the neighbourhood of that pixel. Even though, the median filter preserves the sharp edges, some information at the sharp edges is lost. The intensity gradient at the transition of wrapped regions is the main parameter for unwrapping of the phase (Sect. 6.4). Any decrease in the intensity at the transition regions will affect phase unwrapping adversely. Thus, median filter is not an ideal choice for the present application. For removing the noise without disturbing the sharp edges, the following noise removal algorithm is suggested.

The noise removal algorithm will remove the binary noise based on the difference in the intensities between the neighbourhood pixels along the scanning direction. If the difference in intensities between the current pixel and the neighbourhood pixels along the scanning direction is beyond a threshold, then the current pixel is identified as a noise pixel. The grey level of this pixel is replaced with the average of grey levels of the neighbourhood pixels.

Consider a 3 × 3 pixel mask as shown in Fig. 4.2a, Chap. 4. Let the noise removal gradient threshold be T_n. Scanning the image horizontally, the pixel at (0, 0) is a salt noise pixel when the following condition is satisfied.

$$(g_{00} - g_{-10}) > T_n \text{ and } (g_{10} - g_{00}) < -T_n \tag{6.1}$$

The pixel at $g(0, 0)$ is a pepper noise pixel when the following condition is satisfied.

$$(g_{00} - g_{-10}) > -T_n \text{ and } (g_{10} - g_{00}) < T_n \tag{6.2}$$

The grey level of the noise pixel $g(0, 0)$ is then replaced with the grey levels of the neighbourhood pixels $[(g_{-10} + g_{10})/2]$. The procedure is repeated for scans at 45°, 90° and 135° (*noise.c*, CD). In practice, the noise removal scheme could be merged with the program doing fractional retardation calculation itself.

The value of T_n depends on the maximum grey level gradient between the neighbourhood pixels in the data region and one has to select it by trial and error. A value of 123 (which corresponds to a phase of π) will be sufficient for a variety of problems [1].

6.4 Algorithm for Phase Unwrapping

Figure 6.2a shows the phase map for the case of a circular disc under diametral compression. The phase map is obtained by employing the scheme discussed in Sect. 5.8.3. The calculation of fractional retardation and plotting of phase map is obtained by invoking Eqs. (5.32), (5.33) and (5.37). Salt and pepper noise over the domain has been removed by using the algorithms discussed in Sect. 6.3. Figure 6.2b shows the variation of intensity in the phase map along the horizontal diameter. Since the phase map represents the fractional fringe order as intensities, the plot shows the variation of the fractional fringe order along the horizontal diameter.

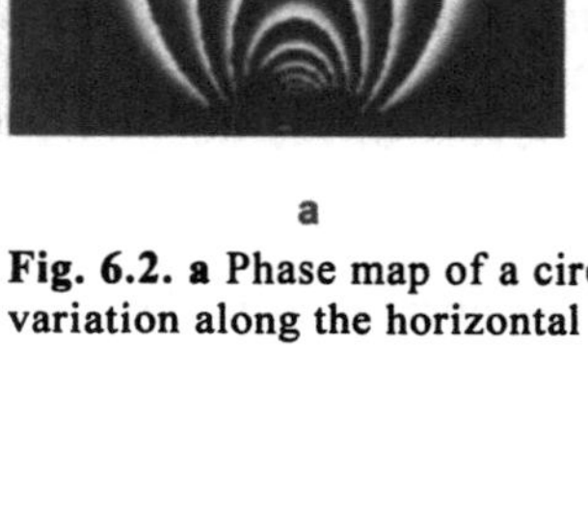

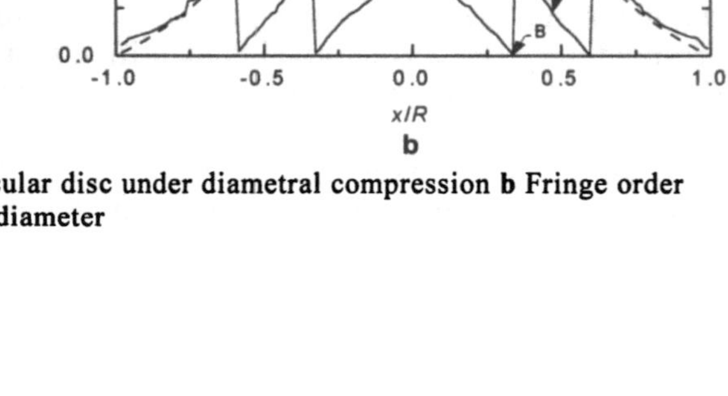

Fig. 6.2. **a** Phase map of a circular disc under diametral compression **b** Fringe order variation along the horizontal diameter

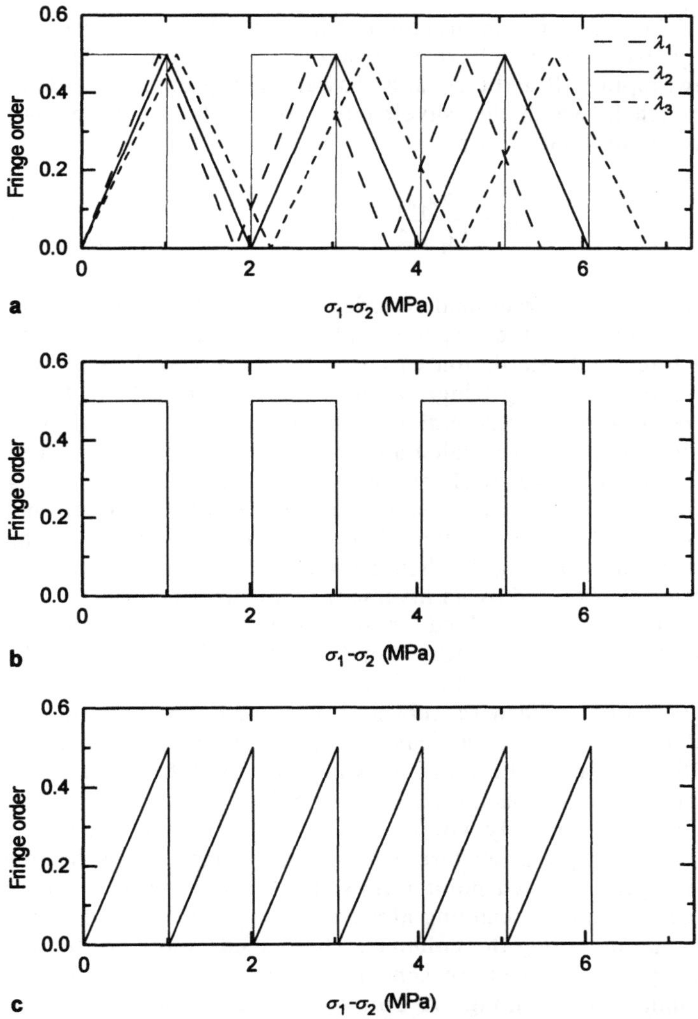

Fig. 6.3. a Fractional retardation plot for three wavelengths satisfying Eq. 6.3 **b** Square wave mask **c** Saw tooth wave form of fractional retardation corresponding to wavelength λ_2

It can be seen from the figure that the fractional fringe order increases in the direction in which the total fringe order increases, and a discontinuity in the slope occurs when the phase retardation is 2π.

Since, the fractional retardation has been calculated using an inverse tangent function, the phase map shown in Fig. 6.2b has a saw tooth waveform. In Chap. 5 it was seen that, while using a plane polariscope for phase shifting, one obtains fractional retardation using an inverse cosine function. In such a case, the sign information of the fractional retardation is lost. The intensity variation will have a triangular shape (Fig. 6.3a). To enable phase unwrapping, Buckberry and Towers [2] proposed a three wavelengths approach in which the wavelengths are so chosen, such that, they follow the following relationship

$$\lambda_2 = \frac{2\lambda_1\lambda_3}{(\lambda_1 + \lambda_3)} \tag{6.3}$$

To effect phase unwrapping, fractional retardation for each of these wavelengths has to be obtained by the respective phase shifting technique. The schematic plot of fractional retardation variation for these wavelengths in a zone of monotonically increasing fringe order is shown in Fig. 6.3a. Let δ_i be the fractional retardation corresponding to the wavelength λ_i. In view of Eq. (6.3), if the sign of $(\delta_1 - \delta_3)$ is calculated over the line of interest, then one can obtain a square wave mask as shown in Fig. 6.3b. Combining Figs. 6.3b and 6.3a, one can get a saw tooth waveform of fractional retardation corresponding to the wavelength λ_2 (Fig. 6.3c). The rest of the procedure for phase unwrapping is then similar to the case where the fractional retardation is obtained by an inverse tangent function.

If an integer fringe order is added or subtracted at the transition of wrapped regions, the total fringe order can be known. The slope of the curves can decide the addition or subtraction. If it is positive, then, one has to add the integer value, otherwise subtract. The methodology however requires the total fringe order for at least one point in the domain to be supplied by auxiliary means. The point at which the total fringe order is interactively supplied is termed as a *primary* seed point.

To unwrap the phase completely within the boundary, the image has to be scanned completely. A simple procedure is to scan the image along the row passing through the primary seed point followed by column-wise scanning of the entire image. The initial scanning along the row provides seed points for effecting the scanning along the columns. These seed points are termed as *secondary* seed points. The entire scheme can be termed as row-wise seeding and column-wise scanning. A software code to perform phase unwrapping is enclosed in the accompanying CD (*unwrap.c*).

Let point 'A' in Fig. 6.2b be the primary seed point at which the total fringe order is to be supplied externally. Once the point is selected, the software (*unwrap.c*) echoes back the fractional fringe order for that point. The user can then appropriately add it with the integer part to specify the full fringe order. The unwrapping is first done along the left direction and then along the right direction. At point 'B', there is a discontinuity in the slope. The discontinuity in the slope is identified by a threshold value T_p. A detailed discussion on the value of T_p influencing the phase unwrapping procedure is discussed in Sect. 6.6. When the slope between the neighbourhood pixels is less than $-T_p$, the total fringe order is increased by one and when the slope between the neighbourhood pixels is greater than T_p, the total fringe order is decreased by one.

Though in principle the fractional fringe order gradually increases in the direction of the overall fringe gradient, there can be kinks in the curve due to random noise, which escaped the noise removal procedure. Since phase unwrapping is basically a scanning procedure, this noise can easily propagate and will adversely affect the final result unless corrective measures are taken. Every effort must be made to select the primary seed point such that, in row-wise seeding along the row containing the seed point, no noise is present.

6.5 Representation of the Unwrapped Phase

The total fringe orders obtained from phase unwrapping have to be stored properly in files. To store the total fringe order (7 digits i.e., 2 digits integer and four digits precision in fractional part) and its location (3-digits each for x and y in pixels) for an image of size 256 × 256, the file size will be approximately 1.0 MB. The file size is too large to handle it conveniently. Hence, use of smaller files with appropriate book keeping is desirable. In the software code enclosed in the CD (*unwrap.c*), for every 15 columns of unwrapping, a new intermediate file (.N*) is created to store the total fringe orders. For a typical image of size 256 × 256, in all, 18 files will be created.

In '*unwrap.c*' unwrapping for any column starts at the seed point and progresses first along the upward direction up to the top window co-ordinate and then from the seed point to the bottom window co-ordinate. The total fringe orders thus obtained are to be sorted, such that, each column in the file corresponds to the total fringe order from top to bottom of the corresponding column of the image. Sorting is done independently for each of the stored files. The total fringe orders from the sorted output is stored into another set of intermediate files (FRI.N*). These files are added side by side to form the final file (*.frn) of total fringe orders. The number of columns and rows in the final file (*.frn) are equal to the number of columns and rows of the selected window of the image. Thus, there exists one to one correspondence between the window chosen and the fringe order file (*.frn) [1].

6.5.1 Three-Dimensional Plots

Three-dimensional plots provide a graphical representation of the results obtained. Several software packages are now available to provide a three-dimensional representation of data. In view of the large file size of the results obtained, it is desirable that a Windows based package be used for the graphical representation. The software *excelplo.c* enclosed in the CD accepts the '*.frn' file as input and provides an output '*.plo' which is

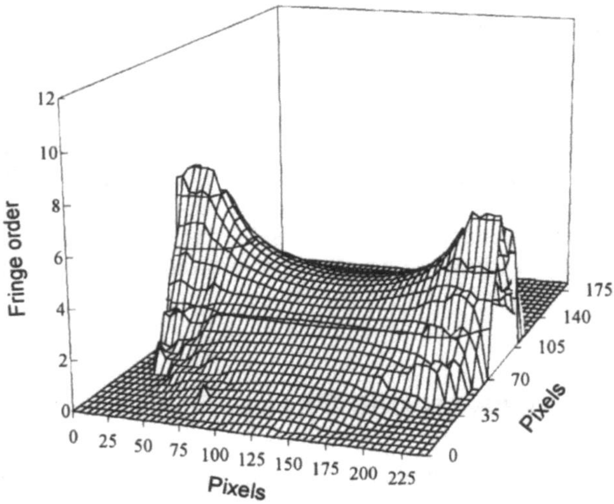

Fig. 6.4. Three-dimensional representation of total fringe order over the model domain

compatible to the MS-Excel package for three-dimensional plotting. To improve clarity of the three-dimensional plots, only every 5th row and 5th column of the '*.frn' file is stored in the '*.plo' file. Figure 6.4 shows the three-dimensional representation of the total fringe order variation for the problem of a circular disc under diametral compression obtained by plotting 'disc.plo' file.

6.5.2 Total Fringe Order Viewing on the Image

In many occasions, one would like to find the value of the total fringe order by moving the cursor over the displayed image. In principle, this can be achieved by combining the file '*.frn' and the cursor utility. However, locating and reading the values of total fringe order from a file will be very difficult to implement. This is because, the searching of a data in the file is in terms of bytes, which is quite cumbersome to implement. On the other hand, the searching would be trivial if the total fringe orders are stored in an array equal to the window size of the image. While doing so, it is enough that one stores the total fringe order and not its location in pixel dimensions. It is important to note here that this will require a very large memory and for an image of size 256 × 256 one would require an array of size 0.26 MB. However, on PC based systems (Dos), the maximum random access memory space available is only 64 kB. In view of this, the file '*.frn' has to be segmented into appropriate number of columns and for each of these, a global array has to be declared in the software. Since, the total fringe order has both integer and fractional part, it is a 'float' variable. A 'float' variable will occupy 4 bytes of memory. For a column of 256, the maximum number of columns that could be stored as an array is around 62. In the implementation of the software (*shown.c*, CD), it is taken as 55 columns. Thus, for an image of size 256 × 256, one would require five different global arrays. The software accepts '*.frn' file as input and generates the appropriate data structure as discussed above to view the fringe orders on-line.

6.6 Parameters Affecting Phase Unwrapping

6.6.1 Influence of the Selection of the Phase Unwrapping Threshold

In Sect. 6.4 it was discussed that using a threshold T_p identifies the transition regions of phase map. In analog data processing one can take T_p as 255. However, in digital data processing even in small stress gradient zones, one has to take a smaller value of T_p than 255 to accommodate the effect of spatial discretization and intensity quantization. This aspect is highly pronounced in high stress gradient zones. The variation of intensity in the phase map along the vertical diameter for the problem of a circular disc under diametral compression is shown in Fig. 6.5. It can be observed from Fig. 6.5, that at stress concentration regions, the intensity in the wrapped regions neither starts with a value of 0 nor ends with a value of 255. This is primarily due to the fact that the fringe gradient is so high that one requires sub-pixel calculations to identify the exact transition zone.

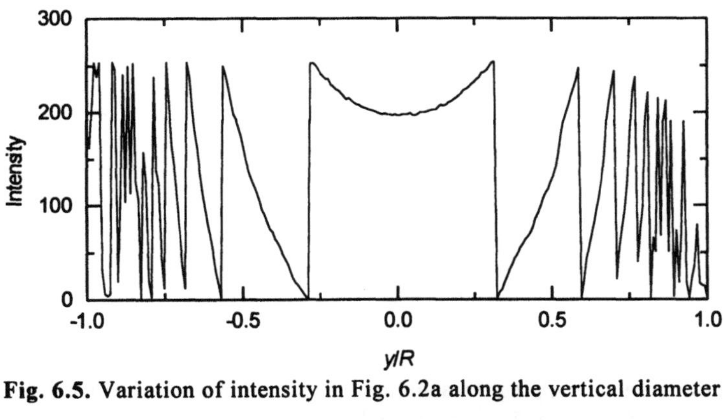

Fig. 6.5. Variation of intensity in Fig. 6.2a along the vertical diameter

The error in mismatch of quarter-wave plates also plays a role in deciding the value of T_p. It is reported in ref. [3], that the error in mismatch of quarter-wave plates has the effect of slightly shifting the point of transition and the intensity does not reach the extremum value of 255.

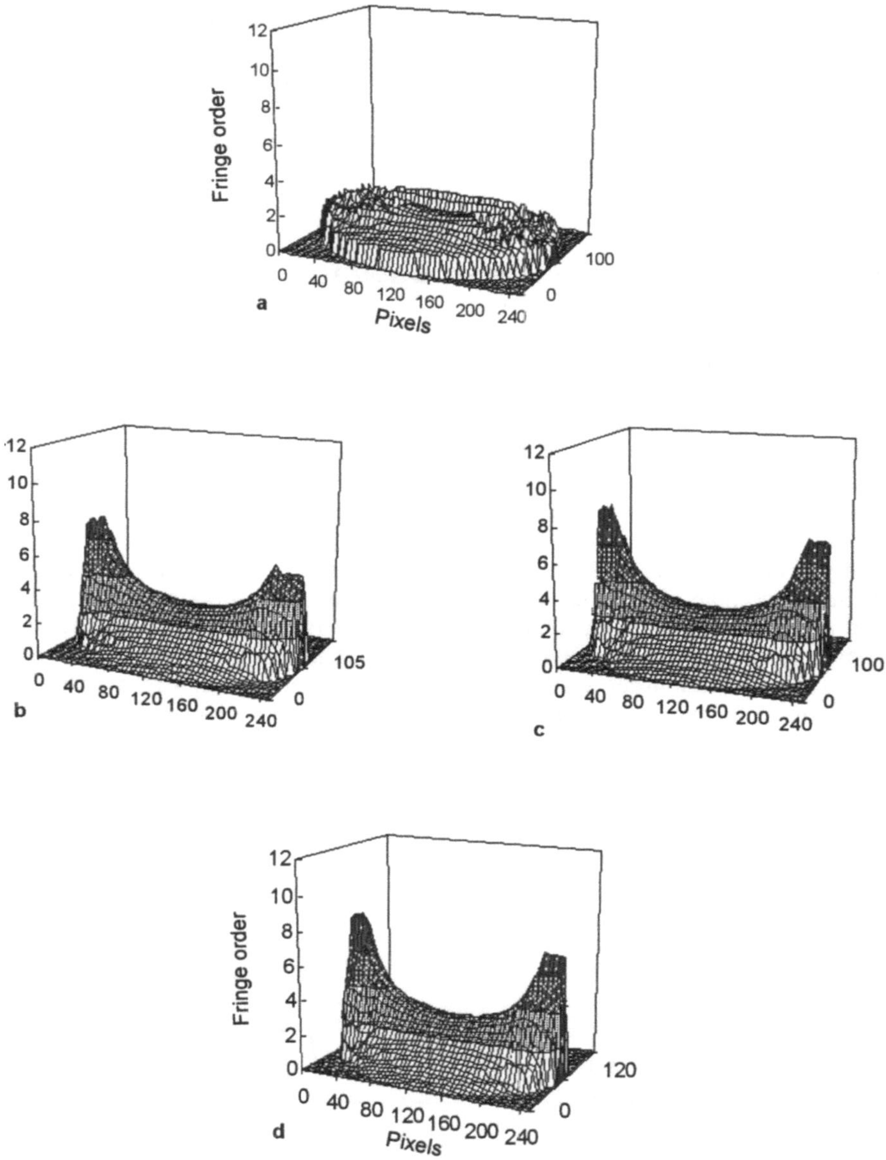

Fig. 6.6. Influence of phase unwrapping threshold on the extraction of total fringe order over the domain: **a** $T_p = 255$ **b** $T_p = 225$ **c** $T_p = 180$ **d** $T_p = 123$

From the above discussion it is clear that T_p should be lower than 255. For identifying noise in the domain, a threshold T_n was used in Sect. 6.3. In view of this in general, T_p should be higher than T_n. Figure 6.6 shows the three-dimensional representation of fringe orders by taking T_p values of 255, 225, 180 and 123 for the problem of a disc under diametral compression. The figure shows the dramatic improvement achieved by lowering the value of T_p.

It can be noted from Table 6.1 that, by decreasing the slope from 255 to 123, the fringe gradient that can be unwrapped is increased from 0 to 0.5176 fringe/pixel and the evaluation of total fringe order is increased up to 9.5. Since T_n has been taken as 123, if T_p is also taken as 123 then any noise not eliminated by the noise removal algorithm may propagate. Thus T_p should be greater than T_n and in most cases a value of 180 would be satisfactory.

Table 6.1. Influence of selection of phase unwrapping threshold

Threshold	Fringe gradient (fringe/pixel)	Maximum unwrapped fringe order
250	-	2
225	0.1176	8
180	0.2940	9
123	0.5176	9.5

6.6.2 Influence of the Location of the Primary Seed Point

It has been mentioned earlier, that to initiate phase unwrapping, the user has to specify the total fringe order for at least one point in the field. Ideally, the location of this point should be arbitrary and the phase unwrapping procedure should be flexible enough to unwrap the phase for the entire domain. However, the phase unwrapping algorithm could be a lot simpler if the user selects the primary seed point appropriately. The choice of the location depends on the specific geometry of the specimen under consideration.

For the problem of a circular disc under diametral compression, while using the row-wise seeding and column-wise scanning scheme to unwrap the entire image, the primary seed point should be chosen along the horizontal diameter. If not, the unwrapping is limited only up to the boundary that is encountered along the row-wise seeding. The problem of limited unwrapping can be overcome by adopting a tiling procedure for phase unwrapping, which is discussed next.

6.7
Use of Tiling Procedure for Phase Unwrapping

Tiling as it is generally understood in image processing literature is that the image can be thought of as an assembly of several windows and one can select process parameters for these tiles individually and later assemble them to get the overall picture. Mostly such operations are meant for intensity manipulations and can be easily accomplished by using the frame buffer memory. However, in phase unwrapping, one deals with *real numbers*, which cannot be stored in the pixel locations of the frame buffer. For storing the fringe order values, only a file structure has to be used. This poses certain restrictions on the selection of the nature of the tiles. Selecting an arbitrary window would be difficult from book keeping point of view. Hence, in phase unwrapping, one has to select tiles either as vertical strips filling the entire window or as horizontal strips filling the entire window. An alternate procedure is to declare a large matrix, which is equal to the size of the image. In such a case, any arbitrary window could be used. However, this is not an efficient approach on Dos based systems.

Consider the problem of a plate with a hole subjected to uniaxial tension. Due to the symmetry of the problem one quarter of the plate could be considered for evaluating the total fringe order. In the previous sections, the use of row-wise seeding and column-wise scanning has been discussed. With this approach, the fringe order for the complete domain could be unwrapped only if the primary seed point lies in the zone 1 of Fig. 6.7a. Similar to row-wise seeding, one can also think of column-wise seeding.

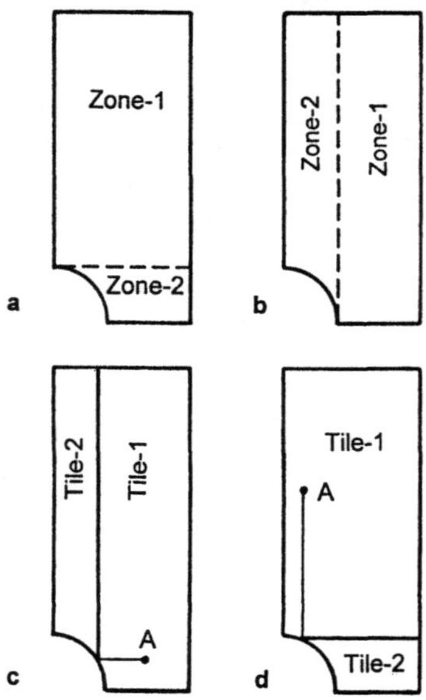

Fig. 6.7. Tiling for phase unwrapping. Identification of zones for locating the primary seed point: **a** row-wise seeding and column-wise scanning and **b** column-wise seeding and row-wise scanning. Tiles required for **c** row-wise seeding and column-wise scanning and **d** column-wise seeding and row-wise scanning

In such a case, the complete image has to be scanned row-wise for phase unwrapping. For column-wise seeding and row-wise scanning, to unwrap the phase for the complete domain, the primary seed point should lie in the zone 1 of Fig. 6.7b. If for some reasons the primary seed points could not be located in these zones, then use of tiling could help in unwrapping the phase. For row-wise seeding and column-wise scanning if the primary seed point is in zone 2, then, the tiles needed for unwrapping are shown in Fig. 6.7c. Similarly for column-wise seeding and row-wise scanning the tiles needed are shown in Fig. 6.7d. Obviously in such cases, the book keeping operations of the phase unwrapping algorithm is quite involved. Implementation of this is left as an exercise to the reader.

6.8 Digital Magnification of High Fringe Density Zones

The algorithms for phase unwrapping discussed so far would work reasonably well in most of the zones of the model domain leaving stress concentration/high fringe density zones. This is because, the spatial resolution in these zones are not satisfactory enough to evaluate the parameters accurately. Magnifying the region of interest can increase the spatial resolution of the stress concentration zone of the image. The magnification of the region of interest of the image can be achieved either digitally or optically. In digital magnification, the new grey levels required for the magnification are obtained using interpolation of the grey levels of the neighbourhood pixels. The various interpolation schemes generally used are as follows.

6.8.1 Replication

In replication, each pixel along a scan line is replaced once and each scan line is repeated. This is equivalent to taking an image of size $M \times N$ and interlacing it by rows and columns of zeros to obtain a matrix of size $2M \times 2N$ and convolving it with a matrix $\mathbf{C}$ defined as

$$\mathbf{C} = \begin{bmatrix} 1 & 1 \\ 1 & 1 \end{bmatrix} \tag{6.4}$$

The block diagram for the interpolation by replication is shown in Fig. 6.8.

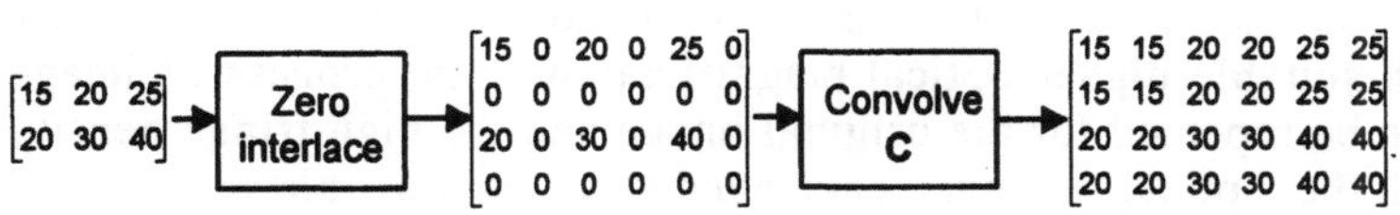

Fig. 6.8. Block diagram of digital magnification by replication

6.8.2 Linear Interpolation

In linear interpolation, a straight line is first fitted in between the pixels along a row. Then the pixels along each column are interpolated along a straight line. The block diagram for the interpolation by replication is shown in Fig. 6.9.

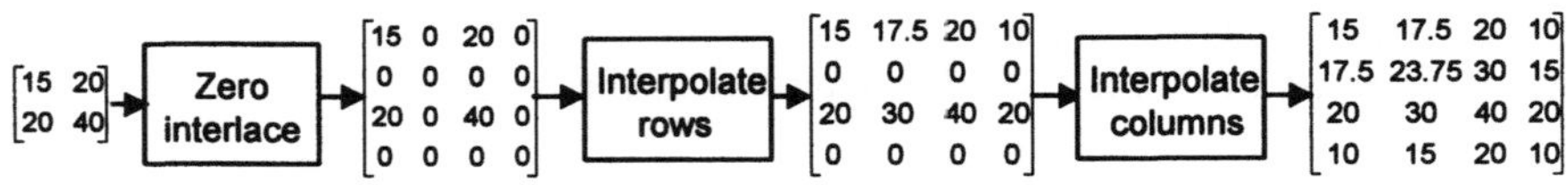

Fig. 6.9. Block diagram of digital magnification by linear interpolation

6.8.3 Higher Order Interpolation

In higher order interpolation, the name implies that a higher order interpolation function is used. In this, each row and each column of the image is padded with p rows and p columns of zeros respectively and convolving it p times with **C**. For $p = 3$ yields a cubic spline interpolation between the pixels [4].

Though these techniques help in magnifying the region of interest, in the process, the intensity information is altered. To employ the methodologies of Chap. 5 this is not desirable.

6.9 Optically Enhanced Tiling (OET)

It was seen before that tiling is nothing but dividing the image into convenient segments (windows) and selecting appropriate process parameters for each of the tiles suitably. In OET, the experimental parameter corresponding to a tile in the original image is extracted from processing the intensity data of the optically magnified tile [5]. As only intensity information is used, unlike tiling in phase unwrapping, here there are no restrictions on the nature of the tile. Any arbitrary window can be selected as a tile.

To effect OET, initially, the overall image is to be captured with the maximum possible optical magnification. Then, the high fringe density zones are to be identified and each of these zones is to be separately captured with suitable higher optical magnifications. The process of image capture has to be repeated for the original image and the high fringe density zones for all the optical arrangements needed to employ phase shifting/ polarization stepping or Fourier transform technique discussed in Chap. 5.

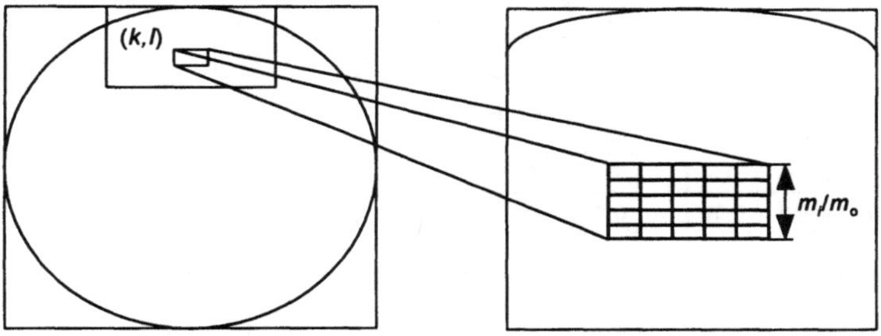

Fig. 6.10. Figure showing the relationship between the overall image and the optically magnified tile

Using this intensity data, the parameters of interest have to be determined by the methodologies discussed in Chap. 5. The experimental parameter (fractional fringe order) at each pixel in high fringe density zones of the original image (Fig. 6.10) is then replaced by the one appropriately obtained from the optically magnified tile (Fig. 6.10).

6.10 Cementing of a Tile

Cementing of the optically magnified tile with the original image involves two steps. First, a one to one correspondence between a pixel in the original image is to be established with that of a block of pixels of the magnified tile. Next, the average value of the total fringe order corresponding to this block of pixels is to be noted as the fringe order for the pixel in the original image.

Let the overall image has a magnification of m_o pixels/mm. Let the tile i has a magnification of m_i pixels/mm. Then, for each pixel (k,l) in the original image, the magnified tile will have a block of m_i/m_o (Fig. 6.10) pixels. Thus, one has to find the average value of the experimental parameter represented by this block of pixels and then to increment it by the value (m_i/m_o) to achieve OET. The process is much simpler and straightforward if this ratio is an integer. However, in general this expression will be a real number with an integer portion and a fractional part. For a typical value of $m_i/m_o = 5.3$, the sequence of weighting required along a row-scan are shown in Fig. 6.11. The important point to note is that the block of pixels involved in OET along a row-scan varies from 6×6 to 6×7 i.e., $\mathrm{Int}(m_i/m_o + 1)$ to $\mathrm{Int}(m_i/m_o + 2)$. Along a column-scan it varies from 6×6 to 7×6. This makes the computer implementation a little involved. The requirement of variable mask size is achieved by taking a matrix of size $\mathrm{Int}(m_i/m_o + 2)$

1	1	1	1	1	0.3
1	1	1	1	1	0.3
1	1	1	1	1	0.3
1	1	1	1	1	0.3
1	1	1	1	1	0.3
0.30	0.3	0.3	0.3	0.3	0.09

0.7	1	1	1	1	0.6
0.7	1	1	1	1	0.6
0.7	1	1	1	1	0.6
0.7	1	1	1	1	0.6
0.7	1	1	1	1	0.6
0.21	0.3	0.3	0.3	0.3	0.18

0.4	1	1	1	1	0.9
0.4	1	1	1	1	0.9
0.4	1	1	1	1	0.9
0.4	1	1	1	1	0.9
0.4	1	1	1	1	0.9
0.12	0.3	0.3	0.3	0.3	0.27

0.1	1	1	1	1	1	0.2
0.1	1	1	1	1	1	0.2
0.1	1	1	1	1	1	0.2
0.1	1	1	1	1	1	0.2
0.1	1	1	1	1	1	0.2
0.03	0.3	0.3	0.3	0.3	0.3	0.06

Fig. 6.11. Weightings of the variable mask for $m_i/m_o = 5.3$ along a row-scan

and making the appropriate rows and columns zeros or non-zeros depending on the size of the mask required.

For a generic situation, let the mask size be $n \times n$. Let the weighting of the mask be w_{pq}. The weighting has to be obtained by using row (a_p) and column (b_q) multipliers as

$$w_{pq} = a_p b_q \quad \text{where } p = 1 \text{ to } n \text{ and } q = 1 \text{ to } n \tag{6.5}$$

It is noted earlier that the mask will have a variable size. This is achieved by declaring a row or a column as zeros appropriately. For the first block a_1, $b_1 = 0$ and a_2 to a_{n-1} and b_2 to b_{n-1} are 1.0. The values of a_n and b_n are fractional part of m_i/m_o. For other blocks, the row multipliers are obtained as

$$\begin{aligned} a_1 &= 1 - a_n \\ a_p &= 1.0 \text{ for } p = 2 \text{ to } n-2 \\ a_{n-1} &= \begin{cases} \dfrac{m_i}{m_o} - \displaystyle\sum_{p=1}^{n-2} a_p & \text{for } a_{n-1} < 1.0 \\ 1.0 & \text{for } a_{n-1} > 1.0 \end{cases} \\ a_n &= \frac{m_i}{m_o} - \sum_{p=1}^{n-1} a_p \end{aligned} \tag{6.6}$$

The column multipliers are also obtained on similar lines.

To cement the tile in the original image, the position of at least one of the block of pixels in the magnified image is to be known with respect to the original image. Using this as a reference, the top left position of the tile with respect to the original image is to be evaluated. To simplify computations, the software relocates the window co-ordinates of the tile such that the ends are at whole number of pixels away from the origin. Cementing of the tile continues first along the horizontal and then along the vertical directions.

6.11 OET Applied to a Circular Disc Under Diametral Compression

Figure 6.12 shows the phase map for the case of a circular disc (diameter = 60.14 mm, thickness = 6.28 mm, diametral load = 669.4 N, materiel stress fringe value F_σ = 9.77 N/mm/fringe) under diametral compression for the overall image (Fig. 6.12a) and the optically magnified stress concentration zones of the top and bottom loading points. Using the 6-step phase shifting method, the fractional retardation for the overall image and the tiles are obtained. The ambiguity in the determination of principal stress direction influencing the phase map is effectively handled by the methodology described in Sect. 5.8.3.

The magnification for the overall image is 2.97 pixels/mm and for the top and bottom tiles, it is 16 pixels/mm. The ratio m_i/m_o is equal to 5.38. The load application point is taken as a reference for cementing the tiles. Figure 6.13a shows a three-dimensional representation of the unwrapped phase for the overall image without OET. Figures 6.13b and 6.13c show the

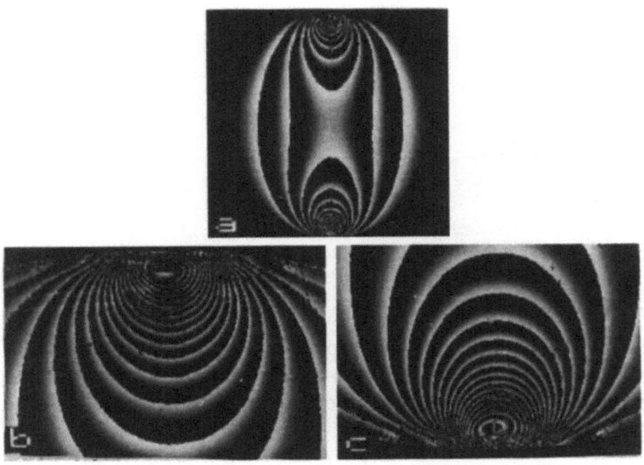

Fig. 6.12. Phase map: **a** disc as a whole **b** top tile and **c** bottom tile

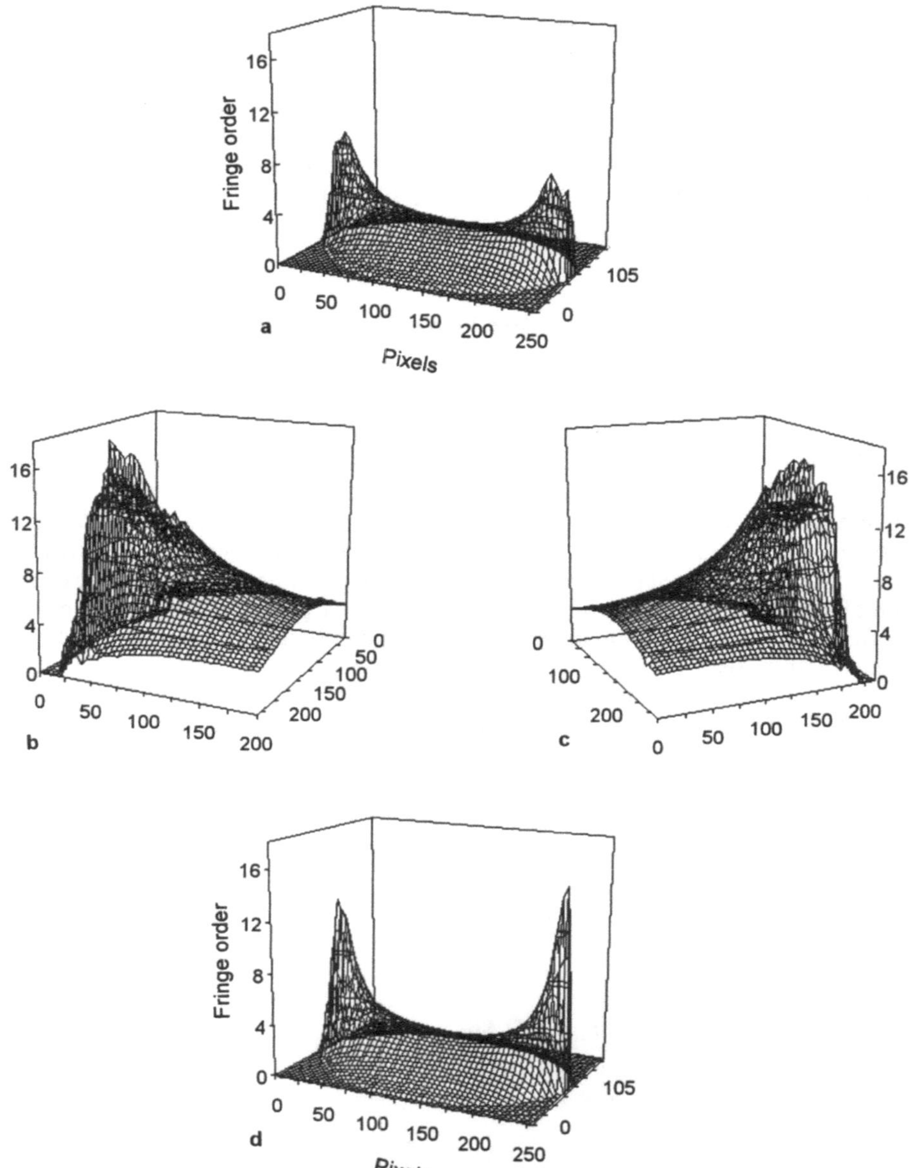

Fig. 6.13. Three-dimensional representation of unwrapped phase for the problem of disc under diametral compression: **a** direct evaluation **b** magnified top tile **c** magnified bottom tile **d** results of top and bottom tiles superposed on the original image by OET

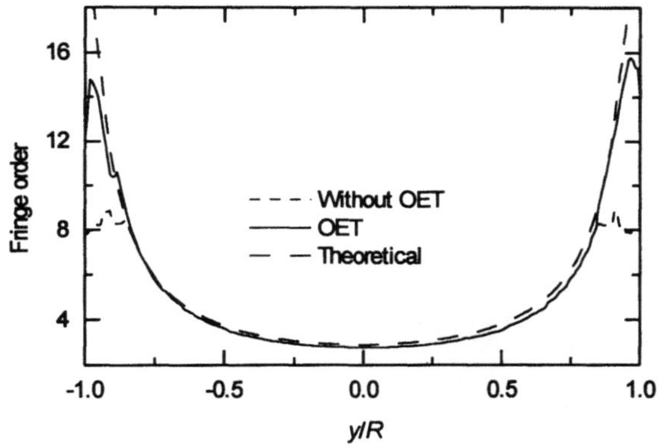

Fig. 6.14. Comparison of fringe order variation along the vertical diameter for the disc under diametral compression obtained by various methods

unwrapped phase for the optically enlarged top and bottom tiles respectively. For all these cases a uniform phase unwrapping threshold of $T_p = 180$ is used [1]. Figure 6.13d shows the unwrapped phase with OET. Figure 6.14 compares the variation of total fringe order along the vertical line passing through the load application points with theory. From the figure it is clear that by direct evaluation, N is limited to a value of 8.8 with a corresponding fringe gradient of 0.238 fringe/pixel. With OET, the fringe order resolvable is up to 15.2 and the fringe gradient is increased to 0.496 fringe/pixel. It is to be noted that in the optically magnified image, the fringe order resolvable is up to 18 and the corresponding fringe gradient is 0.11 fringe/pixel. Thus OET has improved fringe order extraction in stress concentration zones.

6.12 OET Applied to a Ring Under Diametral Compression

In the previous example, no attempt has been made to discuss about what is the minimum optical magnification needed to achieve satisfactory results using OET. OET has been basically proposed, such that, the intensity information in stress concentration zone should be as good as possible for accurate evaluation of the parameters. In Chap. 3, it was shown that the intensity variations in stress concentration zone takes a special form (Fig. 3.14) where the minimum intensity for different fringe orders are different. This was first brought into focus by Friedrich [6] who attributed it to the presence of two different wavelengths of 589 nm and 589.6 nm in the most commonly used sodium vapour source. In fact, the CCD cameras have dramatically recorded this aspect and Fig. 6.15a shows a steep gradient of intensity variation along the horizontal diameter of the dark-field isochromatics of a ring under diametral compression. The question here is,

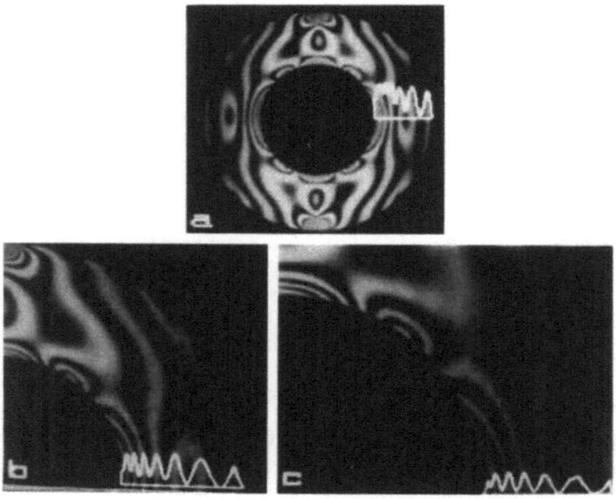

Fig. 6.15. Intensity variation along the horizontal diameter of a ring under diametral compression: **a** ring as a whole (magnification = 1.97 pixels/mm) **b** magnified tile–1 (magnification = 4.09 pixels/mm) **c** magnified tile–2 (magnification = 7.14 pixels/mm) (Courtesy of BSSM [5])

does the variation is only due to the effect of two different wavelengths in sodium vapour or is it also due to the errors in spatial discretization and grey level quantization of the intensity data? Figures 6.15b and 6.15c show the intensity variation of the diametral line of the ring for tow different optical magnifications of 4.09 pixels/mm and 7.14 pixel/mm respectively. The original image is taken with a magnification of 1.97 pixels/mm [7]. The inspection of the figures show that by increasing the spatial resolution, the steep gradient of intensity variation is considerably decreased. This brings out the important role of spatial discretization in evaluating the intensity data in stress concentration/high fringe density zones.

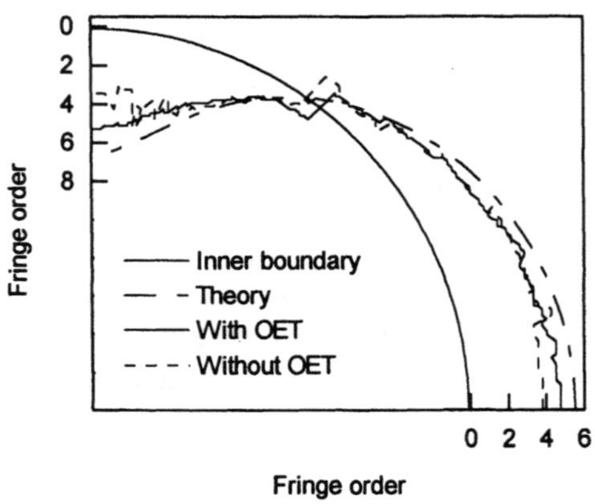

Fig. 6.16. Variation of fringe order along the inner boundary of the ring obtained by various methods

Comparison of Figs 6.15b and 6.15c show that for performing OET, the optical magnification corresponding to Fig. 6.15c would be desirable. Figure 6.16 shows the fringe order variation along the boundary of the hole obtained by various means. From the figure, it is clear, that by direct evaluation the total fringe order is limited to a value of 3.9 along the horizontal diameter. With OET, the fringe order resolvable is 5.0. Along the vertical diameter, the fringe order resolvable is 3.2 and 6.7 respectively [1]. The variation of total fringe order is also smoother with OET. Thus, OET has improved the fringe order extraction in stress concentration zones.

6.13 Closure

In this chapter, a comprehensive presentation of phase unwrapping of fractional retardation has been presented. The need for a noise detection threshold T_n and a phase unwrapping threshold T_p has been shown with examples. The basic procedure for phase unwrapping has been developed for the case of the phase map being obtained by an inverse tangent operation. In situations where a inverse cosine function is used for parameter evaluation, the use of three wavelengths approach has been suggested to convert the phase map to an equivalent one obtainable by an inverse tangent function.

The phase map obtainable in photoelasticity is of high quality and free of speckle noise. The phase maps taken were free of discontinuities. Hence, a simple row-wise scanning or a column-wise scanning has been shown to work for such applications. If the phase map has undue noise, then phase unwrapping algorithms developed for other interferometric techniques may be used [8, 9].

In intensity based measurements, the evaluation of isoclinic parameter θ is in the range $-\pi/4 < \theta < \pi/4$. However, physically θ has to be expressed in the range $-\pi/2 < \theta < \pi/2$. In the literature, expanding the range of θ to $-\pi/2 < \theta < \pi/2$ has been termed as phase unwrapping of isoclinics. It is important to note that while doing so, θ should correspond to either σ_1 or σ_2 direction uniformly over the complete domain. This is important for calculating fractional retardation with correct sign over the domain. Otherwise, discontinuities in the phase map would occur. A detailed discussion on how to overcome this problem interactively for the 6-step phase shifting algorithm has been discussed in Sect. 5.8. Future research should concentrate on automatically evaluating θ such that it represents either σ_1 or σ_2 direction uniformly over the complete domain.

In high stress gradient zones, the poor monochromaticity of the light source as well as the resolution of the digital camera used influence the intensity data recorded. Issues related to these are discussed in detail in this chapter. A novel approach for enhancing data in high fringe density zones using a normal resolution camera has also been presented. The procedure is termed as OET to distinguish it from conventional tiling used in DIP.

Exercises

6.1 How multiple wavelengths are used in unwrapping the retardation calculated by arccosine function? Explain with neat sketches and appropriate equation.

6.2 Mention the role of noise identification threshold T_n and phase unwrapping threshold T_p in effecting phase unwrapping.

6.3 Develop a windows based software to identify the total fringe order on the PC monitor.

6.4 Define OET. How would you select the optimum magnification for performing OET?

References

1. Sreedhar D (1998) Phase unwrapping and optically enhanced tiling in digital photoelasticity. M.Tech thesis, Indian institute of Technology, Kanpur, India
2. Buckberry C, Towers D (1996) New approach to the full-field analysis of photoelastic stress patterns. Optics and Lasers in Engng (24):415–428
3. Ajovalasit A, Barone S, Petrucci G (1998) A method for reducing the influence of the quarter-wave plate error in phase shifting photoelasticity. J Strain Analysis for Engng Des 33(3):207–216
4. Jain AK (1989) Fundamentals of digital image processing, Prentice-Hall, New York
5. Ramesh K, Sreedhar K (1998) Optically enhanced tiling (OET) in digital fringe pattern analysis. Strain 34(4):127–131
6. Friedrich G (1976) Calibration of equidensity lines. Strain 12:140–143
7. Mangal SK, Ramesh K (1998) Factors affecting the intensity distribution in high stress gradient zones in digital photoelasticity. Strain 34(4):133–134
8. Quiroga JA, Bernabeu E (1994) Phase unwrapping algorithm for noisy phase map processing. Appl Optics (33):6725–6731
9. Stetson KA, Wahid J, Gauthier P (1997) Noise-immune phase unwrapping by use of calculated wrap regions. Appl Optics (36):4830–4838

7 Colour Image Processing Techniques

7.1 Introduction

With rapid advancements in computer technology, colour image processing systems are now available at affordable prices. A proper understanding of the colour image processing techniques requires an understanding of the various models for colour representation. Apart from this, one needs to know the spectral response of the colour camera and the light source. The type of colour image processing hardware used for transmission or reflection photoelastic analysis has an influence on the quality of data acquisition. Data interpretation and analysis requires the understanding of the intensity of light transmitted, for various polariscope arrangements, in white light.

Significant achievements in understanding the intensity of light transmitted in white light was possible with the development of spectral content analysis (SCA) [1–3]. Various investigators [1–6] have developed the technique and its extension to digital domain was done by Haake and Patterson [6]. Basically, the technique can provide isochromatic fringe order but a modification of the technique to provide isoclinic fringe order has also been reported [7]. Although, the digital domain SCA uses white light as a source, it requires only the use of a monochrome camera coupled with eight monochrome filters for intensity recording.

In Chapter 5 various techniques based on phase shifting and polarization stepping for determining isoclinic and isochromatic parameters have been discussed. It has also been pointed out that the isoclinic-isochromatic interaction affects the determination of both isoclinics and isochromatics in different ways. To overcome the problems in determining isoclinics, several researchers have resorted to the use of multiple wavelengths that help to shift the points where isochromatic intensity is zero for different wavelengths differently, thereby, achieving a continuous isoclinic. Though these methods were initially aimed at using monochrome cameras [8, 9], the later researchers [10] advocated the use of colour CCD cameras for one step recording of three images corresponding to three wavelengths.

Not all techniques that use multi-wavelengths approach for data reduction can be benefited by using a colour CCD camera or a colour image processing system. In the recording of multi-wavelengths images, one can observe three philosophies. In the first approach, white light is used and images corresponding to different wavelengths are grabbed with the help of narrow-band filters [9, 10]. In the second approach, no intermediate monochrome filters are used but the full colour image is directly recorded by a colour CCD camera [11–17]. In the third approach, the light source is appropriately designed to provide three different wavelengths. Kihara [8] used He-Ne and Ar-ion lasers to produce wavelengths of 632.8 nm, 514.5 nm and 457.9 nm. Yoneyama et al. [18] designed a light source using halogen and mercury lamps to produce wavelengths of 436 nm, 546 nm and 619 nm. The half bandwidth of these wavelengths is quite low and is 3 nm, 3 nm and 5 nm respectively.

In techniques that use monochrome filters in conjunction with a white light source for image recording, no specific advantage could be achieved by using a colour camera, as the images corresponding to different wavelengths cannot be separated from the image recorded by a colour CCD camera. One has to record as many images as required by a particular technique only by using a monochrome camera.

Techniques such as *three fringe photoelasticity* (TFP) [12–14], phase-shifting in colour domain [15, 16], polarization stepping in colour domain [17] use the second philosophy of directly recording full colour image using a colour CCD camera. These techniques use the basic hardware features of a colour image processing system effectively. A colour image processing system can be configured to capture images in RGB mode. In RGB mode, one can identify two main hardware features. In the first, the image is identified as a superposition of the image planes of red, green and blue. The second feature is that the intensity is quantized in the range of 0–255 for each image planes of R, G and B. Three fringe photoelasticity provides total fringe order over the domain. Phase shifting in colour domain can provide isoclinic parameter and fractional fringe order over the domain. Phase unwrapping techniques have to be used to determine the total fringe order. Polarization stepping in colour domain provides isoclinic parameter and total fringe order over the domain.

The light source reported by Yoneyama et al. [18] has three wavelengths corresponding to R-, G- and B-plane detectors of a colour CCD camera. Hence, one step recording and separation into three individual channels is possible by using a colour CCD camera while using their light source. The technique has come to be known as *tricolour photoelastic method* and can provide both isochromatic and isoclinic fringe orders. Although Kihara developed a light source with three wavelengths, in view of the choice of these values, one step recording of images corresponding to three wavelengths is not possible by a colour CCD camera [19]. This is because; the B-filter of the colour camera transmits not only the light corresponding to the wavelength 457.9 nm but also that due to 514.5 nm.

In this chapter, the various techniques that basically require a colour image processing hardware for data acquisition are discussed. Though, SCA does not require a colour image processing hardware, in view of its importance in having contributed to the development of intensity equations in white light, the details of the techniques are discussed. Towards the end of the chapter it will become clear how the various aspects of colour recording have significantly influenced and improved the acquisition of photoelastic data.

7.2 Colour Models

In the electromagnetic energy spectrum, the visible spectrum of light is quite narrow, which is in the range of 380 nm to 780 nm of wavelengths. The colour spectrum can be divided into broad regions consisting of colours in the following sequence: violet, indigo, blue, green, yellow, orange and red. No colour in the spectrum ends abruptly but each colour blends smoothly into the next.

Human eye can only differentiate about twenty-four shades of grey but it can differentiate thousands of colour shades and intensities. The colours that human beings perceive in an object are determined by the nature of the light reflected from the object. A body that reflects light that is relatively balanced in all visible wavelengths appears white to the observer. On the other hand, if the body reflects only a limited range of the visible spectrum, it appears with some shades of colour. Various studies on human vision have concluded that there are three different types of cones in the normal human retina with absorption spectra in the range of 380 nm to 780 nm with peaks at blue, green and yellow-green regions [20]. In view of this, all colours are seen as a mixture of various amounts of three colours that are red (R), green (G) and blue (B).

Colour representation in image processing is based on the classical theory of Thomas Young [21], who stated that any colour could be represented by mixing an appropriate set of three *primary colours*. The choice of *primaries* is such that no two primaries can combine to match the third primary. One obvious choice is the use of red, green and blue as primary colours. For the purpose of standardisation, the CIE (*Commission Internationale de l'Eclairage* – the International Commission on Illumination) designated in 1931, the following specific wavelength values to the three primary colours as, 435.8 nm, 546.1 nm, 700.0 nm for blue, green and red respectively [20]. For true colour rendering, not only the three primaries are to be mixed with various intensity proportions but also their wavelengths be allowed to vary.

Apart from *primary colours*, *secondary colours* are also used for colour representation. The three *secondary colours* are cyan (Cy), magenta (Mg) and yellow (Ye). These are obtained by adding any two primaries. For example, addition of green and blue gives cyan, red and blue gives magenta and red and green gives yellow.

If the light is void of colour (achromatic), its only attribute is its *intensity*. On the other hand, a chromatic light has three characteristics, viz., *hue*, *saturation* and *brightness* or *intensity*. Hue represents the dominant colour as perceived by an observer. Saturation refers to relative purity or

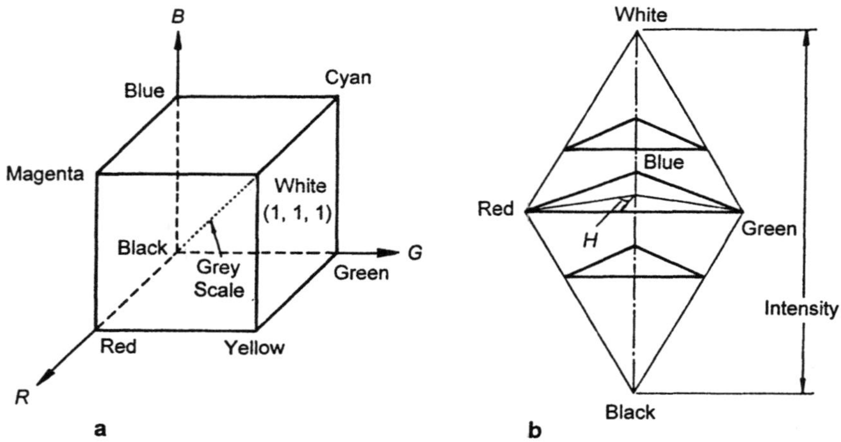

Fig. 7.1. Graphical representation of colour models: **a** RGB model **b** HSI model

the amount of white light mixed with a hue. Brightness is a subjective descriptor and it embodies the chromatic notion of intensity.

A colour model is one that facilitates the specification of colours in a generally accepted way. The colour models are based on either hardware-oriented or application-oriented. Hardware-oriented models are RGB, CMY and YIQ. Most video cameras for colour image capture use RGB (red, green, and blue) model on the other hand, printers use the CMY (cyan, magenta, yellow) model. Colour TV broadcast uses the YIQ (luminance, inphase, quadrature) model. The YIQ model is basically a record of RGB. It is designed such that the information transmitted is compatible to both colour and B&W television sets. The Y component provides all the video information required by a monochrome television set [20]. The colour models most often used for image processing are the RGB and the HSI (hue, saturation, and intensity) models. In some applications YIQ model is also used.

7.2.1 RGB Model

In this model, each image consists of three independent image planes of red, green and blue. When the images are expressed naturally in terms of three colour planes, this model becomes the most ideal one to choose. Most video cameras used for acquiring digital images utilise the RGB format. In view of this, RGB model is very important in colour image processing. RGB model has a special significance in digital photoelasticity. Many techniques use this model for data acquisition in digital photoelasticity.

A graphical representation of this model is shown in Fig. 7.1a. All colour values are normalised such that one gets a unit cube. Colours are points on or inside the cube, defined by vectors extending from the origin. The grey scale extends from black to white along the line joining these two points. For any point along this line, the values of R, G and B are equal.

Noise removal techniques developed for processing black & white images such as time averaging, spatial averaging can be directly used here by applying these to each image planes of red, green and blue. On the other hand, use of image enhancement techniques such as histogram equalisation may result in distorting the colour information, as intensities of three image planes will be altered differently by this operation. This is not a serious handicap to use RGB model in digital photoelasticity, as most of the algorithms for data acquisition do not require any modification of the recorded intensity.

7.2.2 HSI Model

In this model, the colour is represented by its *hue*, *saturation* and *intensity*. Since the intensity information is decoupled from the colour information of the image, no distortion in colour will result while employing image enhancement technique such as histogram equalisation. The *hue* and *saturation* components are closely related to the way in which human beings perceive colour. Thus, HSI is ideally suited for designing image processing applications where the colour sensing properties of the human visual system is used.

A graphical representation of the HSI model is shown in Fig. 7.1b. Hue is the predominant spectral colour of the received light and represents the tint such as red, green, blue, pink, brown etc. The hue H of a colour is determined by its angle with respect to the red axis (Fig. 7.1b). The intensity of a colour is its perpendicular distance from the black point. The farther it is from black, the greater is the intensity of colour. The saturation S, of a colour is the degree to which the colour is undiluted by white and is proportional to the distance from the centre of the triangle. Fully saturated colours lie on the outer boundary of the triangle and have no white component.

As the basic image data recorded by a colour camera is only RGB (Sect. 7.4), the HSI model has to be constructed from the RGB model. After performing any image processing sequence on the HSI model, it has to be converted back to RGB model for display. Conversion from RGB to HSI is obtained as [20]

$$H = \cos^{-1}\left\{\frac{\frac{1}{2}[(R-G)+(R-B)]}{\sqrt{(R-G)^2+(R-B)(G-B)}}\right\} \tag{7.1a}$$

$$S = 1 - \frac{3}{(R+G+B)}[\min R, G, B]$$

$$I = \frac{1}{3}(R + G + B)$$
$$H = 360 - H \quad \text{if} \ (B/I > G/I) \tag{7.1b}$$

It is to be noted that in Eq. (7.1) only normalised values of *R, G* and *B* in the range 0–1 should be used. The values of *S* and *I* obtained from Eq. (7.1) are in the range of 0–1. *H* is normalised by taking $H = H/360^{\circ}$. The conversion of *H, S* and *I* values to *R, G* and *B* is quite involved and one has to use different sets of relations for different zones of *H*. The appropriate expressions are summarised in Table 7.1. From the table one gets only *r, g* and *b* which are related to *R, G* and *B* as

$$r = \frac{R}{R+G+B}$$
$$g = \frac{G}{R+G+B} \tag{7.2}$$
$$b = \frac{B}{R+G+B}$$

Table 7.1. Conversion formulae for converting HSI to RGB values

	Zone-1 $0 < H \le 120^{\circ}$	Zone-2 $120^{\circ} < H \le 240^{\circ}$	Zone-3 $240^{\circ} < H \le 360^{\circ}$
r	$\frac{1}{3}\left[1 + \frac{S\cos H}{\cos(60^{\circ} - H)}\right]$	$\frac{1}{3}(1-S)$	$1-(g+b)$
g	$1-(r+b)$	$\frac{1}{3}\left[1 + \frac{S\cos(H-120^{\circ})}{\cos(180^{\circ} - H)}\right]$	$\frac{1}{3}(1-S)$
b	$\frac{1}{3}(1-S)$	$1-(r+b)$	$\frac{1}{3}\left[1 + \frac{S\cos(H-240^{\circ})}{\cos(300^{\circ} - H)}\right]$

7.3 Colour Image Processing Systems

7.3.1 Hardware

Since, a colour image is a superposition of three image planes of red, green and blue, the size of a colour image is three times that of a monochrome image. In Chap. 3, it was discussed that the spatial resolution and the quantization levels for a particular application are so selected such that the image is cosmetically pleasing and is also easy to handle using the available hardware. In the early use of colour image processing techniques in digital photoelasticity, only 32 quantization levels, unlike 256 quantization levels in B&W images, were used to facilitate easy image handling [11, 12, 17]. However, as the developments in computer hardware are quite rapid, use of 256 quantization levels for each image planes are quite common now [13–16]. In Sect. 7.2, the symbols R, G and B were used to denote the normalised intensity values in the range 0–1 but in the subsequent discussions these are used to denote the intensity variation in the range 0–255 as integers.

7.3.1.1 Transmission Photoelasticity

Portability is not a critical factor in transmission photoelasticity. Thus, one can afford to choose high-resolution systems with ease for transmission photoelasticity. For true colour rendering, one needs to have a frame grabber that can be configured to store a 24-bit image. Systems specially meant for handling colour images are available in the market. They are MVP-AT, Meteor (RGB) of Matrox Corporation. Alternately, one can also have three monochrome cards such as PIP for each image planes of red, green and blue connected to a high-resolution colour monitor. The results shown in this book on transmission photoelasticity are obtained by a TMC-76 RGB PULNiX camera connected to a MVP-AT board. The board can be configured to have four 512×512 pixels $\times$ 8 bits monochrome frame buffers or two 512×512 pixels $\times$ 16 bits colour frame buffers or one 512×512 pixels $\times$ 24 bits colour frame buffer. Overlapping of three 8-bit monochrome frame buffers generates the 24-bit colour frame buffer.

7.3.1.2 Reflection Photoelasticity

Portability is an important criterion while identifying the hardware requirement for photoelastic coating analysis. The systems that have been reported so far, for transmission photoelastic analysis were bulky and consisted of a CCD camera connected directly to a PC based image-processing system. With the introduction of high resolution CCD arrays for portable camcorders, they can easily effect the image capture at the site. However, not all the image processing cards can accept colour images from a VCR (PAL compatible). An intermediate interface is needed to overcome this.

In the applications discussed in this book, Sony Hi8 CCD-TR750E Handycam (PAL compatible) is used to capture images at the site. Matrox-Marvel multimedia system is used as an interface to collect images frame by frame from the Handycam. The colour images thus grabbed could be directly used in a colour image processing system such as MVP-AT or split into three channels of red, green and blue and transferred to a monochrome image processing systems for further processing.

The use of a consumer electronics quality VCR for image recording may be quite cost effective but the quality of the images recorded will not be as good as those recorded directly from a CCD camera connected to a PC based image processing system. In future, it is quite possible to see laptop computers being used for digital reflection photoelasticity for striking a balance between portability and quality.

7.3.2 Software

Use of RGB model is quite common in digital photoelasticity. In this, image is seen as a superposition of image planes of red, green and blue. Each of these planes can be treated as a monochrome image for the purpose of image processing. For example, time averaging, which was used for removing electronic noise in black & white images could be directly extended here by applying these to individual image planes of red, green and blue. Similarly, spatial averaging and median filtering could be applied to individual planes. Any operation involved in altering the intensity values in a major way will lead to distortion in the appearance of colour of the final image. It will be seen in the subsequent discussions, that, in digital photoelasticity, a colour image processing system is basically used as an image acquisition device and no alteration of the recorded intensity is needed.

7.4 Typical Spectral Response of a Colour Camera

The spectral response of a colour camera depends on how the camera is constructed. It has been seen in the earlier sections that colour can be thought of as a combination of either three primaries or secondaries (complements). For high quality colour imagery, ideally three separate detectors each sensitive to a primary or their complements are to be used. However, for consumer applications, a single array is used. In fact many of the early application of colour image processing techniques for digital photoelasticity were based on these due to their easy availability at affordable costs. The cameras that use three detectors for colour recording are labelled as 3CCD in product brochures.

The detectors are usually covered with different filters with the appropriate spectral transmittance to create either the primary colours (R, G, B) or their complements (Ye, Cy, Mg). Complementary filters have higher transmittances than the primary filters. While using a single detector, the filter itself is constructed as an array called coloured filtered array (CFA) to make the camera sensitive to the complete spectrum of colours. The CFAs could be of two types either a *stripe* or a *mosaic* pattern (Fig. 7.2). It is instructive to note that the number of filters devoted to each colour is different and more of these are used near the green wavelength.

R	G	B	G
R	G	B	G
R	G	B	G
R	G	B	G

Ye	G	Cy	G
Ye	G	Cy	G
Ye	G	Cy	G
Ye	G	Cy	G

Stripes

G	Mg	G	Mg
Cy	Ye	Cy	Ye
Mg	G	Mg	G
Cy	Ye	Cy	Ye

R	G	R	G
G	B	G	B
R	G	R	G
G	B	G	B

Mosaics

Fig. 7.2. Various designs of CFA

This is because; the human visual system derives its detail information primarily from the green portion of the spectrum. An array that has 768 horizontal elements may devote 384 to green, 192 to red and 192 to blue [22]. This result in an unequal sampling of the colours and colour aliasing may occur if the filters for various colours are not properly distributed. The layouts of stripes or mosaics depend on the manufacturer's philosophy and cleverness in reducing colour aliasing. The colour video from a single chip CFA must be decoded to produce usable R, G and B signals. Usually, the decoder is part of the camera itself and the cameras, in general, provide an RGB output and a composite video output.

Desired spectral response for the three primaries and their complements are shown in Fig. 7.3 [22, 23]. The spectral response of camera filters is a measure of fidelity of the camera to reproduce actual colours. As the spectral response widens and becomes flat, each of the three channels becomes a mixture of a large number of wavelengths and thus, ceases to act as an optical filter.

The CCD arrays for consumer and industrial applications are within a sealed environment for protection. Light must pass through a glass window to reach the array. To minimise reflection losses at all wavelengths, the glass has to be coated with a broad band anti-reflection coating.

Interline transfer devices use photodiodes as sensors. When CMOS are used as sensors, additional problems occur which are due to poly-silicon effect [22]. To minimise this effect, it is recommended to use a back illuminated arrangement. The typical spectral response of front illuminated and

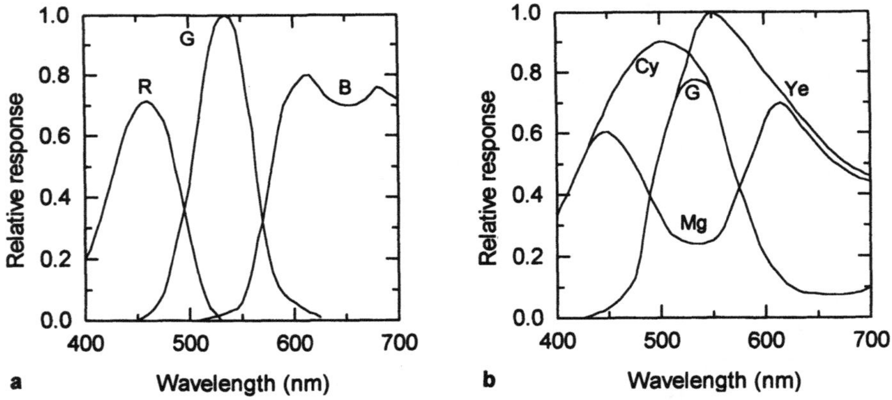

Fig. 7.3. Desired spectral response for the three primaries and their complements: **a** the primaries are red, green, and blue (R, G, B) and **b** their complements are yellow, cyan, and magenta (Ye, Cy, Mg) (Courtesy of refs. [22, 23])

back illuminated CCDs are shown in Fig. 7.4. In Chap. 3 it was mentioned that arrays are cooled to reduce dark current and dark current shot noise has to be accounted for in applications requiring long exposure. However, cooling also reduces the quantum efficiency with the red regions of the spectrum affected the most. These factors have to be taken into account in image analysis.

7.5 Intensity of Light Transmitted in White Light for Various Polariscope Arrangements

Most systems that use colour image processing hardware for data acquisition use white light as the illuminating source. In the previous chapters, intensity equations have been derived for a monochromatic source. It was implicitly assumed that in a circular polariscope arrangement, the quarter-wave plates are matched to the wavelength of the monochromatic source. The monochrome camera was assumed to have its peak spectral response at the wavelength of the light source.

However, while using a white light source, these assumptions have to be re-looked and proper corrections need to be incorporated in the intensity equations. For example, the camera does not have a flat response for all wavelengths and its spectral response $F(\lambda)$ is a function of the wavelength. Further, the light source characteristics also influence the record of intensities at the camera plane. The amplitudes of light corresponding to each of the wavelengths may not be the same for a given source and will be different for different sources [24]. In addition to these effects, one may

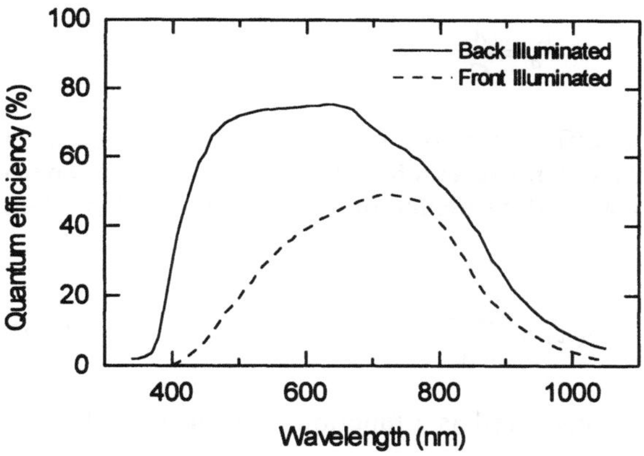

Fig. 7.4. Representative spectral response of front-illuminated and back-side illuminated CCD arrays. The spectral response depends upon the process used to manufacture each device (Courtesy of ref. [22])

have to include a function $T(\lambda)$ which is the sum total of the transmission behaviour of all the optical elements as a function of wavelength forming the particular polariscope arrangement. Finally, in view of a spectrum of colours, the perceived intensity will be an integration of the individual light intensity corresponding to each of the wavelengths in the spectrum. For a plane polariscope, in dark-field arrangement, with the analyzer at an angle β, the Eq. (5.3) could be modified to include these aspects as

$$I_p = I_b + \left[\frac{1}{(\lambda_2 - \lambda_1)} \int_{\lambda_1}^{\lambda_2} T(\lambda) F(\lambda) I_a(\lambda) \sin^2 \frac{\delta(\lambda)}{2} d\lambda \right] \sin^2 2(\theta - \beta) \tag{7.3}$$

where I_b is the total background light intensity for the entire spectrum of colours, the factor $(\lambda_2 - \lambda_1)$ in the denominator is the normalising factor [2, 3] such that the intensity varies in the range 0–1 from dark to bright-field arrangement. Normally, one would like to express the retardation $\delta(\lambda)$ with respect to a reference wavelength λ_{ref}. From stress-optic law one gets

$$N = \frac{\delta}{2\pi} = \frac{t}{F_\sigma}(\sigma_1 - \sigma_2) \tag{7.4}$$

The Eq. (7.4) shows

$$\delta(\lambda) F_\sigma(\lambda) = \text{constant} \tag{7.5}$$

Referring to Chapter 1, the material stress fringe value F_σ is related to the wavelength as

$$F_\sigma = \frac{\lambda}{C} \tag{7.6}$$

where, C is the stress-optic coefficient of the given material. Though C is also a function of wavelength, several researchers have taken C as independent of wavelength due to its weak dependence on wavelength. If C is taken as a constant then

$$\delta(\lambda) = \frac{\delta_{ref}\lambda_{ref}}{\lambda} \tag{7.7}$$

On the other hand if C is also considered as a function of λ, then one has

$$\delta(\lambda) = \frac{\delta_{ref}\lambda_{ref}}{\lambda}\frac{C_\lambda}{C_{ref}} \tag{7.8}$$

where C_λ is the stress-optic coefficient for the wavelength λ and C_{ref} is the stress-optic coefficient at the reference wavelength. Use of Eq. (7.8) becomes important in SCA and also in studies where total fringe order has to be obtained based on fractional retardation measured at various wavelengths [25]. In practice, only F_σ is directly calculated from calibration experiments and Eq. (7.8) can be modified to

$$\delta(\lambda) = \frac{\delta_{ref}F_{\sigma_{ref}}}{F_\sigma(\lambda)} \tag{7.9}$$

The expressions for wavelength dependence of C_λ or F_σ for various materials can be found in references [5, 24, 25]. Substituting Eq. (7.9) in Eq. (7.3) one gets

$$I_p = I_b + \left[\frac{1}{(\lambda_2 - \lambda_1)}\int_{\lambda_1}^{\lambda_2} T(\lambda)F(\lambda)I_a(\lambda)\sin^2\frac{\delta_{ref}F_{\sigma_{ref}}}{2F_\sigma(\lambda)}d\lambda\right]\sin^2 2(\theta-\beta) \tag{7.10}$$

where δ_{ref} is the fractional retardation expressed in radians. In some studies, rather than the fractional retardation, attempts have been made to find the total fringe order from Eq. (7.10). To achieve this, substituting Eq. (7.4) in Eq. (7.10) one gets

$$I_p = I_b + \left[\frac{1}{(\lambda_2 - \lambda_1)}\int_{\lambda_1}^{\lambda_2} T(\lambda)F(\lambda)I_a(\lambda)\sin^2\frac{\pi N_{ref}F_{\sigma_{ref}}}{F_\sigma(\lambda)}d\lambda\right]\sin^2 2(\theta-\beta) \tag{7.11}$$

where, N_{ref} is the total fringe order corresponding to the reference wavelength. To simplify the analysis, in most studies, $T(\lambda)$ is assumed to be unity. This simplification implies that no light energy is absorbed by the optical elements. This makes Eq. (7.11) as

$$I_{\mathrm{p}} = I_{\mathrm{b}} + \left[\frac{1}{(\lambda_2 - \lambda_1)} \int_{\lambda_1}^{\lambda_2} F(\lambda) I_{\mathrm{a}}(\lambda) \sin^2 \frac{\pi N_{ref} F_{\sigma_{ref}}}{F_\sigma(\lambda)} d\lambda \right] \sin^2 2(\theta - \beta) \quad (7.12)$$

The equations of intensity of light transmitted for the circular polariscope (Fig. 1.13, Chap. 1) can be obtained on similar lines. In doing so, the role of quarter-wave plate mismatch needs to be accounted for.

A polariscope is normally constructed with quarter-wave plates meant for a specific wavelength, while white light is used, the quarter-wave plate no longer provides a phase shift of $\pi/2$ for all wavelengths and in general behaves like a retarder. If the reference wavelength for which the quarter-wave plate is matched is say λ_{ref}, the error introduced for all other wavelengths (λ) is

$$\varepsilon = \frac{\pi}{2} \left(\frac{\lambda}{\lambda_{ref}} - 1 \right) \quad (7.13)$$

Considering the quarter-wave plate as a retarder with a retardation of ($\pi/2 + \varepsilon$), the intensity of light transmitted for dark and bright-fields, with crossed quarter-wave plates, for a particular wavelength are respectively as

$$I_{\mathrm{d}}(\lambda) = I_{\mathrm{a}}(\lambda) \sin^2 \left(\frac{\pi N_{ref} F_{\sigma_{ref}}}{F_\sigma(\lambda)} \right) [1 - \cos^2 2\theta \sin^2 \varepsilon] \quad (7.14)$$

$$I_\ell(\lambda) = I_{\mathrm{a}}(\lambda) \left\{ \cos^2 \left(\frac{\pi N_{ref} F_{\sigma_{ref}}}{F_\sigma(\lambda)} \right) (1 - \cos^2 2\theta \sin^2 \varepsilon) + \cos^2 2\theta \sin^2 \varepsilon \right\} \quad (7.15)$$

In Chap. 1, Table 1.1, it was mentioned that dark or bright-field arrangements could also be achieved by keeping the quarter-wave plates parallel to each other and keeping the polarizer and analyzer parallel or crossed respectively. The development of respective intensity equations including the quarter-wave plate error for such arrangements is left as an exercise to the reader. It can be shown [12] that the extinction or minimum intensity condition for such arrangements would be different than that was obtained for Eq. (1.38) for dark-field and Eq. (1.39) for bright-field. However, the extinction or minimum intensity conditions for Eqs. (7.14) and (7.15) are the same as that for Eqs. (1.38) and (1.39) respectively. In other words, in white light, the isochromatic fringes will be distorted for a circular polariscope with quarter-wave plates being kept parallel. Hence, it is always recommended to use the crossed positions of quarter-wave plates for use with multiple wavelengths. Thus, Eqs. (7.14) and (7.15) will be used for subsequent analysis.

Comparing Eq. (7.14) and Eq. (1.38), the role of quarter-wave plate error in a dark-field arrangement, can be modelled as a multiplication term $[1 - \cos^2 2\theta \sin^2 \varepsilon]$. Sanford erroneously extended this to light field arrangement too. However, Eq. (7.15) shows that one also requires an additional term to correctly account for the quarter-wave plate error. This was first brought into focus by Ajovalasit et al. [12].

For the complete spectrum of white light, including the background light intensity, Eqs. (7.14) and (7.15) change to

$$I_d = I_b + \frac{1}{(\lambda_2 - \lambda_1)} \int_{\lambda_1}^{\lambda_2} F(\lambda) I_a(\lambda) \sin^2\left(\frac{\pi N_{ref} F_{\sigma_{ref}}}{F_\sigma(\lambda)}\right)[1 - \cos^2 2\theta \sin^2 \varepsilon]\, d\lambda \quad (7.16)$$

$$I_\ell = I_b + \frac{1}{(\lambda_2 - \lambda_1)} \int_{\lambda_1}^{\lambda_2} F(\lambda) I_a(\lambda) \left\{\cos^2\left(\frac{\pi N_{ref} F_{\sigma_{ref}}}{F_\sigma(\lambda)}\right)(1 - \cos^2 2\theta \sin^2 \varepsilon) + \cos^2 2\theta \sin^2 \varepsilon\right\} d\lambda \quad (7.17)$$

If one has a CCD camera with a single chip to capture a colour image then the above equations can be directly used where λ_1 & λ_2 are the spectral limits of the light source or the camera whichever is lower. If one uses a CCD camera with three detectors, for sensing each primary or its complement, one has to use integration for each of the component light separately using the spectral limits of each detector [17].

Though the intensity equations presented in this section appear quite complex, judicious approximations are invoked while using them for practical applications.

7.6 Three Fringe Photoelasticity (TFP)

Use of a colour code to identify fringe gradient direction and to assign approximately the total fringe order has been in use in conventional photoelasticity. The advent of colour image processing hardware has made this approach easily implementable in digital domain. In the literature one finds the use of both RGB [11, 13] and the HSI models [14] to identify the total fringe order. The basic procedure is to first establish a colour code in terms of *R, G, B* or *H, S, I* by a process of calibration. The total fringe order at a point of interest in the actual model is then established by comparing the values of *R, G, B* or *H, S, I* values at the point of interest with that of the calibration table. In principle, the technique is an extension of half fringe photoelasticity to colour domain. The colours tend to merge beyond fringe order three, and hence the technique is termed as three-fringe photoelasticity (TFP) [13]. Since, R, G and B values of a colour image are used, it is also termed as RGB photoelasticity (RGBP) [11]. An extension of the technique to reflection photoelasticity is also reported [26].

7.6.1 Calibration

In conventional photoelasticity, colour code is basically used in a qualitative sense and the code is available for the dark-field arrangement (Table 1.2, Chap. 1). To use the colour code in a quantitative sense, the calibration has to be systematic and quite detailed to account for practical problems.

In Chap. 3, it was shown how the modern day CCD cameras are quite sensitive to measure even minute variations in intensity values, which a normal human eye cannot detect easily. Colour CCD cameras are equally sensitive to measure intensity variations in RGB planes as monochrome CCD cameras. The only difference is that the illumination levels have to be slightly brighter for colour grabbing than for B&W grabbing.

In view of spatial and intensity digitisation, the values of R, G and B could be sensitive to the fringe gradient. This is because, in zones of high fringe gradient, the colour code of Table 1.2 (Chap. 1) gets compressed and in zones of less fringe gradient, the code is quite stretched. In view of this, the perceived colour could be different. Stretching of the colour code occurs, in the case of reflection photoelasticity and also in the analysis of slices cut from stress frozen models. On the other hand, in 2-D transmission photoelasticity, the fringe gradient could lie between moderate to high. Though, it is impractical to develop a calibration table for every value of the fringe gradient, the calibration scheme should take into account the role of fringe gradient in some manner in the construction of the calibration tables. The quantization levels should be at least 256 to generate a good calibration table.

The calibration table containing RGB values associated with known fringe orders can be prepared using a beam under four point bending. In view of the linear variation of the fringe orders, the total fringe order can be easily calculated for each pixel, which is then used to construct the calibration table. The influence of fringe gradient could be incorporated easily by appropriately selecting the bending load. The loads could be adjusted such that the farthest fringe order seen is 1, 2 or 3. Thus, three calibration tables corresponding to 0–1 (low fringe gradient), 0–2 (medium fringe gradient) and 0–3 (high fringe gradient) fringe orders are to be obtained. Care should be taken while preparing a calibration table, so that the table would not include a noise point with sudden change in the RGB values. This can be ensured by collecting data from the central region of the beam and by taking the average of 40 pixels horizontally for each fringe order. Further, the illumination levels should be so adjusted that no saturation occurs for any of the image planes of R, G and B. Tables 7.2 and 7.3 show the calibration tables for fringe order variation of 0–1 and 0–3 respectively. The models were made by mixing CY230 resin and hardener HY951 in the ratio 100:9 and they have a light yellow tinge.

Table 7.2. The *R*, *G*, and *B* values recorded for 0–1 fringe orders in dark-field arrangement (Courtesy of BSSM [13])

R	*G*	*B*	*N*	*R*	*G*	*B*	*N*
68	91	77	0.00000	169	251	138	0.51163
70	86	82	0.01163	169	250	132	0.52326
67	90	86	0.02326	168	250	127	0.53488
67	93	83	0.03488	179	245	130	0.54651
72	90	95	0.04651	170	243	128	0.55814
73	97	89	0.05814	170	240	120	0.56977
76	101	94	0.06977	169	239	111	0.58140
76	108	93	0.08140	170	236	102	0.59302
81	113	108	0.09302	168	233	99	0.60465
80	122	100	0.10465	165	230	91	0.61628
84	128	117	0.11628	163	228	83	0.62791
89	134	113	0.12791	164	222	73	0.63953
90	142	124	0.13953	161	220	64	0.65116
96	148	120	0.15116	162	231	62	0.66279
98	157	125	0.16279	159	210	58	0.67442
103	161	130	0.17442	156	206	42	0.68605
109	164	149	0.18605	156	200	37	0.69767
110	172	145	0.19767	153	192	52	0.70930
114	180	145	0.20930	150	189	36	0.72093
121	181	157	0.22093	145	185	26	0.73256
124	190	158	0.23256	147	173	45	0.74419
130	192	160	0.24419	143	170	28	0.75581
130	202	161	0.25581	137	166	34	0.76744
138	203	168	0.26744	137	158	34	0.77907
143	208	173	0.27907	138	149	42	0.79070
147	212	172	0.29070	128	146	37	0.80233
147	219	177	0.30233	129	135	63	0.81395
152	221	178	0.31395	123	130	58	0.82558
155	225	180	0.32558	121	123	64	0.83721
158	227	184	0.33721	126	108	99	0.84884
158	235	171	0.34884	112	113	88	0.86047
161	234	185	0.36047	106	108	100	0.87209
160	240	173	0.37209	100	107	106	0.88372
162	242	173	0.38372	106	94	134	0.89535
162	246	166	0.39535	95	98	134	0.90698
163	247	165	0.40698	92	95	147	0.91860
165	247	167	0.41860	94	93	160	0.93023
167	248	166	0.43023	88	94	165	0.94186
166	251	165	0.44186	83	103	169	0.95349
168	251	156	0.45349	84	102	178	0.96512
167	251	158	0.46512	86	103	194	0.97674
169	251	148	0.47674	86	104	208	0.98837
170	251	151	0.48837	85	106	209	1.00000
171	249	146	0.50000				

Table 7.3. The *R*, *G* and *B* values for the range 0–3 fringe orders recorded for dark-field arrangement (Courtesy of BSSM [13])

R	*G*	*B*	*N*
69	104	74	0.00000
76	107	80	0.02586
78	115	96	0.51720
82	127	108	0.07759
88	141	109	0.10345
93	155	117	0.12931
99	168	125	0.15517
111	117	151	0.18103
117	190	153	0.20690
125	201	163	0.23276
135	209	168	0.25862
141	221	170	0.28448
150	226	178	0.31034
155	233	175	0.33621
160	239	172	0.36207
161	244	174	0.38793
166	247	169	0.41379
168	249	162	0.43966
168	251	150	0.46552
171	250	152	0.49138
168	251	139	0.51724
170	246	131	0.54310
167	246	113	0.56897
168	239	110	0.59483
166	234	94	0.62069
166	224	85	0.64655
160	218	67	0.67241
155	209	52	0.69828
152	197	42	0.72414
148	183	40	0.75000
151	164	52	0.77586
137	155	55	0.80172
132	142	62	0.82759
127	125	92	0.85345
119	115	108	0.87931
110	107	133	0.90517
98	106	150	0.93103
89	106	166	0.95690
85	106	195	0.98276
83	112	216	1.00862
81	214	213	1.03448
83	133	225	1.06034
82	147	231	1.08621
82	165	215	1.11207
89	177	213	1.13793
99	184	222	1.16379
103	198	202	1.18966
110	211	189	1.21552
115	221	181	1.24138
124	227	176	1.26724
131	234	165	1.29310
136	241	157	1.31897
142	246	141	1.34483
144	251	127	1.37069
148	252	119	1.39655
153	252	111	1.42241
157	251	107	1.44828
159	250	97	1.47414
160	249	83	1.50000
159	248	77	1.52586
163	241	73	1.55172
166	234	63	1.57759
170	227	52	1.60345
167	219	55	1.62931
166	212	44	1.65517
169	195	62	1.68103
173	182	70	1.70690
164	174	80	1.73276
168	156	113	1.75862
166	143	133	1.78448
165	130	156	1.81034
148	129	167	1.83621
132	131	172	1.86207
132	119	206	1.88793
122	121	214	1.93379
111	125	221	1.93966
99	131	232	1.96552
95	140	219	1.99138
91	148	215	2.01724
87	157	218	2.04310
87	166	204	2.06897
87	177	198	2.09463
88	187	184	2.12069
87	202	162	2.14655
94	206	163	2.17241
98	213	160	2.19828
102	223	142	2.22414
105	229	143	2.25000
109	237	125	2.27586
112	242	118	2.30172
117	248	100	2.32759
119	251	95	2.35345
124	252	96	2.37931
130	251	92	2.40517
134	252	84	2.43103
137	251	83	2.45690
142	248	84	2.48276
141	247	78	2.50862
149	239	82	2.53448
150	234	81	2.56034
154	227	86	2.58621
158	218	90	2.61207
155	212	91	2.63793
159	201	101	2.66379
162	191	107	2.68966
168	173	138	2.71552
164	168	138	2.74138
161	156	155	2.76724
160	149	165	2.79310
146	149	167	2.81897
143	144	179	2.84483
125	151	179	2.87069
116	157	179	2.89655
101	170	171	2.92241
96	182	151	2.94828
90	197	136	2.97414
91	209	117	3.00000

To minimise the effect of non-uniform illumination in actual experimental situations, Ramesh and Deshmukh [13] suggested the following approach. Initially, calibration tables for dark-field and bright-field tables are to be constructed. The final calibration table is one in which the RGB values of dark-field are subtracted from the bright-field intensity values. Tables 7.4 and 7.5 show the calibration tables for (bright – dark) combination for fringe order variation of 0–1 and 0–3.

Table 7.4. The R, G and B values for the range 0–1 fringe orders obtained by subtracting dark-field values from bright-field values (Courtesy of BSSM [13])

R	G	B	N	R	G	B	N
84	133	86	0.00000	−92	−147	17	0.51163
83	138	79	0.01163	−92	−143	26	0.52326
86	136	69	0.02326	−89	−140	40	0.53488
86	132	73	0.03488	−92	−128	32	0.54651
79	135	61	0.04651	−88	−128	56	0.55814
76	129	58	0.05814	−90	−117	56	0.56977
75	121	63	0.06977	−85	−113	75	0.58140
71	114	61	0.08140	−84	−108	95	0.59302
67	106	50	0.09302	−80	−98	95	0.60465
68	95	49	0.10465	−77	−87	97	0.61628
60	89	35	0.11628	−71	−82	119	0.62791
55	80	35	0.12791	−74	−71	132	0.63953
54	67	25	0.13953	−67	−64	137	0.65116
45	60	19	0.15116	−70	−50	141	0.66279
42	48	13	0.16279	−64	−40	134	0.67442
33	41	7	0.17442	−58	−33	161	0.68605
23	34	−13	0.18605	−54	−23	162	0.69767
23	21	−11	0.19767	−53	−6	141	0.70930
13	12	−20	0.20930	−44	−2	167	0.72093
2	8	−34	0.22093	−43	12	164	0.73256
−5	−5	−36	0.23256	−35	23	153	0.74419
−9	−13	−46	0.24419	−33	32	163	0.75581
−15	−23	−60	0.25581	−26	42	154	0.76744
−26	−31	−62	0.26744	−22	53	153	0.77907
−32	−40	−73	0.27907	−19	63	148	0.79070
−33	−54	−64	0.29070	−8	70	150	0.80233
−38	−62	−81	0.30233	−9	86	119	0.81395
−48	−70	−78	0.31395	0	89	130	0.82558
−47	−85	−67	0.32558	6	99	121	0.83721
−57	−86	−91	0.33721	−1	119	71	0.84884
−54	−106	−63	0.34884	14	117	83	0.86047
−65	−105	−82	0.36047	23	121	68	0.87209
−63	−119	−67	0.37209	28	126	55	0.88372
−71	−124	−65	0.38372	29	137	31	0.89535
−74	−130	−61	0.39535	37	137	19	0.90698
−81	−132	−68	0.40698	43	139	7	0.91860
−80	−140	−68	0.41860	42	143	−16	0.93023
−82	−146	−49	0.43023	50	141	−25	0.94186
−87	−147	−43	0.44186	56	132	−32	0.95349
−93	−145	−45	0.45349	56	133	−47	0.96512
−91	−149	−36	0.46512	55	132	−70	0.97674
−91	−153	−6	0.47674	55	130	−91	0.98837
−91	−153	0	0.48837	59	120	−90	1.00000
−94	−147	0	0.50000				

Table 7.5. The *R*, *G* and *B* values for the range 0–3 fringe orders obtained by subtracting dark-field values from the bright-field values (Courtesy of BSSM [13])

R	*G*	*B*	*N*
83	118	97	0.00000
77	114	84	0.02586
73	106	62	0.05172
66	91	43	0.07759
54	74	37	0.10345
43	56	18	0.12931
34	36	3	0.15517
17	20	–35	0.18103
4	1	–41	0.20690
–9	–18	–66	0.23276
–18	–41	–70	0.25862
–28	–65	–75	0.28448
–46	–78	–82	0.31034
–56	–97	–79	0.33621
–70	–109	–87	0.36207
–70	–130	–75	0.38793
–80	–140	–60	0.41379
–89	146	–45	0.43966
–89	–152	–28	0.46552
–95	–149	–15	0.49138
–95	–143	5	0.51724
–92	–134	28	0.54310
–90	–124	56	0.56897
–87	–111	80	0.59483
–83	–92	102	0.62069
–81	–67	104	0.64655
–69	–53	130	0.67241
–58	–33	141	0.69828
–50	–11	152	0.72414
–37	12	149	0.75000
–35	37	138	0.77586
–16	55	124	0.80172
–4	74	109	0.82759
4	96	76	0.85345
18	109	53	0.87931
29	120	20	0.90517
44	125	–5	0.93103
54	125	–37	0.95690
61	124	–70	0.98276
63	116	–103	1.00862
65	102	–105	1.03448
63	91	–133	1.06034
66	70	–144	1.08621
67	44	–137	1.11207
54	30	–145	1.13793
47	11	–159	1.16379
42	–13	–132	1.18966
36	–39	–115	1.21552
24	–56	–101	1.24138
15	–77	–81	1.26724
6	–94	–55	1.29310
–1	–116	–23	1.31897
–18	–123	0	1.34483
–29	–134	34	1.37069
–37	–141	64	1.39655
–54	–138	72	1.42241
–63	–138	96	1.44828
–74	–132	115	1.47414
–76	–130	137	1.50000
–76	–126	157	1.52586
–83	–107	156	1.55172
–86	–92	164	1.57759
–89	–76	172	1.60345
–85	–57	161	1.62931
–84	–38	156	1.65517
–83	–11	129	1.68103
–85	10	114	1.70690
–72	27	91	1.73276
–75	55	42	1.75862
–66	74	9	1.78448
–58	87	–4	1.81034
–38	95	–36	1.83621
–18	97	–46	1.86207
–15	113	–97	1.88793
2	109	–103	1.93379
14	106	–119	1.93966
28	101	–137	1.96552
36	89	–124	1.99138
42	80	–132	2.01724
53	61	–127	2.04310
54	49	–113	2.06897
54	34	–111	2.09483
55	16	–92	2.12069
56	–6	–68	2.14655
55	–22	–55	2.17241
49	–35	–50	2.19828
47	–54	–28	2.22414
41	–68	–11	2.25000
38	–84	17	2.27586
35	–99	39	2.30172
24	–110	76	2.32759
17	–117	91	2.35345
3	–116	89	2.37931
–11	–115	100	2.40517
–18	–119	124	2.43103
–26	–117	129	2.45690
–41	–105	122	2.48276
–45	–101	129	2.50862
–57	–85	116	2.53448
–60	–78	121	2.56034
–64	–64	111	2.58621
–71	–47	93	2.61207
–67	–35	88	2.63793
–71	–18	71	2.66379
–72	0	55	2.68966
–81	26	11	2.71552
–74	38	4	2.74138
–68	52	–8	2.76724
–65	69	–37	2.79310
–46	70	–40	2.81897
–36	76	–56	2.84483
–15	70	–58	2.87069
–2	64	–58	2.89655
21	45	–59	2.92241
33	27	–34	2.94828
43	2	–15	2.97414
47	–19	14	3.00000

Referring to Eqs. (7.16) and (7.17), the intensity of light transmitted is also a function of the isoclinic angle θ due to quarter-wave plate error. This error has the effect of reducing the maximum intensity of light transmitted in dark-field and increasing the minimum intensity of light transmitted in bright-field [12]. To account for the quarter-wave plate error Ajovalasit et al. [12] suggested that calibration table be obtained corresponding to the isoclinic angle of 22.5°. An approximation similar to this is also used in SCA analysis (Sect. 7.9).

The role of quarter-wave plate error could be completely eliminated if a plane polariscope arrangement is used for obtaining the colour code [17]. The determination of colour code is not a one step procedure but requires the recording of four polarization-stepped images. The details on how to find the *R*, *G* and *B* values from these images can be seen in Sect. 7.11.1.

The calibration table thus obtained is a function of the spectral characteristics of the light source, the CCD camera and the transmission characteristics of the model and the optical elements used in the polariscope. Though, this gives an impression that calibration is a tedious task, in fact it is not so. For a particular laboratory situation, since most of these are common from experiment to experiment, a set of calibration tables could be developed for each type of model material used.

7.6.2 Methodology

In TFP/RGBP one has to compare the RGB values of a point with the calibrated RGB values assigned with known fringe orders so as to determine the fringe order at a given data point. Ideally, RGB values have to be unique for any fringe order. However, in view of experimental difficulties, the RGB values corresponding to a data point may not exactly coincide with the RGB values in the calibration table. For any test data point, an error term '*e*' is defined as

$$e=(R_e-R_c)^2+(G_e-G_c)^2+(B_e-B_c)^2 \tag{7.18}$$

where, subscript 'e' refers to the experimentally measured values for the data point and 'c' denotes the values in the calibration table. The calibration table corresponding to 0–1, 0–2 and 0–3 fringe orders are to be searched until the error 'e' is a minimum. For the R_c, G_c and B_c values thus determined, the calibration table provides the total fringe order.

7.6.3 Application to the Problem of a Circular Disc Under Diametral Compression

A circular disc made up of the same material as the beam was loaded to a known load. The disc is digitised in both dark and bright-field

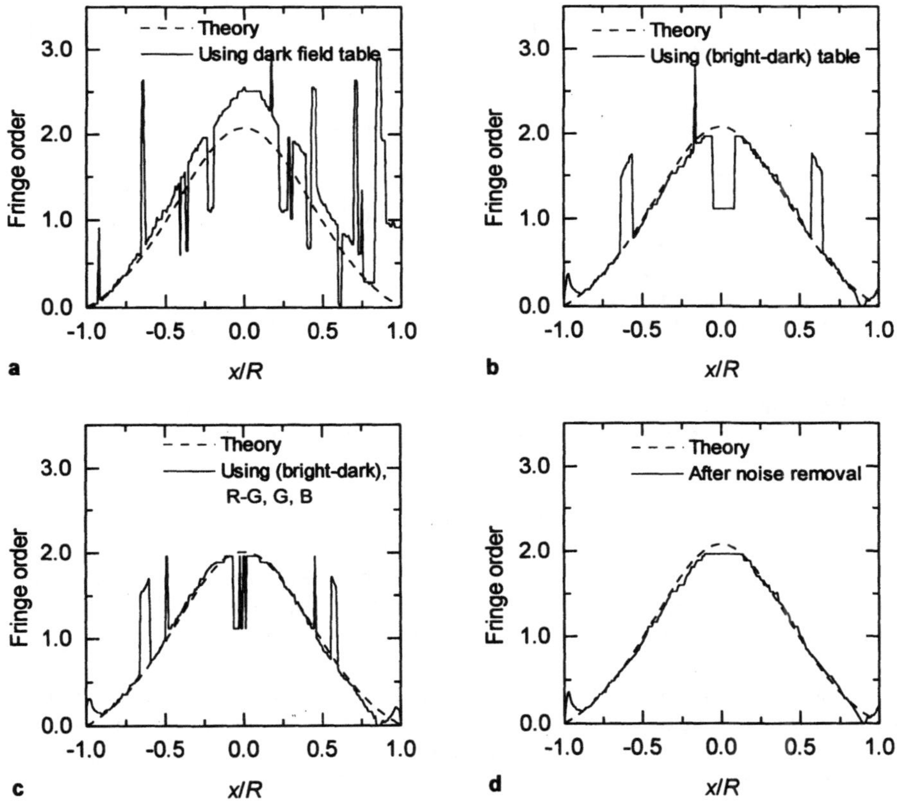

Fig. 7.5. Fringe order variation for disc under diametral compression: **a** using dark-field table **b** using (bright – dark) table **c** (bright – dark) table with combination of $(R-G)$, G, and B and **d** after noise removal (Courtesy of BSSM [13])

arrangements. Figure 7.5a shows the comparison of the fringe orders obtained using the dark-field calibration table with theory for the diametral line. In Fig. 7.5, no smoothening is done while plotting the curves. The curve shows the existence of noise points and these can be attributed to the fact that, the least squares error may be minimum at some other fringe order due to sinusoidal variation of RGB values.

Ramesh and Deshmukh [13] performed a systematic study in reducing the error in the analysis. Figure 7.5b shows the comparison using (bright – dark) table. It is seen that, the number of noise points is less when (bright – dark) combination is used. This is because the technique accounts for non-uniform illumination. It is to be noted here that one of the usual methods to account for non-uniform illumination is that the intensity values of the loaded model is subtracted from those of the intensity values obtained with

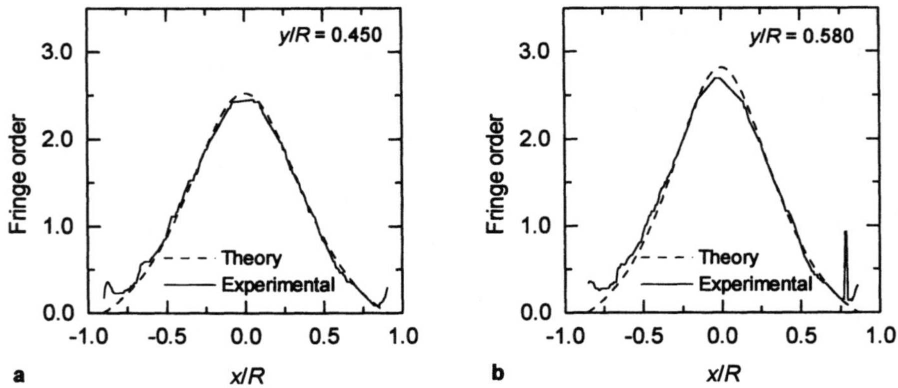

Fig. 7.6. Comparison of variation of total fringe order for two horizontal lines obtained by TFP and theory: **a** $y/R = 0.45$ and **b** $y/R = 0.58$ (Courtesy of BSSM [13])

an unloaded model. However, the models, which are made of epoxies, undergo considerable deformation once the load is applied. Hence, there may be a mismatch between the corresponding pixels of loaded image and unloaded image. However, such a problem does not exist in Fig. 7.5b. The noise points are further reduced if in (bright – dark) combination, $(R - G)$, G and B are used in Eq. (7.18) rather than simply R, G and B.

It can be seen that, in Figs. 7.5a–7.5c, the noise points are distinct. Use of a direction flag to keep track of the fringe gradient direction along with the use of a tolerance value can help in devising a simple and efficient noise removal algorithm. The results are shown in Fig. 7.5d. Using this procedure, the results for two lines with $y/R = 0.45$ and $y/R = 0.58$ respectively are calculated and is shown in Figs. 7.6a and 7.6b.

Unlike SCA [1–5], the computational demand is almost negligible and hence, TFP is easily adaptable for devising automatic polariscopes. Using this method, the least fringe order variation that can be detected is 0.012 fringe order in 0–1 range, 0.017 in 1–2 range and 0.026 in 2–3 range [13]. The accuracy can be further improved by using deep beams coupled with high optical magnification while preparing the calibration tables. To minimise the influence of intensity variation of the light source on the estimation of fringe orders, while using the calibration table constructed from dark-field alone, it is suggested to perform an operation of normalisation of intensities. This is achieved by dividing the R, G and B values of both the model and the calibration table values by their respective sum of $(R + G + B)$ value [17, 27]. It is reported that for intensity variations of the light up to 30 percent, this procedure gives fringe orders with an error lower than 0.03 fringe orders for retardation higher than 0.5 fringe orders.

7.7 Green Image Plane as a Green Filter

It is seen in the previous sections that the colour image could be thought of as a superposition of image planes of red, green and blue. When white light is used as a source, the images corresponding to these individual planes cannot be thought of as monochrome images corresponding to red, green and blue wavelengths as the spectral response of the camera has a large bandwidth (Fig. 7.3). However, if some of these image planes could be approximated to that of a particular wavelength, then data could be extracted easily from photoelasticity point of view. For this, equivalent wavelengths for these image planes need to be computed.

If C, the stress-optic coefficient, is taken independent of wavelength, the dependence of F_σ as a function of wavelength is obtained as

$$F_{\sigma_2} = F_{\sigma_1} \frac{\lambda_2}{\lambda_1} \tag{7.19}$$

The material stress fringe value for the model material could be obtained by analysing the stress field for a model with a known stress distribution (Sect. 1.17) and to evaluate this, the wavelength of light does not come into the equation. If the material stress fringe value corresponding to images in different image planes is obtained by the linear least squares approach involving a large number of data from the fringe field, then by using Eq. (7.19), the equivalent wavelength could be obtained. The equivalent wavelength is obviously the function of the hardware used. Figure 7.7 shows the R, G, and B image planes of a 24-bit colour dark-field isochromatics of a circular disc under diametral compression taken using the TMC 76 PULNiX colour camera.

For two different hardwares, Table 7.6 shows the F_σ values calculated by the least squares method and the wavelength calculated using Eq. (7.19) with the F_σ value corresponding to sodium vapour source as the basis [15, 16]. The table shows that among the three channels, only for the green channel, the experimentally calculated wavelength is very close to the

Table 7.6. Equivalent wavelengths of RGB planes for various systems

S. No.	λ (ref. [20]) (nm)	λ (System 1)†(nm)	λ (System 2)‡(nm)
1.	Red 700.0	598.1	485.66
2.	Green 546.1	551.9	542.34
3.	Blue 435.8	445.7	546.11

† TMC-76 PULNiX CCD camera with Matrox MVP-AT system [15].
‡ Sony Hi8 CCD-TR750E Handycam with Matrox Marvel [16].

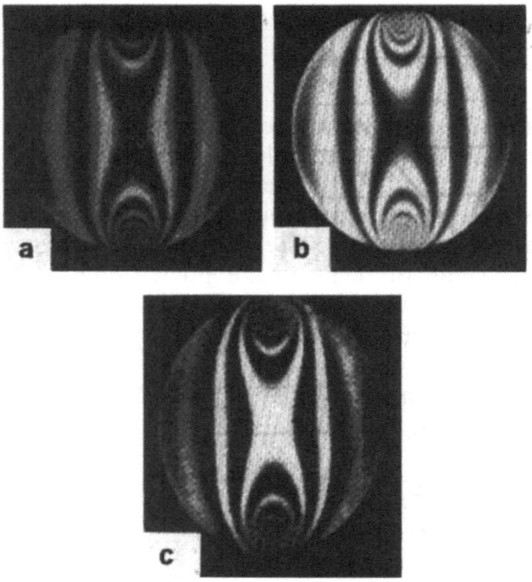

Fig. 7.7. Figure showing the RGB planes of the colour image of a circular disc under diametral compression: **a** R-plane **b** G-plane **c** B-plane

theoretical one. In view of this, the G-plane image can be considered equivalent to an image recorded by a green filter. Further, from Fig. 7.7, the visual quality of the green channel is also good.

7.8 Phase Shifting in Colour Domain

In this technique, the images needed for phase shifting are recorded in colour. For quantitative data extraction, green channel of the colour images is used. Since the images are recorded in colour, the technique is called as phase shifting in colour domain.

7.8.1 Transmission Photoelasticity

The use of phase shifting in colour domain for transmission photoelasticity is simple and straightforward. It is shown in Chap. 5, that among the various phase shifting techniques, the use of a circular polariscope for data extraction is quite rugged and well tested. The procedure requires recording of six images for various positions of the output quarter-wave plates and analyzer (Table 5.3, Chap. 5). Based on these intensity values, the values of isochromatic fringe order and isoclinic values have to be determined as discussed in Sect. 5.8 of Chap. 5. Figure 7.8 shows a simulated dark-field image, bright-field and the phase map using the green channel of Fig. 7.7.

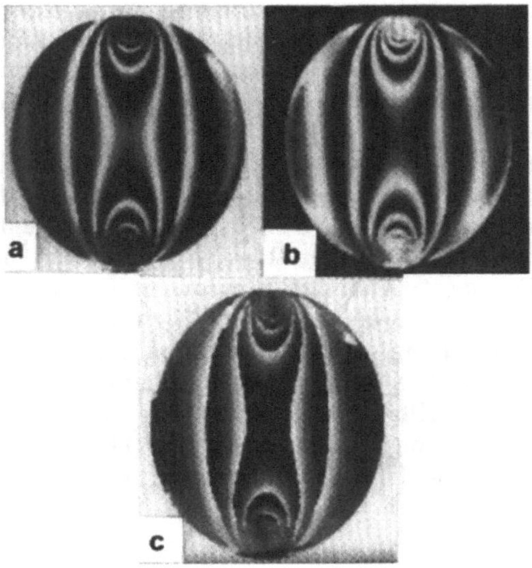

Fig. 7.8. Whole field representation of the fractional retardation: **a** simulated dark-field fringe patterns **b** simulated bright-field fringe patterns **c** phase map of fractional retardation expressed in the range $0–2\pi$

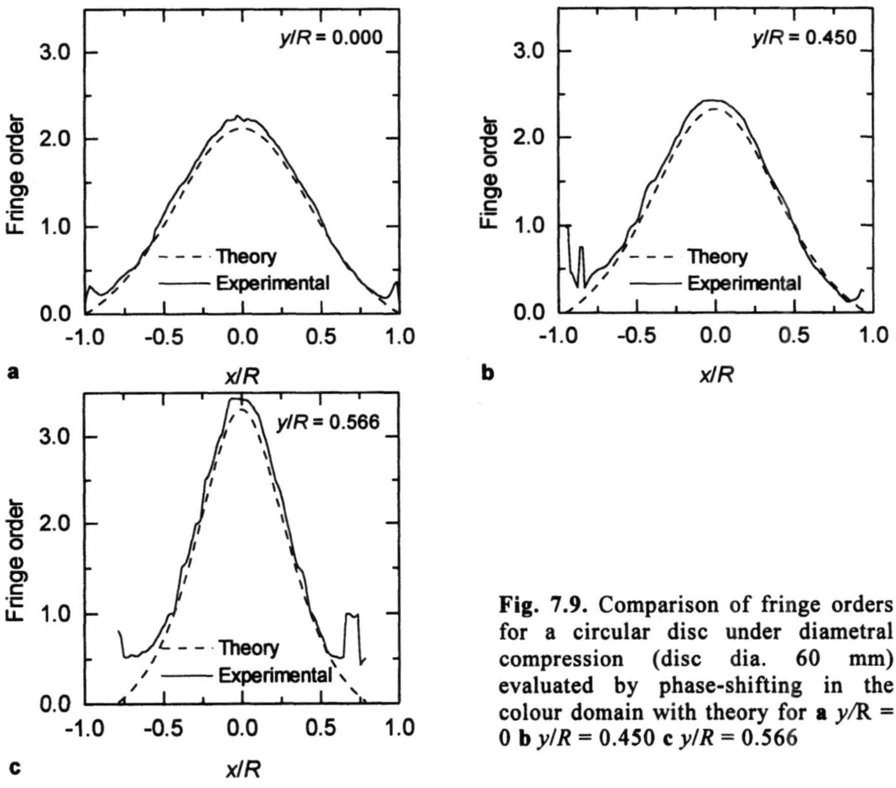

Fig. 7.9. Comparison of fringe orders for a circular disc under diametral compression (disc dia. 60 mm) evaluated by phase-shifting in the colour domain with theory for **a** $y/R = 0$ **b** $y/R = 0.450$ **c** $y/R = 0.566$

A quantitative comparison, for the fringe order variation along three horizontal lines corresponding to $y/R = 0$, $y/R = 0.45$ and $y/R = 0.566$, with disc centre as the origin, are shown in Fig. 7.9 and compared with theory. The comparison is quite good despite using the green plane as equivalent to a green filter.

Ji and Patterson [24] have studied the influence of using multi-spectral light such as white light as a source on the accuracy of evaluation of isochromatic fringe orders and isoclinic angle by PST. They placed medium band filters of 40 nm and 80 nm and recorded the intensity patterns as per Table 5.3, by a monochrome CCD camera. Although, a colour CCD camera should have been used for image capture, the study is nevertheless important. The study showed that by placing various medium band filters in a white light polariscope coupled with a monochrome camera, for an accuracy of ±5 percent, use of a 40 nm filter can measure up to 10 fringes whereas, the 80 nm filter can measure only up to 5 fringes. This result corroborates the use of green channel as a monochrome image for evaluating isochromatic fringe orders by PST.

7.8.2 Reflection Photoelasticity

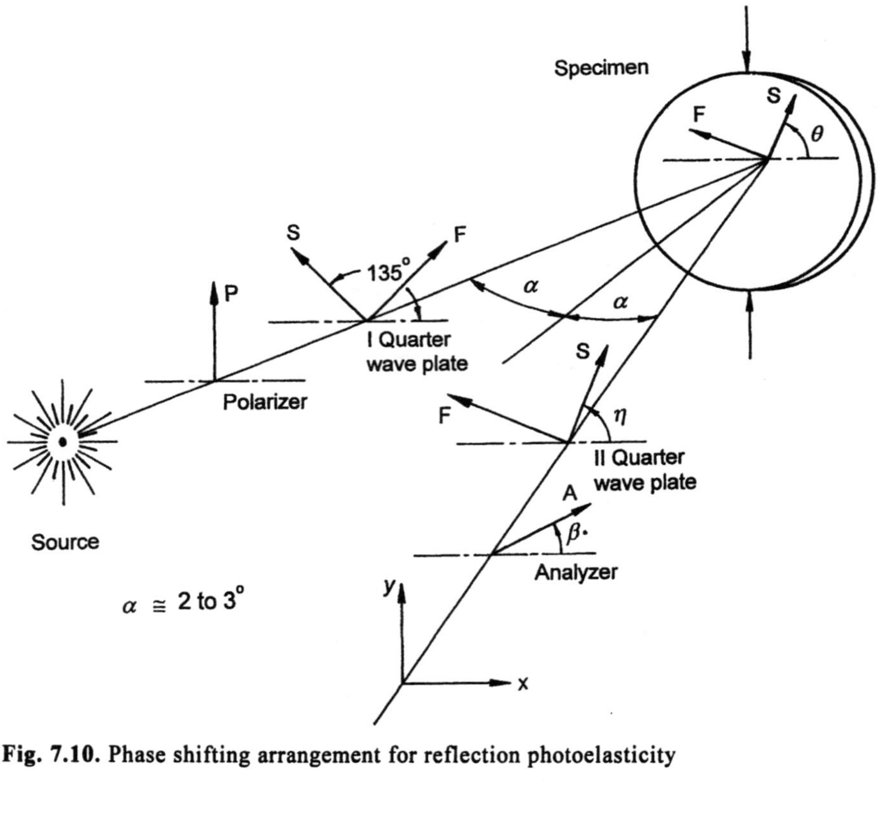

Fig. 7.10. Phase shifting arrangement for reflection photoelasticity

The potential use of phase shifting in colour domain is more appropriate in a reflection photoelasticity arrangement. For image recording, a commercial Handycam could be used. Commercial Handycams, as of available now, have usually a single chip CCD and in view of video recording, the image quality may not be as good as being directly grabbed by a frame grabber attached to a personal computer. The approximation that green plane being considered as monochrome image with an equivalent wavelength may lead to further error. In view of the possible industrial use of the technique, the level of accuracy achievable by this process will be of interest.

The optical arrangement required in reflection photoelasticity to do phase shifting is shown in Fig. 7.10. Using Jones calculus, the components of light vector along the analyzer axis and perpendicular to the analyzer axis are obtained as [16]

$$\begin{Bmatrix} E_\beta \\ E_{\beta+\pi/2} \end{Bmatrix} = \frac{1}{2}\begin{bmatrix} \cos\beta & \sin\beta \\ -\sin\beta & \cos\beta \end{bmatrix}\begin{bmatrix} 1-\mathrm{i}\cos 2\eta & -\mathrm{i}\sin 2\eta \\ -\mathrm{i}\sin 2\eta & 1+\mathrm{i}\cos 2\eta \end{bmatrix}$$
$$\times\begin{bmatrix} \cos\frac{\delta}{4}-\mathrm{i}\sin\frac{\delta}{4}\cos 2\theta & -\mathrm{i}\sin\frac{\delta}{4}\sin 2\theta \\ -\mathrm{i}\sin\frac{\delta}{4}\sin 2\theta & \cos\frac{\delta}{4}+\mathrm{i}\sin\frac{\delta}{4}\cos 2\theta \end{bmatrix} \tag{7.20}$$
$$\times\begin{bmatrix} \cos\frac{\delta}{4}-\mathrm{i}\sin\frac{\delta}{4}\cos 2\theta & -\mathrm{i}\sin\frac{\delta}{4}\sin 2\theta \\ -\mathrm{i}\sin\frac{\delta}{4}\sin 2\theta & \cos\frac{\delta}{4}+\mathrm{i}\sin\frac{\delta}{4}\cos 2\theta \end{bmatrix}\begin{bmatrix} 1 & \mathrm{i} \\ \mathrm{i} & 1 \end{bmatrix}\begin{Bmatrix} 0 \\ 1 \end{Bmatrix} k e^{\mathrm{i}\omega t}$$

In the above equation, δ is used to represent the retardation introduced by the model, θ is the orientation of the principal stress direction with respect to the horizontal, $\mathrm{k}e^{\mathrm{i}\omega t}$ is the incident light vector, E_β and $E_{\beta+\pi/2}$ are the components of light vector along the analyzer axis and perpendicular to the analyzer axis respectively. Since, a reflection arrangement is used, the Jones matrices for the incident light beam and the reflected light beam are separately indicated in Eq. (7.20). For theoretical development of the equation, α (Fig. 7.10) is taken to be zero. The intensity of light transmitted for arbitrary positions of the second quarter-wave plate and the analyzer is obtained as

$$I = \frac{I_{\mathrm{a}}}{2} + \frac{I_{\mathrm{a}}}{2}[\sin 2(\beta-\eta)\cos\delta - \sin 2(\theta-\eta)\cos 2(\beta-\eta)\sin\delta] \tag{7.21}$$

where I_{a} accounts for the amplitude of light. Adding a term I_{b} to denote the background/stray light intensity, Eq. (7.21) becomes

$$I = I_{\mathrm{b}} + \frac{I_{\mathrm{a}}}{2} + \frac{I_{\mathrm{a}}}{2}[\sin 2(\beta-\eta)\cos\delta - \sin 2(\theta-\eta)\cos 2(\beta-\eta)\sin\delta] \tag{7.22}$$

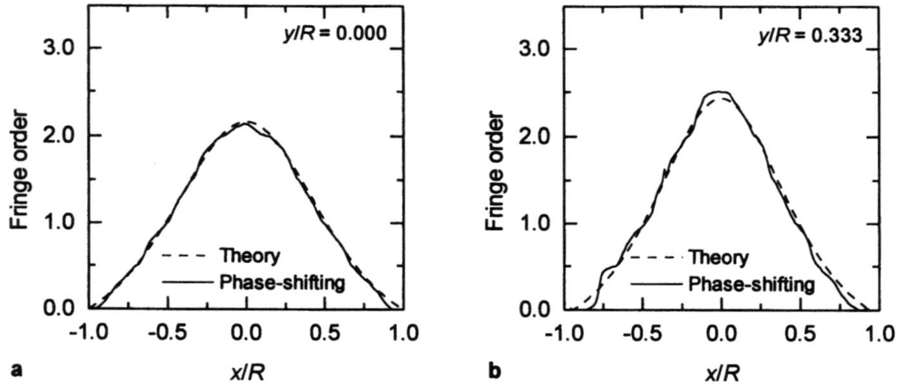

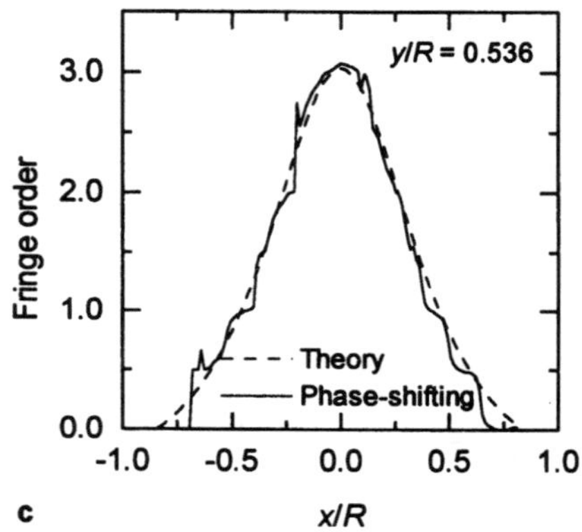

Fig. 7.11. Comparison of fringe order for a circular disc under diametral compression evaluated from spatially averaged image with theory for **a** $y/R = 0.00$ **b** $y/R = 0.333$ **c** $y/R = 0.538$

The Eqs. (7.21) and (7.22) are same as those for a transmission arrangement [Eqs. (5.11) and (5.27), Chap. 5]. In order to achieve this identity, the orientation of the optical elements must conform to Fig. 7.10. Thus, the rest of the procedure for collection of images and processing of them to provide fractional retardation and the isoclinic angle is the same as that for transmission photoelasticity.

A new portable reflection polariscope to do phase shifting has been designed by Ramesh and Mangal [16]. For three lines, in the case of a Polycarbonate disc with a reflective backing under diametral compression, the variation of total fringe order is shown in Fig. 7.11. Despite the use of relatively poor quality images than in transmission photoelasticity, the figure shows a good comparison with theory and demonstrates the usefulness of the technique in an industrial environment. Application of the technique to evaluate stress intensity factors in fracture mechanics is also reported [28].

7.9 Spectral Content Analysis (SCA)

Historically, spectral content analysis (SCA) is the first quantitative approach to use the colour code in photoelasticity for data reduction. It is based on the premise that each fringe order (total fringe order – not just partial fringe order) has a distinct spectral signature. Three fringe photoelasticity or RGB photoelasticity are another variants of using the colour code and they fully utilise the advantage of a colour image processing system hardware for data reduction. The research on SCA has contributed significantly to the development of intensity equations in white light, for various polariscope arrangements, which were discussed in Sect. 7.5.

In SCA, for each data point, the transmission spectrum as a function of wavelength is experimentally obtained. Use of a circular polariscope in dark-field and bright-field arrangements are reported in the literature. The value of total fringe order in Eq. (7.16) or (7.17) is iteratively changed until the theoretical and experimental curves are close in a least squares sense. Thus, the technique is highly demanding computationally.

Redner [1] initially proposed in 1985 the use of spectral content from photoelastic experiments to determine the fringe order. He concluded that SCA couldn't be used to measure small amounts of birefringence and the use of narrow bandwidth detectors yields multiple solutions for the fringe order. Sanford and Iyengar [2] conducted further studies in 1986 and showed that if the effective window of the detector is taken in the analysis, one can evaluate lower level of birefringence uniquely. Sanford [3] had made an experimental verification of his SCA model. Voloshin and Redner [4] came up with commercial equipment based on SCA. Haake and Patterson [5] have applied SCA to find the fringe order in stress frozen photoelastic models containing both high and low fringe orders. The extension of the technique to reflection photoelasticity is reported in ref. [29].

A typical experimental arrangement for conducting SCA is shown in Fig. 7.12. Using a prism or diffraction gratings, the light is split into several rays each corresponding to a narrow band of frequencies. In the literature, one finds several approaches to measure the experimental spectrum with varying degree of precision. The system used by Sanford [3] utilises thermoelectrically cooled detector array with 256 active elements over the range of 390 to 730 nm. Voloshin and Redner [4] used 16 photodiodes to obtain the experimental spectrum, which are obtained at intervals of 21.7 nm. Haake and Patterson [5] developed a highly sensitive instrument wherein a rotating diffraction grating conveyed the spectrum of light to a photomultiplier tube, which served as a detector. The resolution of the experimental spectrum can be varied from 0.25 nm to 10 nm in a range of 200 to 900 nm (i.e., 70–2800 points for representing the experimental

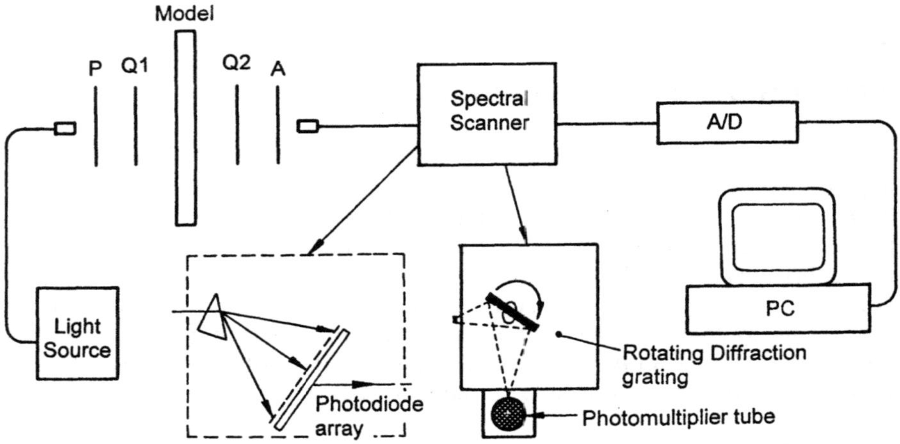

Fig. 7.12. Typical arrangement to perform SCA. Note the different methods to analyse the spectral signature

spectrum). Carazo-Alvarez et al. [30] compared the use of collecting experimental data in steps of 5 nm and also 40 nm. They established that collecting data at 40-nm interval is reasonably accurate for problems where fringe order gradient is less than 1.4 fringes/mm.

Unlike PST or TFP, one does not get whole field information instantly in SCA. For recording a high quality spectrum, one needs to use a fibre optic light source, which has to be moved over the domain for collecting whole field data. The diameter of the fibre-optic bundle is another parameter that needs to be optimised for improved data interpretation. It varies from 0.1 to 1 mm depending on the fringe gradient. Thus, the basic equipment needed for performing conventional SCA requires a new set of hardware involving a spectrophotometer and a cumbersome scanning arrangement for collecting data over the domain.

Though the idea of SCA is conceptually simple, the practical implementation has several problems. Approximations are not only made on recording experimental spectrum but also on data analysis. In view of the quarter-wave plate error, Eqs. (7.16) and (7.17) contain the isoclinic angle θ. Thus theoretically, the fringe order evaluation for a point of interest is influenced by the isoclinic value at the point of interest. The quarter-wave plate error in dark-field arrangement as a function of isoclinic angle with various wavelengths as a parameter is plotted in Fig. 7.13. The reference wavelength is taken as 590 nm [31]. In order to simplify SCA calculations, the value of θ is taken as 22.5°. Thus, the governing equation for SCA in dark-field arrangement for a typical photodiode j (small bandwidth detector) is

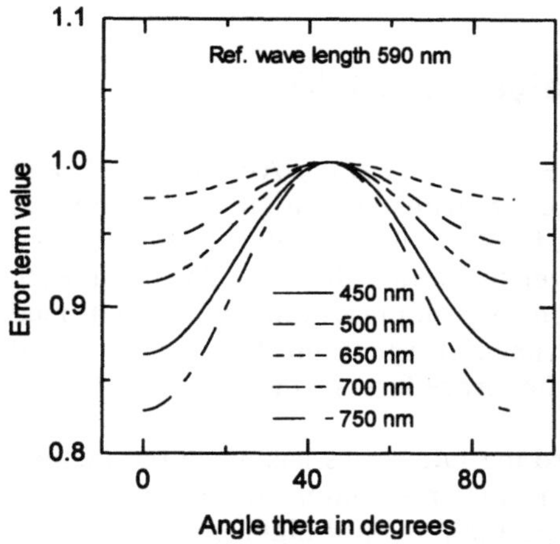

Fig. 7.13. Quarter-wave plate error ($1 - \cos^2 2\theta \sin^2 \varepsilon$) in dark-field arrangement with the reference wave length as 590 nm

$$I_j = I_{b_j} + \frac{1}{(\lambda_{2j} - \lambda_{1j})} \int_{\lambda_{1j}}^{\lambda_{2j}} F(\lambda) I_{a_j}(\lambda) \sin^2\left(\frac{\pi N_{ref} F_{\sigma_{ref}}}{F_\sigma(\lambda)}\right)[1 - 0.5\sin^2\varepsilon]\, d\lambda \quad (7.23)$$

where I_{b_j} represents the background light intensity sensed by the photodiode j and λ_{1j} and λ_{2j} are the lower and upper limit of the spectrum as acquired by the photodiode j. In order to use Eq. (7.23), a calibration has to be performed to find I_{b_j} and I_{a_j} for a generic photodiode j by using an unloaded specimen in the polariscope.

Though the modelling of quarter-wave plate error in SCA literature for circular polariscope in dark-field arrangement was well understood, in earlier studies, the same modelling was erroneously extrapolated for bright-field arrangement too. The correct modelling of quarter-wave plate error in bright-field arrangement was first reported by Ajovalasit et al. [12]. It is worthwhile to point out that references [2, 5, 30] have used light-field arrangement for SCA with incorrectly modelled intensity equation to account for the quarter-wave plate error.

For a point-by-point analysis, SCA is found to give quite accurate results and is estimated to be ±0.005 [3] and ±0.017 [4]. The accuracy can be improved by reducing the beam diameter to 0.2 mm. From an industrial standpoint, it has been found to be quite useful for monitoring the processing of birefringently sensitive polymers [32, 33] and analysis of residual stresses in glasses [34]. The equipment used in ref. [34] employed 512 sensing elements for recording the experimental spectrum. To enhance the accuracy of measurements below 0.5 fringe orders, a plate with a known retardation in series with the specimen is introduced in references [32] and [33] for data recording and interpretation.

7.10 Digital Spectral Content Analysis (DSCA)

Haake and Patterson [6] extended the work of Carazo-Alvarez et al. and used a B&W CCD camera in conjunction with eight high quality optical filters corresponding to wavelengths of 450, 470, 530, 590, 610, 650, 690 and 730 nm to record the intensity over the whole domain. Thus, for each pixel, one gets eight intensity values corresponding to the above mentioned wavelengths. Using these, the experimental spectrum for the point of interest is constructed. The rest of the procedure is the same as that is followed in conventional SCA.

The use of a digital camera to record the intensities has made the method *whole field* and can be termed as Digital Spectral Content Analysis (DSCA). However, the accuracy will not be quite high as only eight points are used to represent the experimental spectrum.

7.11 Hybrid Techniques

The term *hybrid technique* implies that an automated data acquisition system uses a combination of various techniques reported: either in full or in part. Polarization stepping in colour domain is one example wherein, the isoclinics are determined by using four polarization-stepped images in colour domain and for isochromatic fringe order determination, use of TFP is recommended. A detailed description of the technique is discussed in Sect. 7.11.1.

In PST and FT techniques to find the absolute fringe order at any point, one needs the total fringe order for at least one point in the fringe field to be supplied by auxiliary means. One can find the use of two hybrid techniques to achieve this. Carazo-Alvarez et al. [30] used SCA to determine total fringe order at just one point in the fringe field which is used as an input to unwrap the phases obtained by PST. The disadvantage of their approach is that one requires two different hardwares, one for SCA and the other for phase shifting. The other approach is to use PST in colour domain to evaluate the partial fringe order and isoclinic for the whole domain. For determining the total fringe order to initialise phase unwrapping, TFP could be invoked by using the first two full colour images of the six image sequence (Table 5.3) recorded for PST in colour domain [15]. The advantage of the approach is that no specialised hardware other than a general-purpose colour image processing system is required. Further, since the methodology also records the whole-field fringe patterns in colour, one always has sufficient data to verify the results provided by the automated methodology. It has been pointed out in Chap. 5, that isoclinic-isochromatic interaction could adversely affect the unwrapping of fractional fringe order data in some zones. A tile based approach for phase unwrapping was discussed in Chap. 5 to overcome this problem. The method required fringe gradient direction to be supplied for each tile. This can be identified easily based on the full colour image.

7.11.1 Polarization Stepping in Colour Domain

Determination of isoclinic angle by phase shifting has always been a problem when isochromatic intensity in the field goes to zero. This situation occurs when $\delta = 0$, π, 2π etc. To limit such drawbacks, the use of several wavelengths of light [8–10] and acquiring images in white light have been suggested [17, 35]. Of these references [10], [12] and [35] use polarization stepping for isoclinic determination. In polarization stepping, a dark-field plane polariscope is used for intensity measurements. Hecker and Morche [35] proposed the use of three polarization stepped images by keeping the analyzer at 0-, 30- and –30-deg. Let the respective light intensities for these positions be denoted as I_1, I_2 and I_3. The principal stress direction θ is obtained as

$$\theta = \frac{1}{4}\tan^{-1}\left\{\frac{I_2 - I_3}{\sqrt{3}(I_2 + I_3 - 2I_1)}\right\} \tag{7.24}$$

Though, Hecker and Morche proposed the methodology, neither they performed experimental validation of the technique nor they provided any methodology for intensity measurement.

Petrucci [17] used a 4-step approach for isoclinic angle determination, in which he effectively used the wide band characteristics of the detectors of a colour CCD camera used for grabbing full colour image. The method is an extension of the polarization stepping method discussed using a monochromatic source in Chap. 5. With white light as a source, for a plane polariscope in dark-field arrangement, the intensity of light transmitted is given by Eq. (7.12). The second term of Eq. (7.12) can be rewritten as

$$I_w(\lambda)\sin^2 2(\theta - \beta) \tag{7.25}$$

where $I_w(\lambda)$ is defined as

$$I_w(\lambda) = \frac{1}{(\lambda_2 - \lambda_1)}\int_{\lambda_1}^{\lambda_2} F(\lambda) I_a(\lambda)\sin^2\frac{\pi N_{ref} F_{\sigma_{ref}}}{F_\sigma(\lambda)}\, d\lambda \tag{7.26}$$

In view of the wide band characteristics of the detector, the term $I_w(\lambda)$ never goes to zero. Thus, Eq. (7.25), theoretically shows that the problem due to isochromatic intensity becoming zero at some locations in the problem domain, in evaluating the isoclinic value, could be overcome using white light. In practical implementation, one can just use the grey levels recorded by any one of the image planes of red, green or blue for evaluating the isoclinic value. Petrucci [17] observed that instead of using intensity from any one of the image planes, the accuracy of evaluation is improved if the summations of these intensities are used. Thus, the isoclinic angle is determined by

$$\begin{aligned}\theta &= \frac{1}{4}\tan^{-1}\left(\frac{I_4 - I_2}{I_3 - I_1}\right) = \frac{1}{4}\tan^{-1}\left(\frac{I_w(\lambda)\sin 4\theta}{I_w(\lambda)\cos 4\theta}\right) \\ &= \frac{1}{4}\tan^{-1}\left(\frac{(I_{4,R} + I_{4,G} + I_{4,B}) - (I_{2,R} + I_{2,G} + I_{2,B})}{(I_{3,R} + I_{3,G} + I_{3,B}) - (I_{1,R} + I_{1,G} + I_{1,B})}\right)\end{aligned} \tag{7.27}$$

where $I_{i,j}$ ($i = 1, 2, 3, 4$; $j = R, G, B$) correspond to pixel grey levels of R-, G- and B-planes for the analyzer positions of 0-, 22.5-, 45- and 67.5-deg respectively.

Following the approach used for monochromatic source (Sect. 5.11), the isochromatic intensity field could be separated from the isoclinic field for each image planes as

$$I_j = \sqrt{(I_{3,j} - I_{1,j})^2 + (I_{4,j} - I_{2,j})^2} \quad (j = R, G, B) \tag{7.28}$$

These are the R, G and B values to be used in TFP/RGBP to find the total fringe order.

Petrucci performed an experimental validation of the technique by using a colour image processing hardware that provided only 32 quantization levels for each colour planes. He showed that the error in isoclinic determination is quite low of the order of ±0.5 deg. In view of white light being used as a source, no spikes were observed in isoclinic determination. Though he used only 32 quantization levels for each colour plane, he used a three-detector system for image capture, which provided high image fidelity.

The technique of polarization stepping in colour domain does not give satisfactory results while using a single detector CCD camera even with 256 quantization levels for colour image capture and use of three detector CCD is a must for practical applications.

7.12 Tricolour Photoelastic Method

The motivation for this technique emerges from the fact that, whether automated techniques for whole field data reduction could be used for slowly varying dynamic phenomena. It is worthwhile to note that phase shifting and polarization stepping techniques use several optical

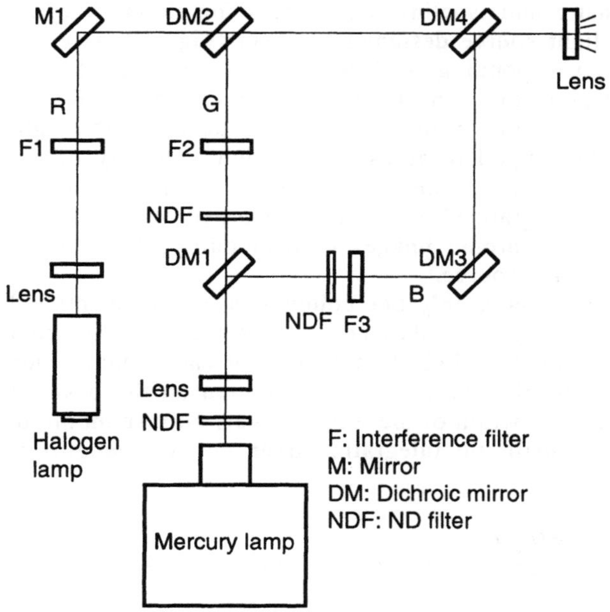

Fig. 7.14. a Design of tricolour light source (Courtesy of ref.[18])

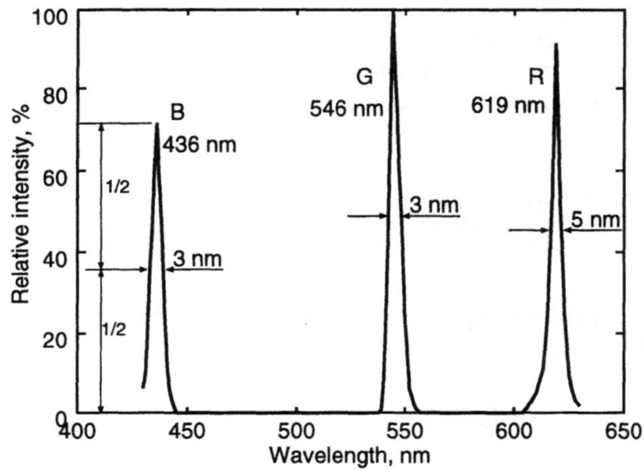

Fig. 7.14. b Spectral distribution of the tricolour light source emittance (Courtesy of ref. [18])

arrangements for data acquisition. Hence, they are not generally applicable for recording time varying phenomenon. Recently, Patterson and Haake [36] reported a technique for simultaneous recording of four phase-stepped images for recording slowly varying dynamic events. The details of their technique are discussed in Chap. 11.

Yoneyama et al. [18] used the colour image processing hardware coupled with an intelligent design of a light source to develop a technique for extracting both isochromatic and isoclinic parameters for slowly varying dynamic situations. The light source designed by them (Fig. 7.14a) emits only three wavelengths corresponding to 436 nm, 536 nm and 619 nm (Fig. 7.14b). These correspond to the peak spectral response of the R-, G- and B-plane detectors of a colour camera. They have designed the light source with ND filters that helped in adjusting the light intensity of the three wavelengths taking into account the spectral sensitivity of the camera. Thus, when a colour image is grabbed, the images recorded by R-, G- and B-planes correspond to monochrome images corresponding to the three distinct wavelengths of the light source.

In tricolour photoelastic method, only one colour image in conventional dark-field arrangement of the plane polariscope using the tricolour light source is recorded. The intensity of light transmitted for each image plane considering the dispersion of birefringence can be written on the basis of Eq. (7.12). Noting that the bandwidth of the wave-packets corresponding to each image plane is very small, no integration over the wavelengths is necessary.

$$I_j = I_{b_j} + S_j \sin^2 \frac{\pi N_{ref} F_{\sigma_{ref}}}{F_\sigma(\lambda_j)} \sin^2 2(\theta - \beta) \quad (j = R, G, B) \tag{7.29}$$

where I_{b_j} is the background light intensity, and $S_j = T(\lambda_j)F(\lambda_j)I_a(\lambda_j)$, which denotes the product of the transmittance of the overall optical system adopted, the spectral responsivity of the camera used and the light source emittance. Let the green wavelength be taken as the reference wavelength. Equation (7.29) then reduces to

$$I_j = I_{b_j} + S_j \sin^2 \frac{\pi N_G F_{\sigma G}}{F_\sigma(\lambda_j)} \sin^2 2(\theta - \beta) \quad (j = R, G, B) \tag{7.30}$$

For simplicity, let the background light intensity be considered equal for different wavelengths. The Eq. (7.30) can be rewritten as

$$I_j = I_b + \psi_j \chi \tag{7.31}$$

where

$$\psi_j = S_j \sin^2 \frac{\pi N_G F_{\sigma G}}{F_{\sigma_j}} \quad (j = R, G, B) \tag{7.32}$$

and

$$\chi = \sin^2 2(\theta - \beta) \tag{7.33}$$

The following set of equations can be constructed from Eq. (7.31).

$$I_j - I_k = (\psi_j - \psi_k)\chi \quad (j,k = R,G,B) \tag{7.34}$$

Eliminating χ in Eq. (7.34) one gets

$$(I_G - I_B)\psi_R + (I_B - I_R)\psi_G + (I_R - I_G)\psi_B = 0 \tag{7.35}$$

Upon substituting Eq. (7.32) in Eq. (7.35) one gets a non-linear equation involving N_G as

$$\begin{aligned}&(I_G - I_B)S_R \sin^2 \frac{\pi N_G F_{\sigma_G}}{F_{\sigma_R}} + (I_B - I_R)S_G \sin^2 \pi N_G + \\ &(I_R - I_G)S_B \sin^2 \frac{\pi N_G F_{\sigma_G}}{F_{\sigma_B}} = 0\end{aligned} \tag{7.36}$$

A numerical solution of Eq. (7.36) provides the fringe order N_G. Though the evaluation of N_G is not restricted to fringe skeletons, for zones where $\chi = 0$ (isoclinic lines) and or $\psi_R = \psi_G = \psi_B$ (i.e., a point where three grey levels I_R, I_G and I_B have same value), the Eq. (7.36) does not have a solution.

For determining the isoclinic value, Eq. (7.34) is recast as

$$|I_G - I_B| + |I_B - I_R| + |I_R - I_G| = (|\psi_G - \psi_B| + |\psi_B - \psi_R| + |\psi_R - \psi_G|)\chi \tag{7.37}$$

The value of χ can be obtained from Eq. (7.37). Invoking Eq. (7.33), the isoclinic value is obtained as

$$\theta = \beta + \frac{1}{4}\cos^{-1}(1 - 2\chi) \quad (\text{when } \chi_R \neq \chi_G \neq \chi_B) \tag{7.38}$$

Figure 7.15 shows the dark-field isochromatics for a disc (dia 50 mm, thickness 6 mm) subjected to a diametral compressive load of 261 N taken using the tricolour light source [18]. The figure also shows the full colour image (printed in black & white) and the individual image planes of R, G and B. The variation of R, G and B values for an arbitrary line A–B indicated in Fig. 7.15 is shown in Fig 7.16a. Figures 7.16b and 7.16c show the variation of isochromatic fringe order calculated by Eqs. (7.36) and (7.38). The figure shows that the estimation of isochromatic fringe order is very good in comparison to isoclinic estimation. However, one could observe that below fringe order 0.3, the extraction of isochromatic fringe order is not satisfactory. Future research should concentrate on this aspect to improve the accuracy of evaluation.

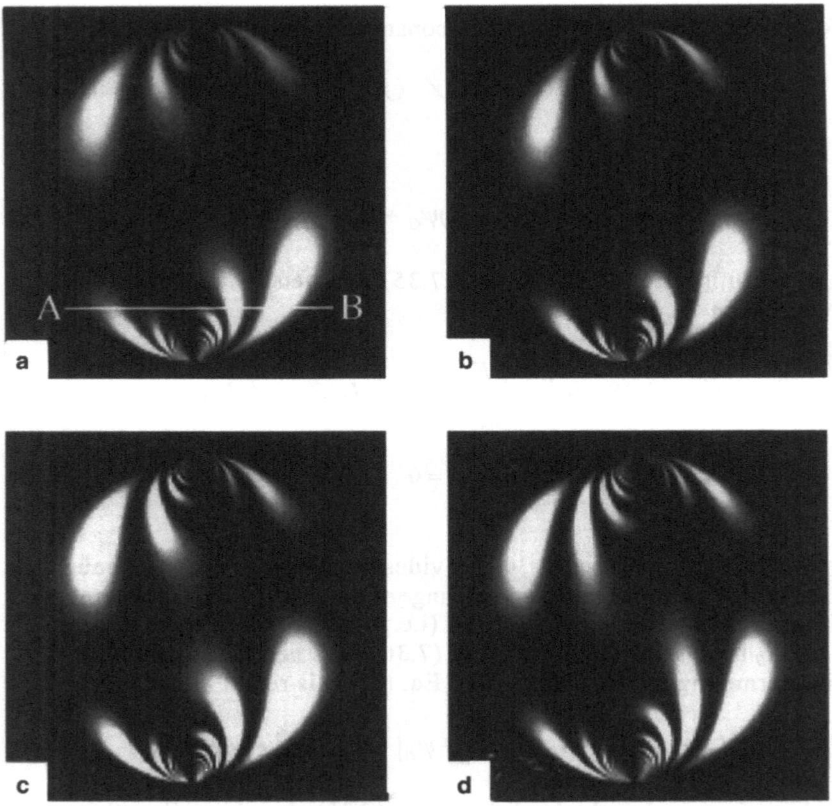

Fig. 7.15. One-step recording of photoelastic image data for three wavelengths by tricolour photoelasticity (15-deg isoclinic): **a** 24-bit image **b** R-plane image (619 nm) **c** G-plane image (546 nm) **d** B-plane image (436 nm) (Courtesy of ref. [18])

7.13 Closure

Among the various optical techniques, photoelasticity has the unique advantage of providing stress information in the form of coloured fringes. Use of colour code for identifying fringe gradient direction and estimation of approximate fringe orders was in vogue in conventional photoelasticity. The first attempt to quantitatively use this colour information came into existence with the development of SCA. The method is basically a point by point technique and involves the determination of isochromatic fringe order in a least squares sense and thus demanding computationally. Nevertheless, the technique has contributed significantly to the development of a proper understanding of intensity of light transmitted in white light.

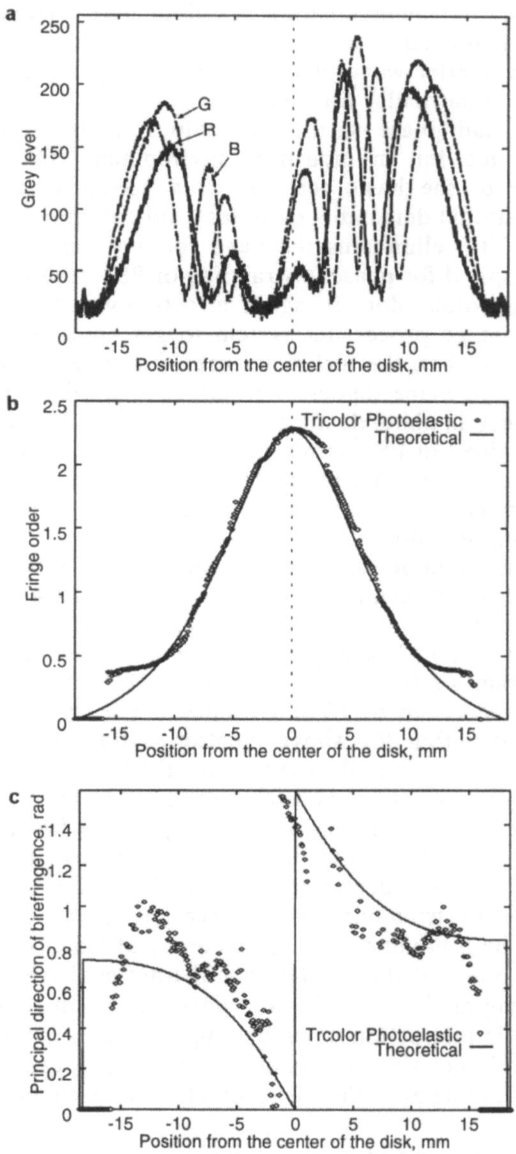

Fig. 7.16. a Variation of *R*, *G* and *B* values along line A–B shown in Fig. 7.15. Results of **b** total fringe order and **c** isoclinic angle along the same line (Courtesy of ref. [18])

Though, the implementation and usage of the techniques such as TFP, PST in colour domain and polarization stepping in colour domain are simple, the theoretical verification of these techniques were possible only with the development of equations for intensity of light transmitted in white light. In future, it is expected that image acquisition in colour will become more of a routine in automated photoelastic analysis. The analysis could be greatly simplified if achromatic quarter-wave plates are used [37] in place of quarter-wave plates meant for a particular wavelength.

Three fringe photoelasticity and phase shifting in colour domain are simple to employ and well suited for an industrial environment. These techniques take the advantage of the basic features of a colour image processing system. The computational demand is quite less and combination of PST in colour domain and TFP eliminates the need for any auxiliary information being externally supplied for phase unwrapping of PST results.

The polarization stepping in colour domain again effectively uses the hardware features of a colour image processing system to get continuous isoclinic map over the domain. In fact, the technique uses the broad band characteristics of the colour image plane detectors to advantage, which is otherwise considered as a nuisance in photoelastic analysis. It is well known that the quality of isoclinic obtained in PST is not quite good. TFP on the other hand may not work beyond fringe order three. Thus, a better approach for high quality data reduction is possible by combining PST in colour domain with polarization stepping in colour domain.

The development of tricolour photoelastic method has added a new dimension to colour image processing techniques, viz., that of a capability for using automated methods for analysing slowly varying dynamic situations [38]. It was seen in Chap. 3 that general purpose solid state cameras can record either 25 frames/s (PAL) or 30 frames/s (NTSC). With electronic triggering they can record up to 10,000 frames/s. For dynamic events which could be analysed within the above mentioned time frame, tricolour photoelastic method could be readily used. A detailed discussion on time scale requirement for dynamic studies is discussed in Chap. 11.

The development of tricolour light source has also given promise for the development of new generation of automated techniques [39]. The range of TFP is extended beyond three fringe orders by replacing the white light source with a tricolour light. Use of tricolour light source in conjunction with PST in colour domain can help in improving the accuracy of isochromatic fringe order evaluation. In view of the availability of fractional fringe order for three different wavelengths, in principle, it is possible to solve the problem of isoclinic-isochromatic interaction in evaluating the isochromatic fringe order. Particularly, the interactive approach to solve this aspect discussed in Chap. 5 could be automated. Future research should concentrate on this.

Conceptually, polarization stepping in colour domain can provide the best value of isoclinics for the whole domain. For the technique to be effective only a multi spectral light source is needed. Use of polariscopes with white light and tricolour light may become common in future. The use of polarization stepping in colour domain for isoclinic determination and PST by using tricolour light source for isochromatic determination may prove to be the best combination for analyses of static and three-dimensional problems using slicing and stress freezing approach.

For improving data acquisition in stress concentration zones, one can use the technique of OET [40] in colour domain too. Recently, the use of a colour scanner rather than a colour camera to enhance information in stress concentration zones has been proposed [41]. However, the data interpretation and analysis requires further development.

Exercises

7.1 What are the different philosophies of recording multi-wavelength images? How does this affect the choice of processing these in digital photoelasticity?

7.2 Compare the RGB and HSI models for colour representation.

7.3 Develop the equation for intensity of light transmitted for a plane polariscope with a white light source. Take into account the spectral characteristics of the light source and the spectral response of the camera. If the light source is changed, list the modifications that you need to do?

7.4 Develop a software to evaluate total fringe orders by three-fringe photoelasticity. Use a modular approach in designing the software and provide a feature so that calibration can be verified as and when required.

7.5 What is the motivation for doing phase shifting in colour domain? How does this help in automating reflection photoelasticity?

7.6 Explain the methodology of spectral content analysis. What are the approximations made in both recording the spectrum and its analysis?

7.7 Explain with appropriate equations how isochromatic interaction on the determination of isoclinics is handled in the methodology of polarization stepping in colour domain.

7.8 What is the advantage of a tricolour light source? Explain how it can improve the technique of phase shifting in colour domain.

7.9 In which of the techniques discussed in this chapter a pre-calibration is required? Summarise this in the form of a table.

References

1. Redner AS (1985) Photoelastic measurements by means of computer-assisted spectral content analysis. Exp Mech 25(2):148–153
2. Sanford RJ, Iyengar V (1985) The measurement of the complete photoelastic fringe order using a spectral scanner. Proc SEM Spring Conference on Experimental Mechanics:160–168
3. Sanford RJ (1986) On the range and accuracy of spectrally scanned white light photoelasticity. Proc SEM Spring Conference on Experimental Mechanics:901–908
4. Voloshin AS, Redner AS (1989) Automated measurement of birefringence: development and experimental evaluation of the techniques. Exp Mech 29(3):252–257

5. Haake SJ, Patterson EA (1992) Photoelastic analysis of frozen stressed specimens using spectral-content analysis. Exp Mech 32(3):266–272
6. Haake SJ, Patterson EA (1995) Photoelastic analysis using automated polariscopes. Proc of the Int conference on Mechanics of Solids and Materials Engineering (MSME 95) Singapore:884–888
7. Haake SJ, Patterson EA (1995) Photoelastic analysis using a full field spectral contents analyser. Proc of SEM Conference on Experimental Mechanics, Michigan:342–345
8. Kihara T (1994) Automatic whole field measurement of principal stress directions using three wavelengths. Proc of X Int conference on Experimental Mechanics, Lisban 1:95–99
9. Buckberry C, Towers D (1996) New approaches to the full-field analysis of photoelastic stress patterns. Optics and Lasers in Engng 24:415–428
10. Nurse AD (1997) Full-field automated photoelasticity using a three-wavelength approach to phase shifting. Appl Optics 36:5781–5786
11. Ajovalasit A, Barone S, Petrucci G (1995) Towards RGB photoelasticity: full-field automated photoelasticity in white light. Exp Mech 35(3):193–200
12. Ajovalasit A, Barone S, Petrucci G (1995) Automated photoelasticity in white light: influence of quarter-wave plates. J Strain Analysis for Engng Des 30:29–34
13. Ramesh K, Deshmukh SS (1996) Three fringe photoelasticity – use of colour image processing hardware to automate ordering of isochromatics. Strain 32(3):79–86
14. Hoy DEP, Yu F (1996) Fuzzy logic approach for analysis of white light isochromatic fringes. Post conference Proc of SEM VIII Int Cong:279–284
15. Ramesh K, Deshmukh SS (1997) Automation of white light photoelasticity by phase shifting technique using colour image processing hardware. Optics and Lasers in Engng 28(1):47–60 Also in: Chen TY (ed) (1999) Selected papers on photoelasticity. SPIE Milestone Series, Vol MS158, pp 290–303
16. Ramesh K, Mangal SK (1997) Automation of data acquisition in reflection photoelasticity by phase shifting methodology. Strain 33:95–100
17. Petrucci G (1997) Full-field automatic evaluation of an isoclinic parameter in white light. Exp Mech 37(4):420–426
18. Yoneyama S, Shimizu M, Gotoh J, Takashi M (1998) Photoelastic analysis with a single tricolour image. Optics and Lasers in Engng 29:423–435
19. Kihara T (1998) A study of automatic whole field measurement of photoelasticity using three wavelengths with a colour camera. Proc of XI Int Conference on Experimental Mechanics, University of Oxford, UK:515–519
20. Gonzalez RC, Woods RE (1993) Digital image Processing. Addison-Wesley Publishing Co, California
21. Young T (1802) On the theory of light and colours. Phil Trans Royal Soc London 92:20–71
22. Holst GC (1996) CCD arrays, cameras and displays. SPIE Optical Engineering Press, Bellingham, Washington
23. TMC-76 colour camera. PULNiX Brochure
24. Ji W, Patterson EA (1998) Simulation of error in automated photoelasticity. Exp Mech 38(2):132–139
25. Srinath LS, Sarma AVSSSR (1973) Determination of Integral fringe orders in photoelasticity. Exp Mech 13(3):138–141
26. Barone S, Petrucci G (1994) Automated photoelasticity in white light: application to birefringent coatings. Proc of XIII Imeko world congress, Torino 3:1829–1834
27. Ajovalasit A, Barone S, Petrucci G (1998) A review of automated methods for the collection and analysis of photoelastic data. J Strain Analysis for Engng Des 33(2):75–91

28. Mangal SK, Ramesh K (1997) Automation of evaluation of stress field parameters in fracture mechanics by reflection photoelasticity using phase-shifting technique. Proc of II conference on Engng Applications of Solid Mechnaics IGCAR, Kalpakkam:11–12
29. Marwitz, Kzler, Schuster (1990) Improved efficiency in photoelastic coatings. Fast detection of fringe orders using computer controlled spectrometry. Proc of IX Int Conference on Experimental Mechanics, Copenhagen 2:828–838
30. Carazo-Alvarez J, Haake SJ, Patterson EA (1994) Completely automated photoelastic fringe analysis. Optics and Lasers in Engng 21:133–149
31. Ramesh K, Mangal SK (1998) Data acquisition techniques in digital photoelasticity: a review. Optics and lasers in Engng 30:53–75 Errata (1999) 31:85
32. Sanford RJ, McGinnis A (1990) Automatic birefringence measurement for inspection and quality control. Proc of 36th Int Instrumentation Symposium:799–809
33. Sanford RJ, McGinnis (1991) A New method for measuring low birefringence using a spectral scanner. Proc of 37th Int Instrumentation Symposium, San Diego 1029–1041
34. Ivanova L, Nechev G (1990) A general method for investigation of the residual stresses in glasses with spectral polariscope. Proc of IX Int Conference on Experimental Mechanics 2:876–883
35. Hecker FW, Morche B (1986) Computer-aided measurement of relative retardations in plane photoelasticity. In: Wieringa H (ed) Experimental stress analysis. Martinus Nijhoff, Dordrecht, The Netherlands, pp 535–542
36. Patterson EA, Wang ZF (1998) Simultaneous observation of phase-stepped images for automated photoelasticity. J Strain analysis for Engng Des 33(1):1–15
37. Hariharan P (1996) Achromatic and apochromatic half-wave and quarter-wave retarders. Opt Engng 35(11):3335–3337
38. Yoneyama S, Ayame D, Gotoh J, Takashi M (1997) Photoviscoelastic stress and strain analysis around a surface groove under rolling contact load. JSME Int J Sc A40: 44–50
39. Yoneyama S, Shimizu M, Takashi M (1998) Higher retardation analysis in automated white light photoelasticity. Proc of XI Int Conference on Experimental Mechanics University of Oxford, UK:527–532
40. Ramesh K, Sreedhar D (1998) Optically enhanced tiling (OET) in digital fringe pattern analysis. Strain 34(4):127–131
41. Hoy DEP (1998) A colour scanner as a digital recording polariscope. Exp Tech 22:26–27

8 Evaluation of Contact Stress and Fracture Parameters

8.1 Introduction

In many mechanical devices, there exist contact between two or more parts such as in gears, rolling element bearings, locomotive wheels and rails etc. The stresses caused by the pressure distribution between the bodies in contact are of importance in the design of these parts. Hertz [1] pointed out that in the absence of friction, the maximum shear stress occurs beneath the surface of contacting bodies. This leads to *pitting* of the contacting surfaces. The material lost from the surface due to *pitting* may get trapped into the contacting surfaces, causing abrasive wear. Smith and Liu [2] studied the effect of friction between the contacting surfaces. They reported that under certain conditions, the point of maximum shear stress could also occur at the surface of contacting bodies. The knowledge of contact zone and coefficient of friction between the contacting bodies is essential for evaluating the design of such contacting elements.

Fracture mechanics deals with the mechanics of cracked bodies. The loading modes at the crack-tip are classified as the opening mode (mode-I), the in-plane shear mode (mode-II) and the out-of-plane shear mode (mode-III). Out of these three modes, the mode-I plays a dominant role and the shear modes tend to alter the crack growth direction. The various theories developed in fracture mechanics studies over the years can provide answers to the questions such as what is the residual strength of a cracked body or how long the crack will take to grow to a critical level etc. [3, 4]. In linear elastic fracture mechanics studies, the stress field at the crack-tip is characterised by a parameter called stress intensity factor (SIF). Several handbooks are available which give SIFs for various geometry and types of loading conditions [5, 6]. Nevertheless, experimental evaluation is needed for creating such handbooks or for new geometry not listed in them. Though in fracture mechanics studies, very near stress field equations involving only SIFs are in use quite widely, for experimental evaluation of SIFs for practical problems involving finite geometry, a multi-parameter stress field solution is essential. Use of such equations increases the reliability in evaluating the SIFs. However, the resulting equations are very complex and the use of digital computers is a must for determining the results.

The role of digital photoelasticity for data acquisition and the role of digital computers for processing these data to get the parameters of interest in these high stress gradient problems are discussed in this chapter. The use of computer graphics to validate the solution procedure is also discussed in this chapter.

8.2 Basic Data Required and its Digital Acquisition

For photoelastic analysis, in general, one requires the value of both isochromatic and isoclinic fringe orders at a point of interest. However, a large class of practically important problems can be solved even if only isochromatic fringe orders are known. For the evaluation of contact stress parameters and fracture parameters, isochromatic fringe order and the corresponding positional co-ordinates are needed.

In conventional analysis, a photograph of the fringe patterns is recorded. By knowing the magnification of the image and selecting a suitable origin, the data can be expressed in a quantitative sense. For proper interpretation of the fringe patterns, sufficient care needs to be exercised in orienting the camera normal to the model to record the fringe patterns.

8.2.1 Conversion of Pixel Co-ordinates to Model Co-ordinates

In digital photoelasticity, a CCD camera replaces the human observer and the fringe patterns are digitised. As in conventional photoelasticity, sufficient care needs to be exercised in aligning the camera with the model for proper interpretation of the results. In a digitised image, the image elements can be easily accessed through frame-buffer (or monitor) pixel co-ordinates. In general, the pixels are rectangular although in modern systems, square pixels are becoming the standards. To convert the image in pixel co-ordinates to real world co-ordinates, one needs to know the physical dimensions of the pixel in frame-buffer co-ordinates. It has been pointed out in Chap. 3 that the monitor pixels are not in general the same as the camera pixels. By knowing the physical dimensions of the pixel of the CCD sensor and also the sampling frequencies used by the CCD sensor (f_c) and the frame grabber (f_f) it is possible to establish a relationship between the frame-buffer pixel co-ordinates (x_f, y_f) to the camera pixel co-ordinates (x_c, y_c) as [7]

$$x_f = \frac{f_f}{f_c} x_c, \quad y_f = y_c \tag{8.1}$$

Let d_x and d_y denote the CCD sensor pixel dimensions in mms along the x and y directions. For an optical magnification of m, the real world co-ordinates (x_w, y_w) are related to (x_f, y_f) as

$$x_w = m\frac{f_c}{f_f}x_f d_x,\quad y_w = m y_f d_y \tag{8.2}$$

However, in practice, it is desirable to establish the frame-buffer pixel dimensions by direct experimentation. This has to be done in two stages. For a given camera frame grabber combination, the aspect ratio (y_f/x_f) of the frame buffer pixels remains a constant and needs to be determined first. This is easily achieved by taking an image containing two mutually perpendicular scales aligned to the horizontal and vertical directions of the monitor. Then for each experiment, the optical magnification needs to be obtained either by knowing a particular physical dimension of the specimen in the field of view or by keeping a scale in the field of view while recording the image. It is common practice to express the magnification as m_x pixels per mm in the x direction or m_y pixels per mm in the y direction. The real world co-ordinates are then related as

$$x_w = \frac{x_f}{m_x},\quad y_w = a_r\frac{y_f}{m_x} \tag{8.3}$$

or

$$x_w = \frac{x_f}{a_r m_y},\quad y_w = \frac{y_f}{m_y} \tag{8.4}$$

where a_r is the aspect ratio of the camera – frame-grabber combination. The use of Eq. (8.3) or Eq. (8.4) is recommended for conversion of pixel co-ordinates to model co-ordinates. In subsequent discussions, the word pixel refers to the frame-buffer pixel.

8.2.2
Rotational Transformation

The axes of reference are problem dependent and will not be parallel to the monitor horizontal and vertical directions. Further, the origin of the monitor is the left top most corner (Fig. 3.1, Chap. 3) and for most problems, the origin will be different from that. In the problems of circular disc or ring under diametral compression, the centre of the disc or ring is the origin and the loading direction coincides with the y-axis. For problems in fracture mechanics, the tip of the crack is taken as the origin and the crack axis is taken as the x-axis. In contact stress problems, the point of contact is the origin and the tangential and normal directions at the point of contact serve as the reference axes. Hence, in general, the frame-buffer co-ordinates must be applied with a translation to fix the origin of interest and a rotational transformation such that the reference axes are as per the requirement of the problem under consideration. The determination of the origin can be done either interactively or by using geometric primitives.

Although, translation can be done in digital domain after selecting the origin, rotational transformation has to be done only in the real world co-ordinate system. The real world co-ordinates are to be determined as per Sect. 8.2.1, with respect to the origin for the problem under consideration. Then these co-ordinates are to be rotated by using the rotation matrix (Sect. 3.13.3). The amount of rotation that needs to be applied has to be evaluated from the digital image using geometric primitives or from obvious references such as the orientation of the crack axis as the case may be.

8.3 Stresses in Terms of Contact Length and Geometrical and Elastic Properties of the Bodies in Contact

The stresses for the combined normal and tangential loading in a two-dimensional plane stress problem (Fig. 8.1) in terms of the contact length ($2a$) and coefficient of friction (μ) at the contact for $z > 0$ (beneath the surface), and $z = 0$ (on the surface) can be respectively expressed as [2]

$$\begin{Bmatrix}\sigma_x\\ \sigma_z\\ \tau_{xz}\end{Bmatrix}_{z>0} = \begin{Bmatrix} -\dfrac{a}{\pi\zeta}\left[(a^2-2x^2+2z^2)\dfrac{z\psi_1}{a}-\dfrac{2\pi z}{a}-3xz\psi_2+\mu\left\{(2x^2-2a^2-3z^2)\psi_2+2\pi x+2(a^2-x^2-z^2)\dfrac{x\psi_1}{a}\right\}\right] \\ -\dfrac{a}{\pi\zeta}(a\psi_1-x\psi_2+\mu z\psi_2) \\ -\dfrac{a}{\pi\zeta}\left[z^2\psi_2+\mu\left\{(a^2+2x^2+2z^2)\dfrac{z\psi_1}{a}-\dfrac{2\pi z}{a}-3xz\psi_2\right\}\right] \end{Bmatrix} \quad (8.5)$$

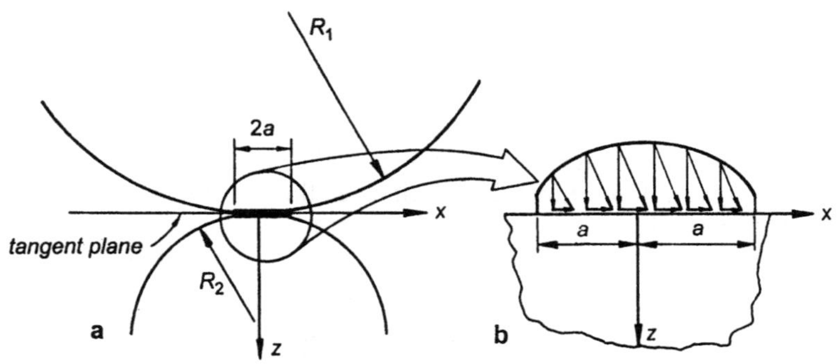

Fig. 8.1 **a** Two cylindrical bodies in contact **b** Loading at the contact region

$$\sigma_x|_{z=0} = \begin{cases} -\dfrac{2a\mu}{\pi\zeta}\left(\dfrac{x}{a} - \sqrt{\dfrac{x^2}{a^2} - 1}\right) & \text{for } x \geq a \\ -\dfrac{2a\mu}{\pi\zeta}\left(\dfrac{x}{a} + \sqrt{\dfrac{x^2}{a^2} - 1}\right) & \text{for } x \leq -a \\ -\dfrac{a}{\pi\zeta}\left(\sqrt{1 - \dfrac{x^2}{a^2}} + \dfrac{2x\mu}{a}\right) & \text{for } |x| \leq a \end{cases} \tag{8.6}$$

$$\sigma_z|_{z=0} = \begin{cases} -\dfrac{a}{\pi\zeta}\sqrt{1 - \dfrac{x^2}{a^2}} & \text{for } |x| \leq a \\ 0 & \text{for } |x| > a \end{cases} \tag{8.7}$$

$$\tau_{xz}|_{z=0} = \begin{cases} -\dfrac{a\mu}{\pi\zeta}\sqrt{1 - \dfrac{x^2}{a^2}} & \text{for } |x| \leq a \\ 0 & \text{for } |x| > a \end{cases} \tag{8.8}$$

In the above set of equations, a represents the semi-contact length and μ denotes the coefficient of friction at the contact. ζ is a function of the geometrical and elastic properties of the bodies in contact and is given as

$$\zeta = \frac{1}{\dfrac{1}{2}\left(\dfrac{1}{R_1} + \dfrac{1}{R_2}\right)}\left(\frac{1-\nu_1^2}{E_1} + \frac{1-\nu_2^2}{E_2}\right) \tag{8.9}$$

The parameters ψ_1, ψ_2 are defined as

$$\psi_1 = \frac{\pi(r_1 + r_2)}{r_1 r_2\sqrt{2r_1 r_2 + 2x^2 + 2z^2 - 2a^2}}$$

$$\psi_2 = \frac{\pi(r_1 - r_2)}{r_1 r_2\sqrt{2r_1 r_2 + 2x^2 + 2z^2 - 2a^2}} \tag{8.10}$$

$$r_1 = \sqrt{(a+x)^2 + z^2}, \quad r_2 = \sqrt{(a-x)^2 + z^2}$$

8.4
Evaluation of Contact Stress Parameters by Least Squares Analysis

In Chap. 1, the use of least squares approach for evaluating the material stress fringe value was discussed. The governing equation was linear and hence the use of Gaussian elimination procedure directly provided the solution. Referring to Eqs. (8.5)–(8.10), the expression for $(\sigma_1 - \sigma_2)$ for the contact stress problem is non-linear in terms of a and μ. This necessitates an iterative procedure for evaluating these parameters [8, 9]. The principal stresses in terms of the stress components are

$$\sigma_1;\sigma_2 = \frac{\sigma_x + \sigma_z}{2} \pm \sqrt{\frac{(\sigma_x - \sigma_z)^2}{4} + (\tau_{xz})^2} \tag{8.11}$$

Combining stress-optic law (Eq. (1.15), Chap. 1) and Eq. (8.11), a function g is defined for the mth data point as follows:

$$g_m = \left\{\frac{\sigma_x - \sigma_z}{2}\right\}_m^2 + (\tau_{xz})_m^2 - \left\{\frac{N_m F_\sigma}{2h}\right\}^2 \tag{8.12}$$

If initial estimates are made for a and μ and substituted in the above equation, it is possible that $g_m \neq 0$ since the estimates may not be accurate. To correct the estimates, a series of iterative equations based on a Taylor series expansion of g_m are written as

$$(g_m)_{i+1} = (g_m)_i + \left(\frac{\partial g_m}{\partial a}\right)_i (\Delta a)_i + \left(\frac{\partial g_m}{\partial \mu}\right)_i (\Delta \mu)_i \tag{8.13}$$

The corrections are determined such that $(g_m)_{i+1} = 0$ and thus the above equation gives

$$-(g_m)_i = \left(\frac{\partial g_m}{\partial a}\right)_i (\Delta a)_i + \left(\frac{\partial g_m}{\partial \mu}\right)_i (\Delta \mu)_i \tag{8.14}$$

Here the subscript i refers to the evaluation of g_m and its derivative at the estimates a_i and μ_i for the ith iteration and $(\Delta a)_i$ and $(\Delta \mu)_i$ denote the corrections to this estimates for $(i+1)^{\text{th}}$ iteration. Application of the above iteration scheme to M data points results in an over determined set of linear equations in terms of the unknown corrections $(\Delta a)_i$ and $(\Delta \mu)_i$ given in the matrix form by

$$\begin{Bmatrix} g_1 \\ g_2 \\ \vdots \\ g_m \\ \vdots \\ g_M \end{Bmatrix}_i = -\begin{bmatrix} \dfrac{\partial g_1}{\partial a} & \dfrac{\partial g_1}{\partial \mu} \\ \dfrac{\partial g_2}{\partial a} & \dfrac{\partial g_2}{\partial \mu} \\ \vdots & \vdots \\ \dfrac{\partial g_M}{\partial a} & \dfrac{\partial g_M}{\partial \mu} \end{bmatrix}_i \begin{Bmatrix} \Delta a \\ \Delta \mu \end{Bmatrix}_i \tag{8.15}$$

or in a concise form by

$$\{g\}_i = -[b]_i \{\Delta A\}_i \tag{8.16}$$

Now $\{\Delta A\}_i$ can be determined in a least squares sense from the above equation as

$$\{\Delta A_i\} = -[c]_i^{-1}\{d\}_i \tag{8.17}$$

where

$$[c]_i = [b]_i^T [b]_i \tag{8.18}$$

and

$$\{d\}_i = [b]_i^T \{g\}_i \tag{8.19}$$

The solution of Eq. (8.17) gives $(\Delta a)_i$ and $(\Delta\mu)_i$ which are used to improve the estimates of a and μ for the next iteration by

$$\begin{aligned} a_{i+1} &= a_i + (\Delta a)_i \\ \mu_{i+1} &= \mu_i + (\Delta\mu)_i \end{aligned} \tag{8.20}$$

The iteration has to be stopped by appropriate convergence criteria. One can use either of the following two criteria:

1. The parameter error minimisation and
2. The fringe order error minimisation.

In the parameter error minimisation, the iterations are stopped when the values of $\{\Delta A\}_i$ become reasonably small (say of the order of 10^{-6}). In the fringe order error minimisation, using the newly calculated values of $(a)_{i+1}$ and $(\mu)_{i+1}$, the fringe orders corresponding to the selected data points are calculated theoretically during every iteration step and are compared with the experimentally obtained fringe orders. The convergence criteria is satisfied if

$$\frac{\sum\left|N_{\text{theory}} - N_{\text{exp}}\right|}{\text{total no. of data points}} \leq \text{convergence error} \tag{8.21}$$

The use of fringe-order-error minimisation criteria generally gives better results and is recommended for general use. Best solution is obtained when the convergence error is less than 0.1.

To make the solution not to be influenced significantly by the choice of data points, it is desirable to implement a sampled least squares approach as discussed in Chap. 1. To get accurate results, in problems involving contact studies and fracture mechanics, it is recommended that in sampled least squares analysis, the total number of data points to be collected be kept about 75 to 100 and the data points per set be kept as 35.

8.4.1 Validation for Hertzian and Non-Hertzian Contact

Using four circular discs in one loading arrangement (Fig. 1.25, Chap. 1), both the normal contact and the combined normal and tangential loading are simulated [9]. Figures 8.2a and 8.3a show the fringe patterns with the skeletons superimposed in white for the contact zone experiencing the Hertzian and non-Hertzian contacts respectively. In Fig. 8.3a, due to the presence of friction, one can clearly see that the eye of the fringe packet is pulled towards the surface indicating that the maximum shear stress can occur at the surface. Using the non-linear least squares method the contact stress parameters are evaluated. Figures 8.2b and 8.3b show the theoretically reconstructed fringe patterns with the data points echoed back for each of these cases. The comparison is very good and this validates the methodology.

a b

Fig. 8.2. Isochromatics for Hertzian contact: **a** experimental isochromatics with the fringe skeleton superimposed **b** theoretically reconstructed isochromatics with data points echoed back (a = 1.15 mm, μ = 0.004, E = 3.3 GPa, ν = 0.34, F_σ = 14.25 N/mm/fr, R_i = 29.45 mm, thickness = 6.26 mm)

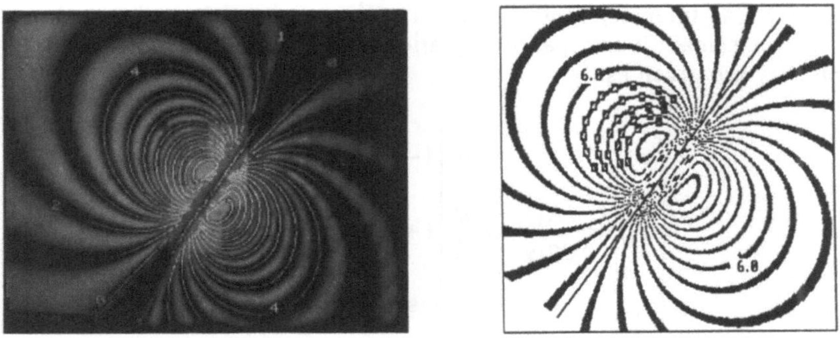

a b

Fig. 8.3. Isochromatics for Non-Hertzian contact: **a** experimental isochromatics with the fringe skeleton superimposed **b** theoretically reconstructed isochromatics with data points echoed back (a = 0.76 mm, μ = – 0.08, E = 3.3 GPa, ν = 0.34, F_σ = 14.25 N/mm/fr, R_i = 29.45 mm, thickness = 6.26 mm)

8.5 Developments in the Description of the Stress Field Equations in the Neighbourhood of a Crack-tip

8.5.1 Mode-I Stress Field Equations

The Westergaard [10] complex stress function technique for solving opening mode crack problems has played an important role in the development of linear elastic fracture mechanics. He gave the stress functions for the case of a central crack and a series of equally spaced straight cracks of length $2a$ in an infinite plate subjected to a biaxial field of tension σ. Irwin [11] in 1957 following the semi-inverse procedure suggested by Westergaard added three additional examples. They are

1. Single crack along the x-axis extending from $-a$ to a with a wedge action applied to produce a pair of *splitting force* of magnitude P located at $x = b$.
2. The situation of the above example with an additional pair of forces of magnitudes P at $x = -b$.
3. Example (1) repeated along the x-axis at intervals ℓ and with the wedge action centred so that b is zero.

Irwin [11] showed that in all the above five cases, the stress distribution near the end of the crack can be expressed independent of the type of loading if the region of interest is very close to the crack-tip (i.e., r/a and $r/(a - b)$ may be neglected in comparison to unity) and introduced the concept of stress intensity factor (SIF). The expressions thus obtained are known as classical Westergaard equations and for a mode-I loading situation they are given as

$$\begin{Bmatrix} \sigma_x \\ \sigma_y \\ \tau_{xy} \end{Bmatrix} = \frac{K_{\mathrm{I}}}{\sqrt{2\pi r}} \cos\frac{\theta}{2} \begin{Bmatrix} 1-\sin\frac{\theta}{2}\sin\frac{3\theta}{2} \\ 1+\sin\frac{\theta}{2}\sin\frac{3\theta}{2} \\ \sin\frac{\theta}{2}\cos\frac{3\theta}{2} \end{Bmatrix} \tag{8.22}$$

When the fringes are theoretically simulated using Eq. (8.22), one observes that the fringes are symmetric with respect to both x and y-axes. Even for relatively large bodies, with cracks far removed from the free surface, experimentally one observes that the fringes are not symmetrical about y-axis and they have a prominent tilt. The angle of tilt is a function of the fringe order and for fringes very close to the crack-tip, this angle approaches 90° [12]. Thus, the practical utility of Eq. (8.22) for the evaluation of SIF in experimental mechanics is limited, as the region of its validity is very small. Irwin added a constant stress term σ_{ox} to the σ_x term to account for the use of finite geometry specimens used in the experiments. The addition of this term also explained satisfactorily the fringe tilt observed in experiments [13]. Thus, the modified Westergaard equations become

$$\begin{Bmatrix} \sigma_x \\ \sigma_y \\ \tau_{xy} \end{Bmatrix} = \frac{K_{\mathrm{I}}}{\sqrt{2\pi r}} \cos\frac{\theta}{2} \begin{Bmatrix} 1-\sin\frac{\theta}{2}\sin\frac{3\theta}{2} \\ 1+\sin\frac{\theta}{2}\sin\frac{3\theta}{2} \\ \sin\frac{\theta}{2}\cos\frac{3\theta}{2} \end{Bmatrix} + \begin{Bmatrix} -\sigma_{ox} \\ 0 \\ 0 \end{Bmatrix} \tag{8.23}$$

Though Irwin introduced σ_{ox} term intuitively, analytical justifications for this additional term in the near-field equations of fracture mechanics came much later [14–17].

In order to account for finite bodies, Tada, Paris and Irwin [6] introduced a series form of Westergaard stress function. This predicted only a constant fringe order along the crack axis. However, for cracks approaching a boundary, several fringe orders cross the crack axis. This has been observed for various specimen geometries. Chona et al. [18] observed this experimentally for a rectangular double cantilever beam (RDCB) specimen at large crack lengths (a/w = 0.9, w is the width of the specimen). A finite element analysis [19] of the single edge notched (SEN) specimen using quarter-point singular elements for modelling the crack-tip shows that for larger crack

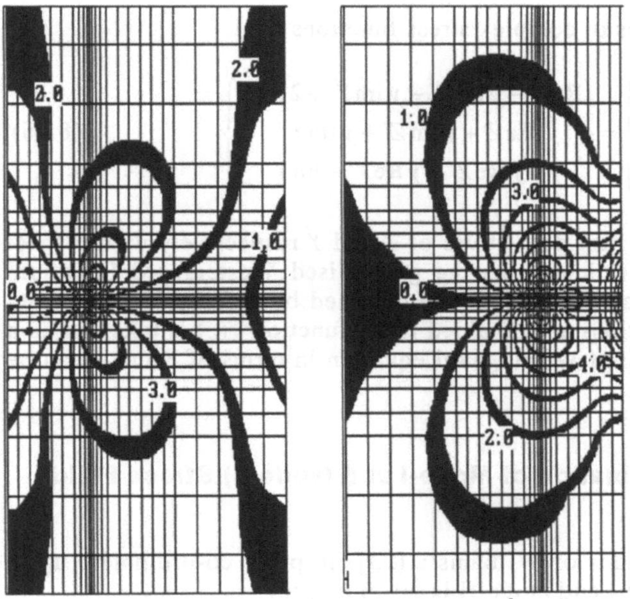

a b

Fig. 8.4. Isochromatic fringe patterns in the neighborhood of a crack in a SEN specimen obtained using the finite element method: **a** a/w = 0.3 **b** a/w = 0.7. Note that for larger crack length, several fringes cross the crack-axis

lengths, several fringes cross the crack-axis (Fig. 8.4). The numerically evaluated fringe contours also simulate the thickness variation of fringes observed in the experiments. This has been made possible through the use of a novel software for fringe plotting from FEM results [20]. The program code is included in the CD under the name *femfrn.c*.

To explain the crossing of several fringes along the crack axis, Sanford reported [21] that an additional stress function $Y(z)$ should be added to the series form of Westergaard stress function $Z(z)$ where, z is $x + iy$ and i is $\sqrt{-1}$. For a single ended crack, with the origin of co-ordinates at the crack tip and the negative x-axis coinciding with the crack faces, the functions $Z(z)$ and $Y(z)$ are expressed as

$$Z(z) = \sum_{j=0}^{J} C_{2j} z^{j-\frac{1}{2}} \tag{8.24}$$

$$Y(z) = \sum_{j=0}^{J} C_{2j+1} z^{j} \tag{8.25}$$

where C_{2j} and C_{2j+1} are real coefficients. The function $Y(z)$ is selected, such that it satisfies the condition $\text{Im}(Y(z)) = 0$ on $y = 0$. This form of $Y(z)$ ensures the condition $\tau_{xy} = 0$ on $y = 0$ and is a less restrictive assumption than used by Westergaard.

The stress field in terms of complex stress functions is

$$\begin{Bmatrix} \sigma_x \\ \sigma_y \\ \tau_{xy} \end{Bmatrix} = \begin{Bmatrix} \operatorname{Re} Z - y \operatorname{Im} Z' - y \operatorname{Im} Y' + 2 \operatorname{Re} Y \\ \operatorname{Re} Z + y \operatorname{Im} Z' + y \operatorname{Im} Y' \\ - y \operatorname{Re} Z' - y \operatorname{Re} Y' - \operatorname{Im} Y \end{Bmatrix} \tag{8.26}$$

where Z' and Y' are the first derivatives of Z and Y respectively with respect to z. Equation (8.26) is known as the generalised Westergaard Equation. The classical Westergaard equations are obtained by making $Y(z)$ zero and Irwin's modified equations are obtained if the function $Y(z)$ is taken to be a real constant. Generalised Westergaard equation in terms of r and θ can be found in ref. [22].

8.5.2 Mixed-Mode (Combination of Mode-I and Mode-II) Stress Field Equations

The stress field equations of Williams's [23] (in polar co-ordinates) up to six terms each in mode-I and mode-II is given by [22]

$$\begin{Bmatrix} \sigma_r \\ \sigma_\theta \\ \tau_{r\theta} \end{Bmatrix} = \begin{Bmatrix} \left\{ r^{-\frac{1}{2}} c_{11} \left(\frac{5}{4}\cos\frac{\theta}{2} - \frac{1}{4}\cos\frac{3\theta}{2} \right) + 4c_{12}(\cos^2\theta) + r^{\frac{1}{2}} 3c_{13} \right. \\ \left(\frac{3}{4}\cos\frac{\theta}{2} + \frac{1}{4}\cos\frac{5\theta}{2} \right) + 2rc_{14}(\cos\theta + 3\cos 3\theta) + r^{\frac{3}{2}} c_{15} \\ \left. \left(\frac{5}{4}\cos\frac{3\theta}{2} + \frac{15}{4}\cos\frac{7\theta}{2} \right) + r^2 c_{16}(12\cos 4\theta) \right\} \\ \left\{ r^{-\frac{1}{2}} \frac{3}{4} c_{11} \left(\cos\frac{\theta}{2} + \frac{1}{3}\cos\frac{3\theta}{2} \right) + 4c_{12}(\sin^2\theta) + r^{\frac{1}{2}} \frac{15}{4} c_{13} \right. \\ \left(\cos\frac{\theta}{2} - \frac{1}{5}\cos\frac{5\theta}{2} \right) + 6rc_{14}(\cos\theta - \cos 3\theta) + \frac{35}{4} r^{\frac{3}{2}} c_{15} \\ \left. \left(\cos\frac{3\theta}{2} - \frac{3}{7}\cos\frac{7\theta}{2} \right) + 12r^2 c_{16}(\cos 2\theta - \cos 4\theta) \right\} \\ \left\{ r^{-\frac{1}{2}} \frac{1}{2} c_{11} \left(\frac{1}{2}\sin\frac{\theta}{2} + \frac{1}{2}\sin\frac{3\theta}{2} \right) - 2c_{12}(\sin 2\theta) + r^{\frac{1}{2}} \frac{3}{2} c_{13} \right. \\ \left(\frac{1}{2}\sin\frac{\theta}{2} - \frac{1}{2}\sin\frac{5\theta}{2} \right) + 2rc_{14}(\sin\theta - 3\sin 3\theta) + r^{\frac{3}{2}} \frac{5}{2} c_{15} \\ \left. \left(\frac{3}{2}\sin\frac{3\theta}{2} - \frac{3}{2}\sin\frac{7\theta}{2} \right) + 3r^2 c_{16}(2\sin 2\theta - 4\sin 4\theta) \right\} \end{Bmatrix}$$

$$
+\left\{\begin{array}{l}
\left\{-r^{-\frac{1}{2}}c_{21}\left(\frac{5}{4}\sin\frac{\theta}{2}-\frac{3}{4}\sin\frac{3\theta}{2}\right)+r^{\frac{1}{2}}3c_{23}\left(\frac{3}{4}\sin\frac{\theta}{2}+\frac{5}{4}\sin\frac{5\theta}{2}\right)\right.\\
+2rc_{24}(\sin\theta+\sin 3\theta)+r^{\frac{3}{2}}c_{25}\left(\frac{5}{4}\sin\frac{3\theta}{2}+\frac{35}{4}\sin\frac{7\theta}{2}\right)+\\
\left.\qquad r^{2}c_{26}\left(6\sin 4\theta\right)\right\}\\
\left\{-r^{-\frac{1}{2}}\frac{3}{4}c_{21}\left(\sin\frac{\theta}{2}+\sin\frac{3\theta}{2}\right)+r^{\frac{1}{2}}\frac{15}{4}c_{23}\left(\sin\frac{\theta}{2}-\sin\frac{5\theta}{2}\right)\right.\\
+6rc_{24}(\sin\theta-\frac{1}{3}\sin 3\theta)+\frac{35}{4}r^{\frac{3}{2}}c_{25}\left(\sin\frac{3\theta}{2}-\sin\frac{7\theta}{2}\right)\\
\left.\qquad 12r^{2}c_{26}\left(\sin 2\theta-\frac{1}{2}\sin 4\theta\right)\right\}\\
\left\{r^{-\frac{1}{2}}\frac{1}{2}c_{21}\left(\frac{1}{2}\cos\frac{\theta}{2}+\frac{3}{2}\cos\frac{3\theta}{2}\right)-r^{\frac{1}{2}}\frac{3}{2}c_{23}\left(\frac{1}{2}\cos\frac{\theta}{2}-\frac{5}{2}\cos\frac{5\theta}{2}\right)\right.\\
-2rc_{24}(\cos\theta-\cos 3\theta)-r^{\frac{3}{2}}\frac{5}{2}c_{25}\left(\frac{3}{2}\cos\frac{3\theta}{2}-\frac{7}{2}\cos\frac{7\theta}{2}\right)-\\
\left.\qquad 3r^{2}c_{26}(2\cos 2\theta-2\cos 4\theta)\right\}
\end{array}\right\} \tag{8.27}
$$

Atluri and Kobayashi [24] also reported the stress field equations for the mixed mode case (Fig. 8.5) in a general form. They are described as follows:

$$
\left\{\begin{array}{c}\sigma_x\\ \\ \sigma_y\\ \\ \tau_{xy}\end{array}\right\}=\sum_{n=1}^{\infty}\frac{n}{2}A_{\mathrm{I}n}r^{\frac{n-2}{2}}\left\{\begin{array}{l}
\left\{2+(-1)^n+\frac{n}{2}\right\}\cos\left(\frac{n}{2}-1\right)\theta-\left(\frac{n}{2}-1\right)\cos\left(\frac{n}{2}-3\right)\theta\\
\left\{2-(-1)^n-\frac{n}{2}\right\}\cos\left(\frac{n}{2}-1\right)\theta+\left(\frac{n}{2}-1\right)\cos\left(\frac{n}{2}-3\right)\theta\\
-\left\{(-1)^n+\frac{n}{2}\right\}\sin\left(\frac{n}{2}-1\right)\theta+\left(\frac{n}{2}-1\right)\sin\left(\frac{n}{2}-3\right)\theta
\end{array}\right\}
$$

$$
-\sum_{n=1}^{\infty}\frac{n}{2}A_{\mathrm{II}n}r^{\frac{n-2}{2}}\left\{\begin{array}{l}
\left\{2-(-1)^n+\frac{n}{2}\right\}\sin\left(\frac{n}{2}-1\right)\theta-\left(\frac{n}{2}-1\right)\sin\left(\frac{n}{2}-3\right)\theta\\
\left\{2+(-1)^n-\frac{n}{2}\right\}\sin\left(\frac{n}{2}-1\right)\theta+\left(\frac{n}{2}-1\right)\sin\left(\frac{n}{2}-3\right)\theta\\
-\left\{(-1)^n-\frac{n}{2}\right\}\cos\left(\frac{n}{2}-1\right)\theta-\left(\frac{n}{2}-1\right)\cos\left(\frac{n}{2}-3\right)\theta
\end{array}\right\} \tag{8.28}
$$

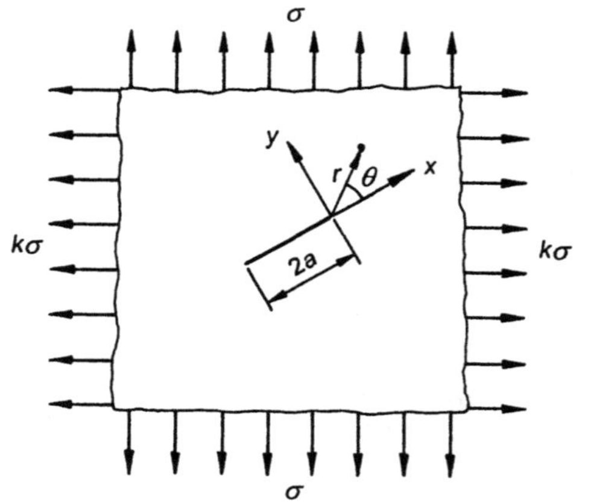

Fig. 8.5. Figure showing a crack in a biaxial stress field. The figure also shows the co-ordinate system followed in fracture mechanics studies

where $A_{I1} = K_I/\sqrt{2\pi}$ and $A_{II1} = -K_{II}/\sqrt{2\pi}$. The σ_{ox} stress term has played a very important role in fracture mechanics and the second term in the Mode-I series is related to σ_{ox} as $4A_{12} = -\sigma_{ox}$.

8.5.3 Equivalence Between the Multi-Parameter Stress Field Equations

Though Eqs. (8.26)–(8.28) appear different, there exists an inter-relationship between them [25]. This relationship can be established, if in all of them, the rectangular stress components are expressed in terms of r and θ. The relationship between the mode-I series coefficients of Eqs. (8.28) and (8.26) is

$$A_{I(i+1)} = \frac{C_i}{i+1} \tag{8.29}$$

Similarly, when the Williams stress field equations are expressed in terms of rectangular stress components using standard transformation law and compared with Eq. (8.28) one gets

$$A_{Ii} = c_{1i} \quad \text{and} \qquad A_{IIi} = -c_{2i} \tag{8.30}$$

8.6 Developments in SIF Evaluation Methodology

The Irwin's method for SIF extraction from photoelastic patterns was the accepted method for many years. In this method, two parameters, namely K_I

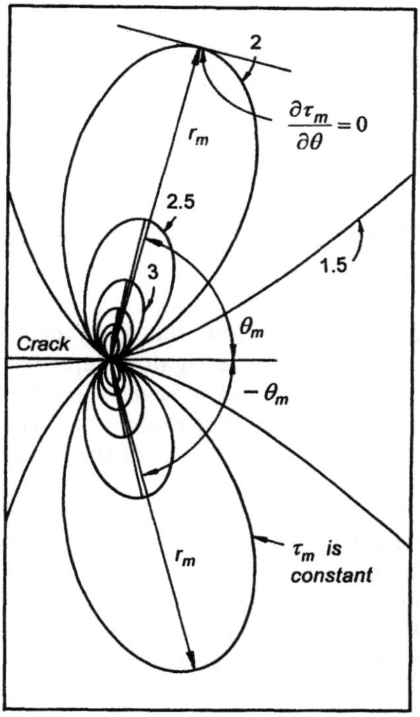

Fig. 8.6. Geometric features of the isochromatics observed near short-cracks in SEN specimens. Note that the fringes are forward tilted and the angle of tilt approaches 90-deg as one approaches the crack-tip

and σ_{ox} are to be evaluated. One needs two equations to solve them. One of the equations comes from the definition of isochromatic fringe contour itself, which is given as

$$(2\tau_m)^2 = \frac{K_I^2}{2\pi r}\sin^2\theta + \frac{2\sigma_{ox}K_I}{\sqrt{2\pi r}}\sin\theta\sin(3\theta/2) + \sigma_{ox}^2 \tag{8.31}$$

where,

$$\tau_m = \frac{NF_\sigma}{2h} \tag{8.32}$$

Referring to Fig. 8.6, for the point $r = r_m$, $\theta = \theta_m$,

$$\left.\frac{\partial \tau_m}{\partial \theta}\right|_{r=r_m,\theta=\theta_m} = 0 \tag{8.33}$$

Application of the above condition gives another equation. The two unknown parameters K_I and σ_{ox} are determined as

$$K_{\rm I} = \frac{2\tau_m(2\pi r_m)^{1/2}}{\sin\theta_m}\left[1+\left(\frac{2}{3\tan\theta_m}\right)^2\right]^{-1/2}\left[1+\frac{2\tan(3\theta_m/2)}{3\tan\theta_m}\right] \tag{8.34}$$

and

$$\sigma_{ox} = \frac{-2\tau_m\cos\theta_m}{\cos(3\theta_m/2)[\cos^2\theta_m + (9/4)\sin^2\theta_m]^{1/2}} \tag{8.35}$$

Bradley and Kobayashi [26] modified the formulation, but fundamentally, the approach was unaltered. Since then, several investigators reported various methods and Etheridge and Dally [27] reported a review on the relative accuracy of the various two-parameter methods. The review brought out the fact that the zone of data collection has to be very small of the order of $r_m/a < 0.03$ and $73^\circ < \theta_m < 139^\circ$.

Though photoelasticity gives whole field information, the previous reluctance to use information from the entire field stemmed primarily from practical limitations on data reduction. The positional co-ordinates of each data point used had to be measured manually and the results are calculated with the aid of a hand held calculator. With the introduction of PC based image processing systems, the drudgery of data reduction is removed [19, 28].

Improvements in data collection methodology have lead to the development of several techniques to process the data from the field rather than restricting the analysis to a point (as in Irwin's method) or a few points along a specified line. Among the several methods [28] the over-deterministic approach of Sanford and Dally [29] has come to stay as the most useful technique for evaluating the SIFs and σ_{ox}.

The approaches described above use the near field equations of fracture mechanics to extract the SIF from the full field fringe patterns. In principle, the accuracy of these methods should improve as the region of data acquisition is reduced to small regions around the crack-tip. It has been independently verified by several researchers [30, 31] that the plane stress assumption ceases to be valid in a small region at the crack-tip. In this region the stress field is three-dimensional and experimental observations are influenced by stress gradients through the thickness. In addition, localised crack-tip blunting alters the stress field. On the other hand, if this region is excluded from the analysis, the region of data acquisition may not lie fully within the domain in which the inverse square root singularity dominates the fringe pattern (Fig. 8.7). The size and shape of the singularity-dominated zone is an important factor to be determined and such studies [18, 31] emphasised the fact that the singularity-dominated zone is very small.

This has lead to the development of multi-parameter solution by experimentalists. The zone of data collection in such a case can be larger

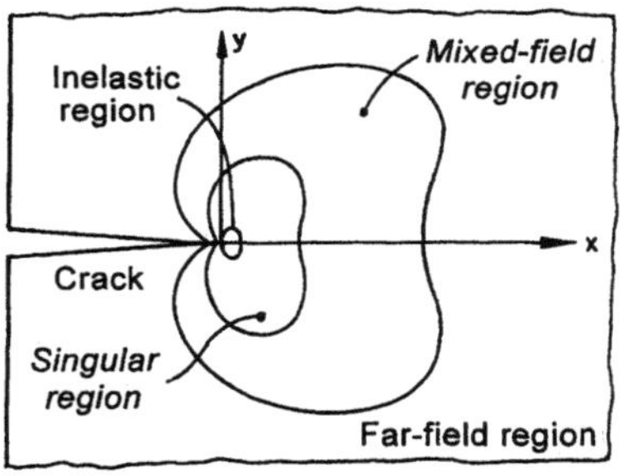

Fig. 8.7. Qualitative representation of various regions near the crack-tip

and one can collect data from the mixed field zone also (Fig. 8.7). It is noteworthy to point out here that in FEM analysis, the use of quarter-point element coupled with a fine discretization of the domain has faithfully captured the fringe field features for both short and long cracks (Fig. 8.4). In view of numerical modelling one can extract data from a region very close to the crack-tip for SIF evaluation in both these cases. Hence in numerical studies, the researchers did not bother about the higher order solution for a long time. However, experimentalists cannot collect data in a zone very close to the crack-tip due to practical difficulties and a multi-parameter solution is needed to interpret the experimental data.

8.7 Evaluation of Mixed-Mode Stress Field Parameters Using Least Squares Technique

Following the discussion in Sect. 8.5.3, it is elegant to use the multi-parameter solution reported by Atluri and Kobayashi for modelling the stress field for experimental studies. The evaluation of the multi-parameters governing the stress field needs to be done in an iterative sense as the expression for the maximum shear stress is non-linear in terms of the parameters A_{I1}, A_{I2} , ... , A_{Ik}, A_{II1}, A_{II2}, ... , $A_{II\ell}$ where k is the number of mode-I parameters and ℓ is the number of mode-II parameters being considered. Apart from these unknowns, *a priori* one also does not know the values of k and ℓ and they have to be found as part of the solution. Following the steps discussed in Sect. 8.3, the unknown parameters can be obtained for known values of k and ℓ by solving the equation

$$\{g\}_i = -[b]_i\{\Delta A\}_i \tag{8.36}$$

where

$$\{g\}_i = \begin{Bmatrix} g_1 \\ g_2 \\ \vdots \\ g_m \\ \vdots \\ g_M \end{Bmatrix}_i ,$$

$$[b]_i = \begin{bmatrix} \frac{\partial g_1}{\partial A_{\mathrm{I}1}} & \frac{\partial g_1}{\partial A_{\mathrm{I}2}} & \cdots & \frac{\partial g_1}{\partial A_{\mathrm{I}k}} & \frac{\partial g_1}{\partial A_{\mathrm{II}1}} & \frac{\partial g_1}{\partial A_{\mathrm{II}2}} & \cdots & \frac{\partial g_1}{\partial A_{\mathrm{II}\ell}} \\ \frac{\partial g_2}{\partial A_{\mathrm{I}1}} & \frac{\partial g_2}{\partial A_{\mathrm{I}2}} & \cdots & \frac{\partial g_2}{\partial A_{\mathrm{I}k}} & \frac{\partial g_2}{\partial A_{\mathrm{II}1}} & \frac{\partial g_2}{\partial A_{\mathrm{II}2}} & \cdots & \frac{\partial g_2}{\partial A_{\mathrm{II}\ell}} \\ \vdots & \vdots & \vdots & \vdots & \vdots & \vdots & \vdots & \vdots \\ \frac{\partial g_M}{\partial A_{\mathrm{I}1}} & \frac{\partial g_M}{\partial A_{\mathrm{I}2}} & \cdots & \frac{\partial g_M}{\partial A_{\mathrm{I}k}} & \frac{\partial g_M}{\partial A_{\mathrm{II}1}} & \frac{\partial g_M}{\partial A_{\mathrm{II}2}} & \cdots & \frac{\partial g_M}{\partial A_{\mathrm{II}\ell}} \end{bmatrix}_i , \tag{8.37}$$

and

$$\{\Delta A\}_i = \begin{Bmatrix} \Delta A_{\mathrm{I}1} \\ \Delta A_{\mathrm{I}2} \\ \vdots \\ \Delta A_{\mathrm{I}k} \\ \Delta A_{\mathrm{II}1} \\ \Delta A_{\mathrm{II}2} \\ \vdots \\ \Delta A_{\mathrm{II}\ell} \end{Bmatrix}_i$$

In forming the Eq. (8.37) one needs to calculate $(\partial g_1/\partial A_{\mathrm{I}1})$, $(\partial g_1/\partial A_{\mathrm{I}2})$ ... etc., and at first it appears that it may be difficult to do it in closed-form. However, it can be determined very easily. A typical derivative for the mth data point with respect to $A_{\mathrm{I}n}$ is obtained by taking the derivative of Eq. (8.12) (Note: the stress components are expressed in the *xy*-plane).

$$\frac{\partial g_m}{\partial A_{\mathrm{I}n}} = \frac{1}{2}(\sigma_x - \sigma_y)_m \left(\frac{\partial \sigma_x}{\partial A_{\mathrm{I}n}} - \frac{\partial \sigma_y}{\partial A_{\mathrm{I}n}} \right)_m + 2\left(\tau_{xy} \frac{\partial \tau_{xy}}{\partial A_{\mathrm{I}n}} \right)_m \tag{8.38}$$

Since, $A_{\mathrm{I}n}$ is a linear coefficient, the terms $(\partial \sigma_x/\partial A_{\mathrm{I}n})$ etc. in Eq. (8.38) are simply obtained as

$$\begin{Bmatrix} \dfrac{\partial \sigma_x}{\partial A_{\mathrm{I}n}} \\ \dfrac{\partial \sigma_y}{\partial A_{\mathrm{I}n}} \\ \dfrac{\partial \tau_{xy}}{\partial A_{\mathrm{I}n}} \end{Bmatrix} = \frac{n}{2} r^{\frac{n-2}{2}} \begin{Bmatrix} \left\{2+(-1)^n+\dfrac{n}{2}\right\}\cos\left(\dfrac{n}{2}-1\right)\theta - \left(\dfrac{n}{2}-1\right)\cos\left(\dfrac{n}{2}-3\right)\theta \\ \left\{2-(-1)^n-\dfrac{n}{2}\right\}\cos\left(\dfrac{n}{2}-1\right)\theta + \left(\dfrac{n}{2}-1\right)\cos\left(\dfrac{n}{2}-3\right)\theta \\ -\left\{(-1)^n+\dfrac{n}{2}\right\}\sin\left(\dfrac{n}{2}-1\right)\theta + \left(\dfrac{n}{2}-1\right)\sin\left(\dfrac{n}{2}-3\right)\theta \end{Bmatrix} \quad (8.39)$$

The solution to Eq. (8.36) gives $\{\Delta A\}_i$ which can be used to find $\{A\}_{i+1}$ for the next iteration as

$$\{A\}_{i+1} = \{A\}_i + \{\Delta A\}_i \quad (8.40)$$

As mentioned before, the iteration needs to be stopped by fringe-order-error minimisation criterion.

A *priori* one does not know how many terms are required to model the stress field. It is always recommended to start from a minimum number of parameters and to progressively increment them by one as the problem demands. If the problem demands say six parameters, but the solution process is started with two parameters, then convergence will not be achieved unless a large convergence error of the order of 0.5 is specified in Eq. (8.21). The converged solution at this stage does not model the stress field correctly. However, it does provide a good starting guess for the two parameters for the subsequent iterations. The number of parameters is then incremented by one and the iterations are initiated with a less convergence error. The process is to be repeated until one could achieve a convergence error of 0.1 or less.

A PC based software to evaluate the fracture parameters using up to 13 terms for each of mode I and mode II of the Atluri and Kobayashi [24] form of the stress field equations is available [32]. The software requires a data file containing the x and y co-ordinates and the corresponding fringe order as the input. The software can accept the input from manual data entry or directly from DIP data collection software. The least squares procedure is invoked and the results of the parameters are written to another data file. The software has a provision to increase the number of stress field parameters and to supply the convergence error required at every stage of the iteration interactively. Thus the number of terms to describe the stress field can be increased progressively until one gets a convergence error less than 0.1.

Using the parameters thus obtained, the program recalculates $(\sigma_1 - \sigma_2)$ values, hence the fringe order at every point in the data field and the fringes are then plotted. The reconstruction routine of fringe patterns is written such that the crack line can be rotated to suit the experimental situation so as to give a realistic representation of the fringe field for easy comparison. At the end of the reconstruction, the data points are echoed back. A cursor routine in-built in the software is used to label the fringes.

8.8 Experimental Validation of the Methodology

8.8.1 Mode-I Loading

The problem chosen is that of analysing the stress field in the neighbourhood of radial cracks in an internally pressurised thick cylinder. The model is cut from an epoxy sheet made of CY230 resin and HY951 hardener mixed in the proportion 100:9 by weight. An Aluminium disc slightly larger than the internal diameter of the epoxy disc was inserted to simulate the internal pressure. Using a specialised cutter, thin radial slits with thickness of the order of 0.1 mm were made to simulate the cracks.

Figure 8.8 shows the dark-field fringe patterns with fringe skeletons superimposed for the case of a radial crack emanating from the outer boundary of an internally pressurised thick ring [33]. It also shows the fringe skeletons obtained by the various fringe-thinning algorithms and brings out the superiority of the global fringe-thinning algorithm in comparison to other fringe thinning methodologies. It is interesting to note that in the present example, ahead of the crack-tip, several fringes cross the crack axis. This can be attributed to the fact that the crack is located in a non-uniform stress field and the crack-tip is closer to the boundary.

Using the methodology discussed in Sect. 8.7, the parameters governing the stress field are determined in a least squares sense. These are given

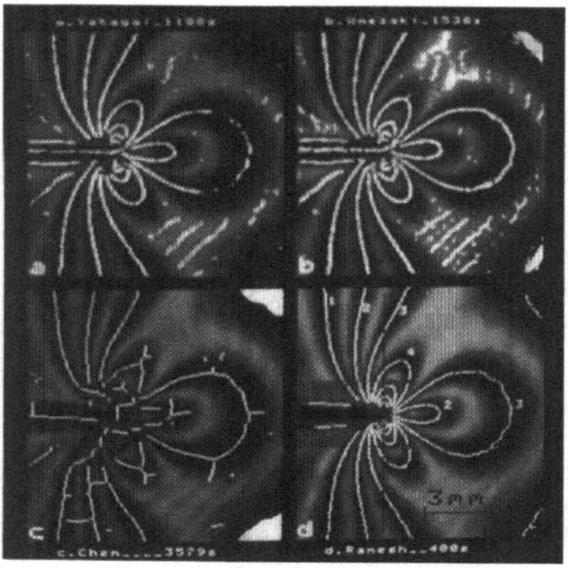

Fig. 8.8. Dark-field isochromatics observed for a radial crack emanating from the outer boundary of an internally pressurised thick ring. The figure also shows the comparative performance evaluation of various fringe thinning algorithms. (Thinning algorithm processing time on a PC-Xt based system): **a** mask-based algorithm-I, 1188 s **b** mask-based algorithm-II, 1530 s **c** binary-based algorithm, 3579 s **d** global fringe thinning, 400 s

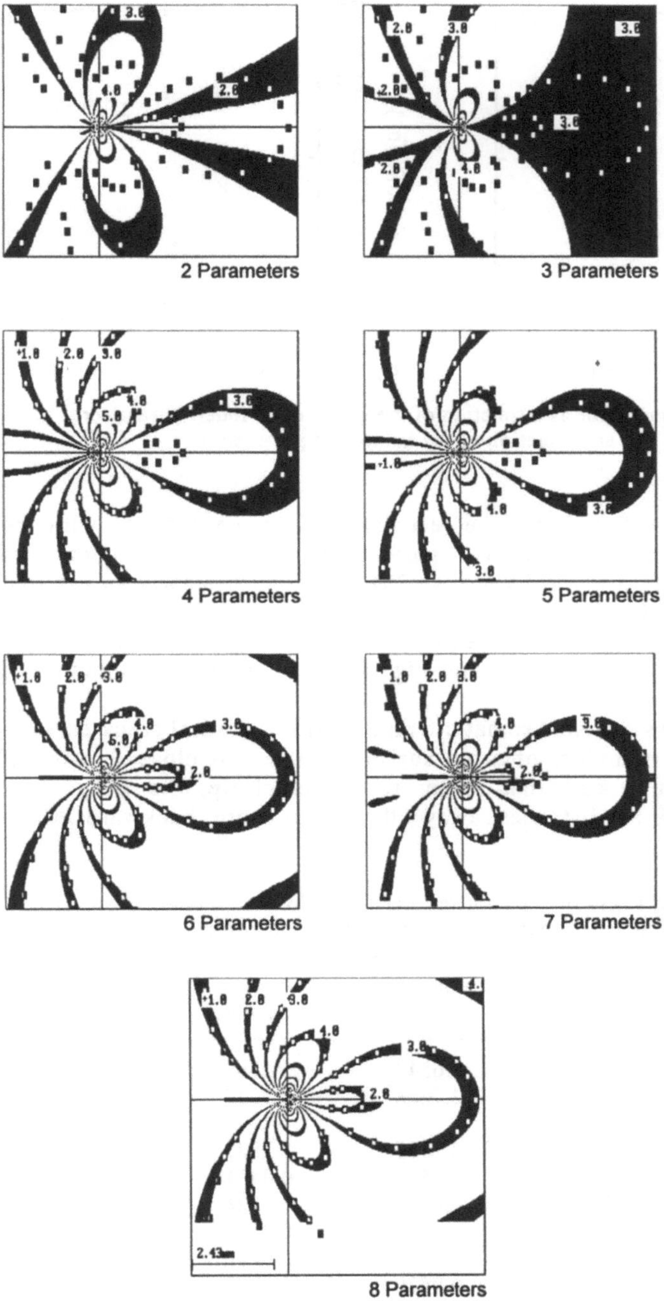

Fig. 8.9. Theoretically reconstructed fringe patterns with data points echoed for various number of mode-I stress-field parameters for the fringe patterns shown in Fig.8.8

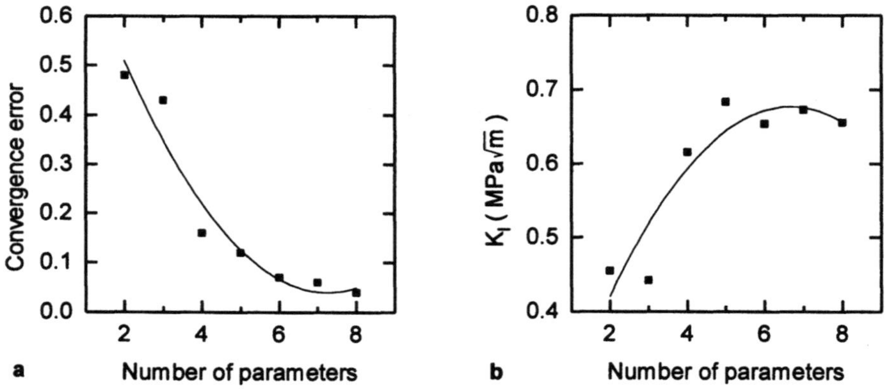

Fig. 8.10. a Convergence error as a function of the number of Mode-I parameters **b** Variation of Mode-I SIF as a function of the number of parameters

in Table 8.1. Figure 8.9 shows the reconstructed fringe patterns with the data points echoed starting from two parameters up to eight parameters of the mode-I stress field given by Eq. (8.28). The figure clearly shows that only a multi-parameter solution beyond six parameters actually models the stress field better. Figure 8.10a shows the percentage error in N as a function of number of parameters and Fig. 8.10b shows K_I as a function of number of parameters.

Figure 8.11 shows the isochromatic fringe field for a radial crack emanating from the inner boundary of a pressurised cylinder. A close scrutiny of the fringe patterns reveals that closer to the crack-tip, the fringes are forward tilted and at distances away, the fringes are backward tilted.

Table 8.1. Summary of stress field parameters for the problem of a radial crack emanating from the outer boundary of a pressurised thick ring

	2-parameter solution	6-parameter solution	8-parameter solution
K_I (MPa$\sqrt{m}$)	0.4550	0.6540	0.6560
σ_{ox} (MPa)	4.2636	2.9668	3.3640
A_{I1} (MPa(mm)$^{1/2}$)	5.7380	8.2581	8.2833
A_{I2} (MPa)	−1.0659	−0.7417	−0.8411
A_{I3} (MPa(mm)$^{-1/2}$)		0.0305	−0.2228
A_{I4} (MPa(mm)$^{-1}$)		−0.0513	−0.0169
A_{I5} (MPa(mm)$^{-3/2}$)		0.0080	−0.0186
A_{I6} (MPa(mm)$^{-2}$)		−0.0020	−0.0023
A_{I7} (MPa(mm)$^{-5/2}$)			0.0015
A_{I8} (MPa(mm)$^{-3}$)			−0.0003

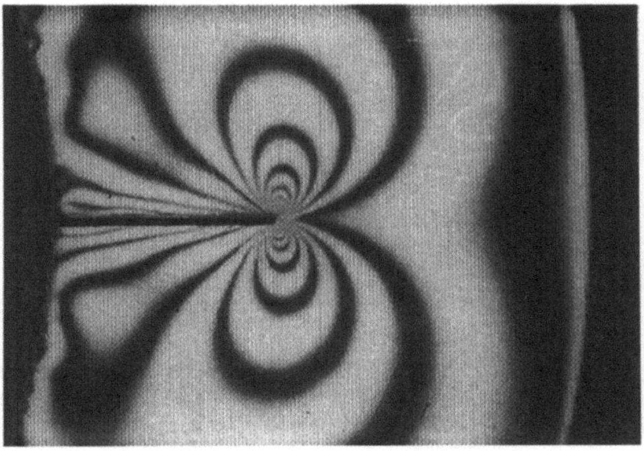

Fig. 8.11. Dark-field isochromatics observed for a radial crack emanating from the inner boundary of an internally pressurised thick ring. Note that the direction of the fringe tilt changes as one moves towards the crack-tip (Courtesy of ref. [32])

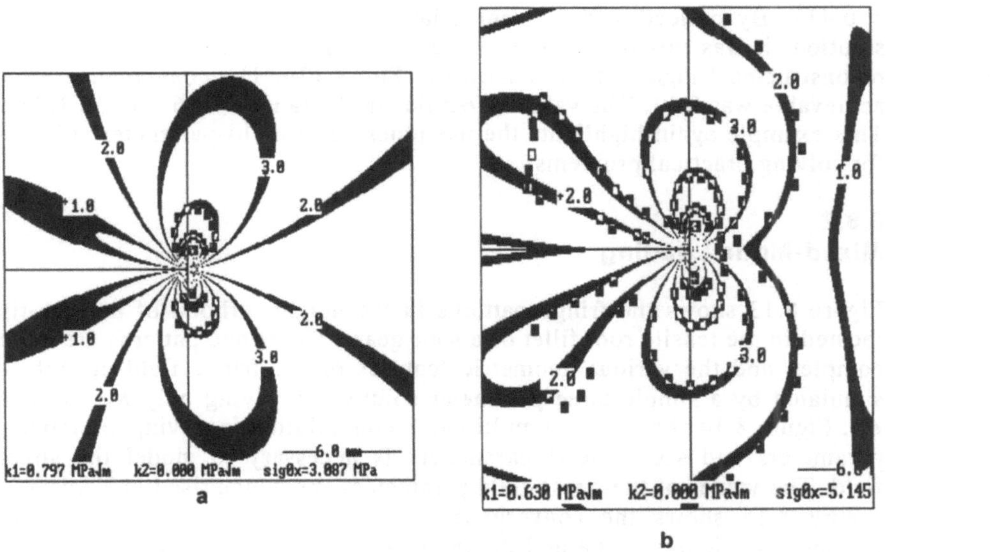

Fig. 8.12. Theoretically reconstructed fringe patterns: **a** two-parameter solution **b** eight-parameter solution. Note that the zone of data collection is substantially increased and the fringes are better modelled in this case

In principle, by collecting data only from the forward tilted loops, it is possible to evaluate K_I and σ_{ox} using a two-parameter over-deterministic

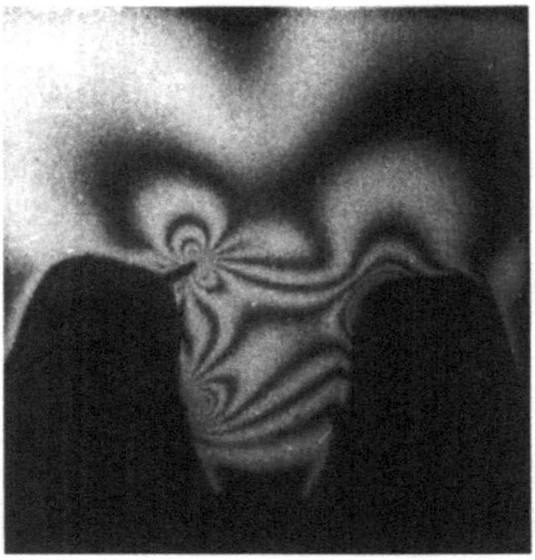

Fig. 8.13. Dark-field isochromatics in the neighbourhood of a crack emanating from the tensile-root fillet of a spur gear

solution [28, 29]. The resulting reconstructed fringe patterns are shown in Fig. 8.12a. The reconstructed fringe patterns clearly brings out the fact that the solution models only a small field closer to the crack-tip. The convergence error achievable was 0.15 and r/a lies in the range $0.116 < r/a < 0.412$. By collecting data from a larger zone, with an eight-parameter solution, it was possible to model the stress field better and the resulting reconstructed fringe pattern is shown in Fig. 8.12b. The convergence error achievable was 0.11. The value of r/a lies in the range $0.116 < r/a < 1.349$. This example again highlights the usefulness of a multi-parameter solution for solving practical problems.

8.8.2 Mixed-Mode Loading

Figure 8.13 shows the fringe patterns in the neighbourhood of a crack-tip located in the tensile root fillet of a spur gear. The fringe patterns are quite complex and the various geometric features of the fringe field cannot be simulated by a simple three-parameter solution involving only K_I, K_{II}, and σ_{ox}. Figure 8.14 shows that a multi-parameter solution involving six mode-I parameters and six mode-II parameters is necessary to model the stress field. The values of the stress-field parameters are summarised in Table 8.2. Figure 8.15 shows the convergence error as a function of number of parameters. Figures 8.16a and 8.16b show the variation of K_I and K_{II} as a function of number of parameters.

In view of the proximity of the crack location to the contact region, the contact stress field due to the tooth pairs in contact also influences the stress field in the neighbourhood of the crack. It is encouraging to note that even for such complex problem situations, the multi-parameter solution has provided accurate results. The experimental evaluation is self-consistent as the experimental fringe patterns are faithfully re-checked by theoretical reconstruction.

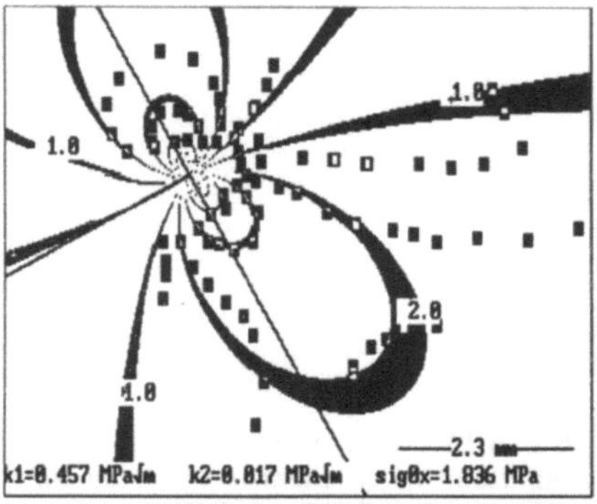

2 Mode I and 2 Mode II Parameters

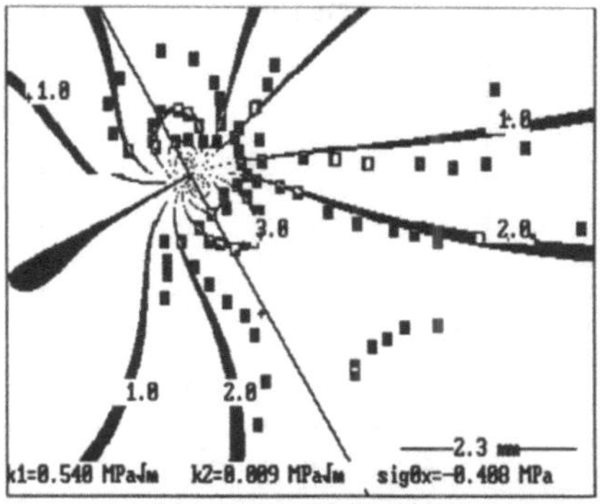

3 Mode I and 3 Mode II Parameters

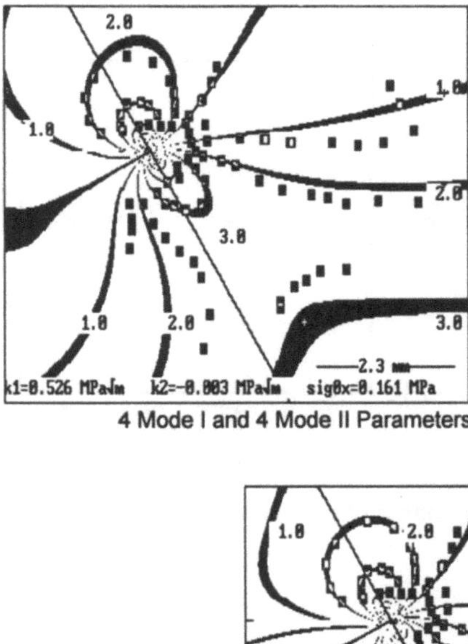

4 Mode I and 4 Mode II Parameters

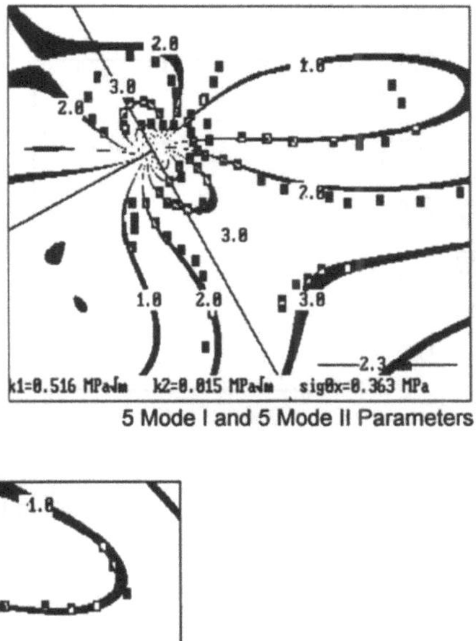

5 Mode I and 5 Mode II Parameters

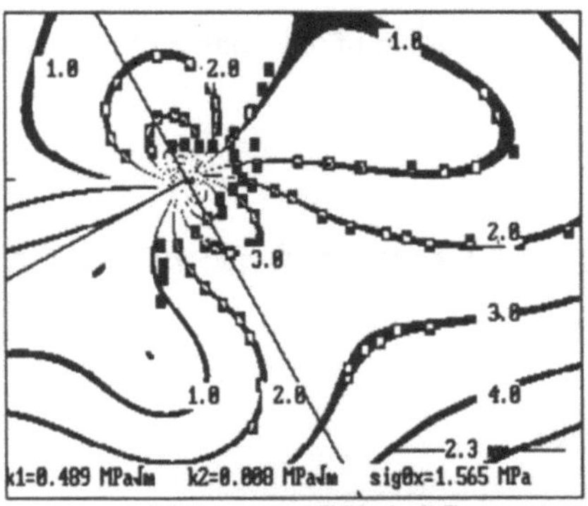

6 Mode I and 6 Mode II Parameters

Fig. 8.14. Theoretically reconstructed fringe patterns with data points echoed for various parameters for the fringe patterns shown in Fig. 8.13

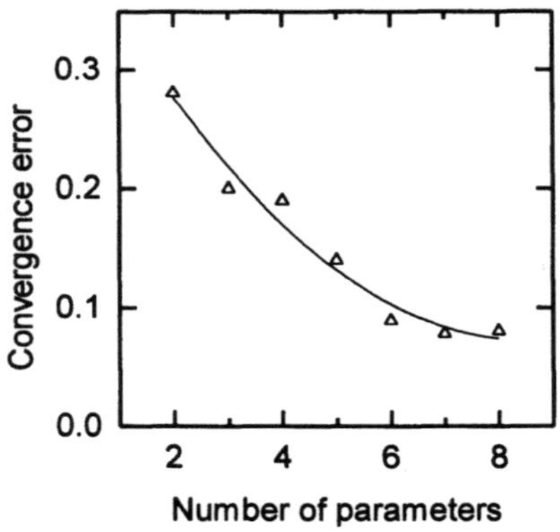

Fig. 8.15. Convergence error as a function of number of stress-field parameters

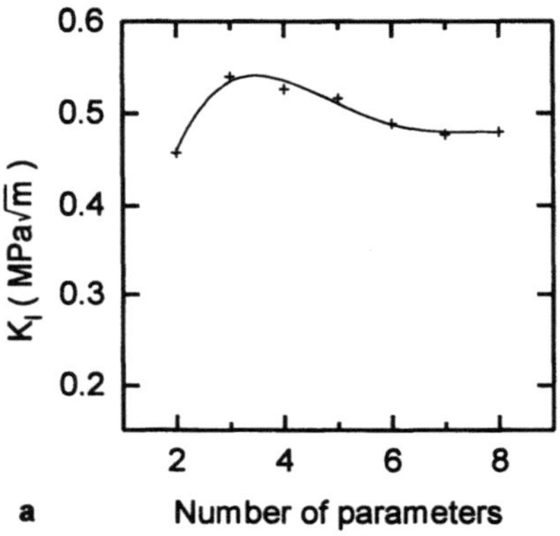

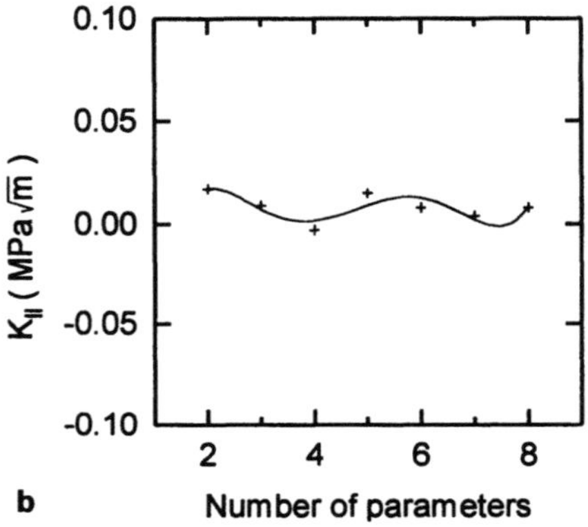

Fig. 8.16. Variation of SIF as a function of the number of parameters: **a** mode-I SIF **b** mode-II SIF

8.9 Contact Stress and Fracture Analysis of a Spur Gear

8.9.1 Loading Frame Design

A simple loading system to load spur gears, consisting of two shaft assemblies, which can be mounted to a standard loading frame of a photoelastic bench [34] is shown in Fig. 8.17. The upper shaft assembly is fixed to a crossbeam, which in turn is fixed to the stationary frame of the loading frame. The lower shaft assembly is fixed to the sliding frame of the loading frame. On the front end of each shaft, the gears are mounted which mesh-in

Table 8.2. Summary of stress field parameters for the problem of a crack emanating from the tensile root fillet of a spur gear

	2-parameter solution	4-parameter solution	6-parameter solution
K_I (MPa$\sqrt{m}$)	0.457	0.526	0.489
σ_{ox} (MPa)	1.836	0.161	1.565
K_{II} (MPa$\sqrt{m}$)	0.017	−0.003	0.008
A_{I1} (MPa(mm)$^{1/2}$)	5.7604	6.6411	6.1642
A_{I2} (MPa)	−0.4590	−0.0402	−0.3913
A_{I3} (MPa(mm)$^{-1/2}$)		−0.4978	−0.0515
A_{I4} (MPa(mm)$^{-1}$)		0.0002	0.0328
A_{I5} (MPa(mm)$^{-3/2}$)			−0.0687
A_{I6} (MPa(mm)$^{-2}$)			0.0113
A_{II1} (MPa(mm)$^{1/2}$)	0.2174	−0.0413	0.1029
A_{II2} (MPa)	0.0000	0.0000	0.0000
A_{II3} (MPa(mm)$^{-1/2}$)		0.0058	−0.0619
A_{II4} (MPa(mm)$^{-1}$)		−0.0717	0.0456
A_{II5} (MPa(mm)$^{-3/2}$)			−0.0219
A_{II6} (MPa(mm)$^{-2}$)			0.0026

Fig. 8.17. Spur gear loading arrangement

when both the shafts are brought in by vertical movement of the sliding frame. On the rear end of the upper shaft, a friction ring is mounted which is keyed to the shaft. When this friction ring is tightened against the upper shaft housing, it provides the resisting torque by locking the angular movement of the shafts. Between the upper gear and housing, a protractor scale is mounted on the shaft collar, which can be locked to the shaft by means of a locking screw. Different angular positions of the upper shaft and the lower shaft can be measured with respect to a reference position by means of this protractor scale.

The lower shaft is free to rotate in its housing. On the rear end of the lower shaft, a pulley is mounted which is rigidly keyed to the shaft. An end of high tensile strength steel wire cable is fixed to the pulley and the cable is wound on it. The free end of the cable passes over an idler pulley and a dead weight load hook is attached to it. The idler pulley bracket assembly is rigidly clamped to the loading frame table. Parallelism of 0.1 mm over the length of the shafts is obtained when the gears are loaded by careful sequencing of the manufacturing operations of the loading frame [34]. Unlike the simplified loading systems that have been usually designed to load a gear tooth assuming the hub being rigid, here the flexibility of the hub also influences the loading of the gear tooth.

For the example problems taken up for discussion, a spur gear of 14.5-deg pressure angle, diametral pitch 6 and having twenty four teeth cut from a sheet of 5.26 mm thickness made of CY230 resin and HY951 hardener mixed in the proportion 100:9 is used. The properties of the gear material are, E is 3.3 GPa, Poisson's ratio is 0.34 and the material stress fringe value is 10.95 N/mm/fringe.

8.9.2 Evaluation of Contact Parameters

During a mesh cycle, a pair of contacting gear teeth undergoes an inherent complex relative motion of rolling and sliding at the contact. Surface fatigue failures in gears are attributed to contact stresses and are classified as sub-surface origin fatigue and surface origin fatigue. Sub-surface origin fatigue in gears essentially occurs near the pitch line where the maximum shear stress is located beneath the surface. The surface origin fatigue failure occurs elsewhere on the surface where sliding occurs. The evaluation of contact stress parameters is thus important to understand the failure behaviour of gears. In order to understand the contact behaviour, the contact stress parameters need to be obtained for representative points along the tooth profile. This could be at pitch point contact, addendum and dedundum contacts. To use the methodology of Sect. 8.3, the radius of curvature at the point of contact needs to be evaluated first.

8.9.2.1 Measurement of Radius of Curvature at the Point of Contact

The radius of curvature at the point of contact varies as the contact moves along the involute surface. Also the curvature of the involute vary across the contact area. It is normal to ignore the later variation and take the contact of spur gear teeth as equivalent to the contact of parallel cylinders with the radii same as the radii of curvature of the mating spur gear teeth at the point of contact.

For the determination of radii of curvatures of the gears at the point of contact, the contact at the pitch point can be chosen as the reference. For getting above or below pitch point contact, one has to rotate the upper gear through a known angle. Then for the known angle of rotation, the radius of curvature of the gears can be found from standard formulae on spur gears.

8.9.2.2
Experimental Results

Figure 8.18(i) shows the experimental isochromatic fringe patterns for the pitch point, above pitch point and below pitch point contacts for an applied torque of 7.116 Nm. The contact region is taken with a magnification of 55 pixels per mm in the x direction. The aspect ratio of the camera-frame grabber system is 0.683. Using the principles discussed in Sect. 8.4, the contact stress parameters are evaluated. Theoretically reconstructed fringe patterns are shown in Fig. 8.18(ii). The figure shows a reasonable comparison between the experimentally recorded data points and the theoretically simulated fringe patterns. Table 8.3 summarises the results of the contact stress parameters. The table shows that the direction of friction is reversed for above and below pitch point contacts. Though for pitch point contact, one should anticipate a frictionless contact, some level of friction exists due to errors in involute profile generation.

The results show the applicability of the methodology discussed in Sect. 8.4 for solving practical problems. However, the modelling can be improved further by incorporating the role of photoelastic material non-linearity at higher stress levels. Further, in contact stress problems, identifying the origin for data reduction is usually subjective and to minimise such errors, apart from considering a and μ as unknowns, the origin also has to be evaluated as part of the solution procedure. Evaluation of contact stress parameters for multiple-contact problem is discussed in ref. [35].

Table 8.3. Summary of contact stress parameters for the problem of a spur gear for various points of contact

	Above pitch point	Pitch point	Below pitch point
Contact length (mm)	0.234	0.246	0.243
Friction coefficient	−0.110	0.134	0.112
Magnification (pixels/mm)	55	55	55
Radius of curvature R_1 (mm)	8.50	13.14	17.99
Radius of curvature R_2 (mm)	17.99	13.14	8.50
Convergence error	0.095	0.146	0.088

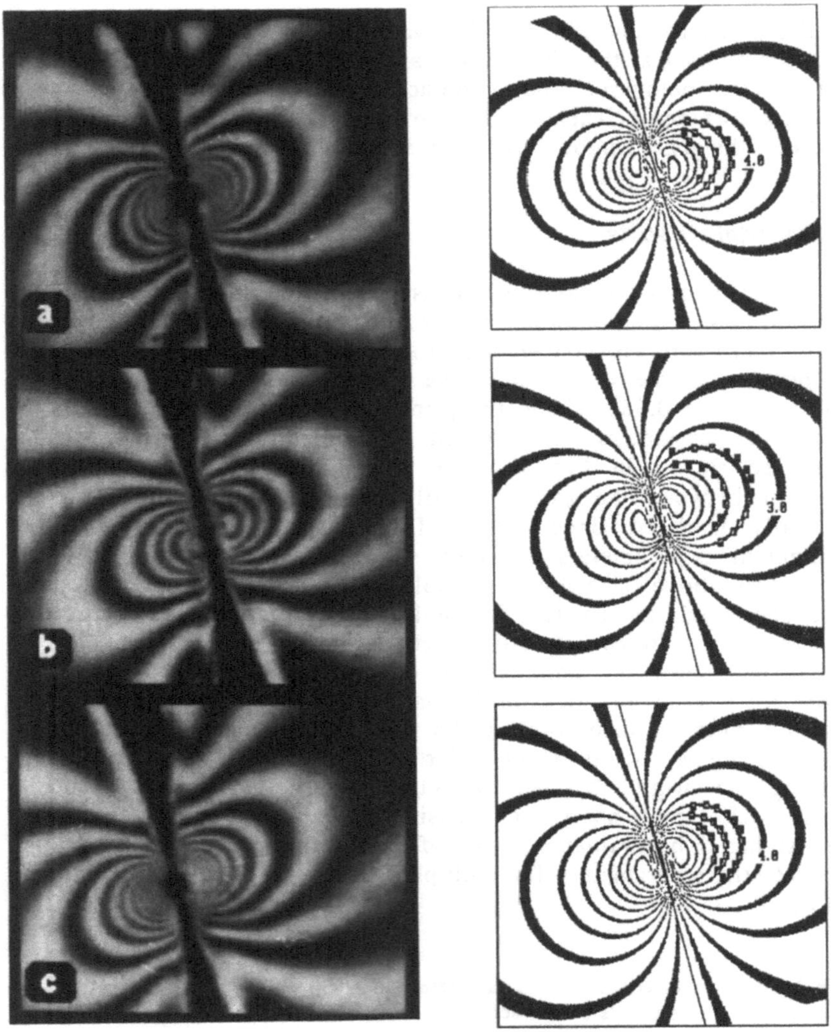

(i) Experimental (ii) Theoretically reconstructed

Fig. 8.18. Experimental and theoretically reconstructed fringe patterns of the contact region of the two spur gears in contact: **a** above pitch point **b** pitch point and **c** below pitch point

8.9.3
Evaluation of Fracture Parameters

During a mesh cycle, the root fillet stresses of a gear pulsate between a minimum and a maximum value. The tensile root fillet, which is on the loaded side of the gear tooth, is the most favourable location for the fatigue crack to initiate. The bending fatigue failure presents itself in the form of a single crack propagating through the base of the tooth. Fatigue crack propagation studies [36] have shown that crack propagation account for 30–40% of the total fatigue lives in carborised gears and 15–30% for shot-peened gears.

Evaluation of fracture parameters helps one to evaluate the allowable load and also to compute the residual life of a gear having a crack in the tensile root fillet [37]. A slit having a width of about 0.1 mm is cut, which simulates the crack at the tensile root fillet of the gear teeth. The cracked gear is meshed with another gear having the same specifications in the loading frame described earlier. Experiments are carried out for different points of contact along the involute profile.

Figure 8.19(i) shows the experimental fringe patterns of stress field near the crack-tip for a torque of 6.84 Nm for three different positions – above pitch point, pitch point and below pitch point contact respectively. The angular positions of above and below pitch point contact are 4-deg clockwise and anti-clockwise directions respectively from pitch point contact of the upper gear.

The zone of data collection has to be very carefully selected for data interpretation. It was found that using a higher magnification of the order of 55 pixels per mm, as was used in contact stress problems, was not suitable in the case of evaluating fracture parameters. The optical magnification has to be selected keeping in mind that one uses a multi-parameter solution for data interpretation. This necessitates that one does not go very close to the crack-tip or very far away from the crack-tip. For the problems discussed in this section, a magnification of 44 pixels per mm in the x-direction was found to be optimum. From the experimental images, 70–90 data points were collected to represent every geometric feature of the fringe patterns.

The least squares analysis is carried out as discussed in Sect. 8.7 by taking 35 data points randomly. Fringe order error minimisation criteria was used to stop the iteration. The number of parameters to model the stress field is increased step by step until the convergence error is less than 0.1. By using the experimentally determined values of the parameters, the isochromatic fringe pattern is reconstructed. To pictorially visualise the accuracy of the analysis, the data points are echoed back [Fig. 8.19(ii)]. The reconstructed fringes are matching very well with the experimental fringes. Table 8.4 summarises the results. It is interesting to note from Table 8.4 that the sign of K_{II} changes from above pitch point contact to below pitch point contact, which was numerically predicted by Honda and Conway [37]. They also observed that the crack would take a curved path. In the above example, the crack length was 2.95 mm which is quite long. In order to experimentally verify the observation of Honda and Conway, another example is taken with an initial crack of length 1.3 mm. The procedure is repeated for evaluating the stress field parameters. The results are summarised in Table 8.5. By knowing the stress field parameters, the direction of crack propagation can be found as described next.

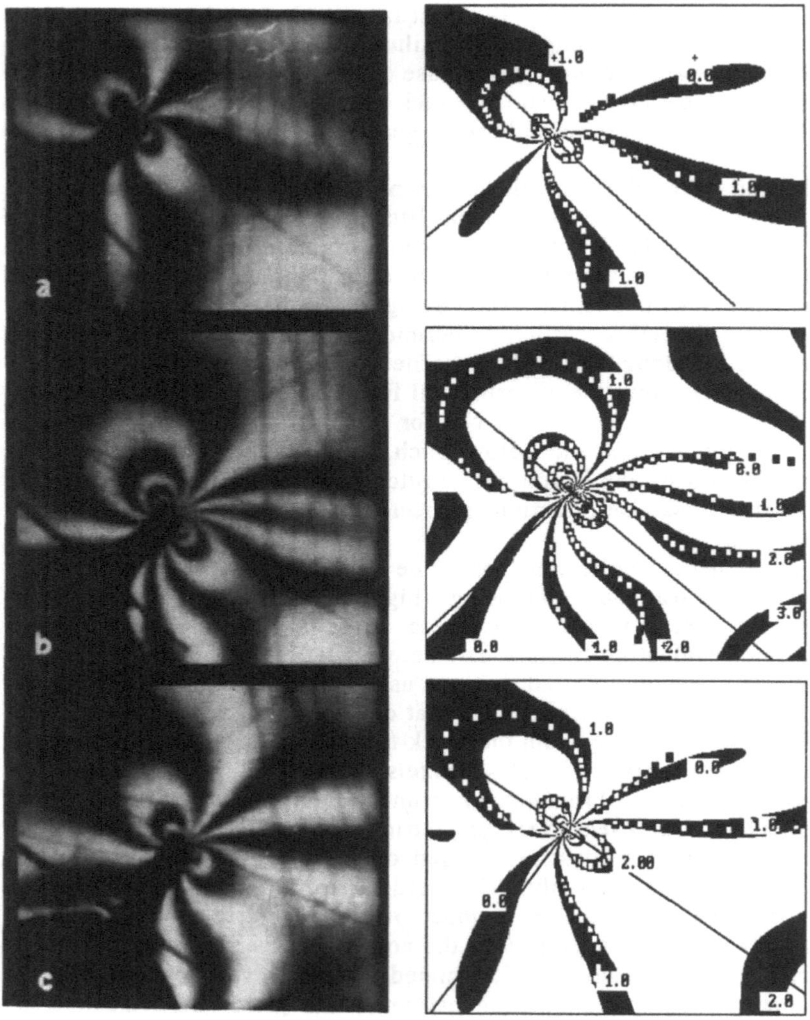

(i) Experimental (ii) Theoretically reconstructed

Fig. 8.19. Experimental and theoretically reconstructed fringe patterns in the neighborhood of a crack emanating from the tensile root fillet of a gear for various positions of contact: **a** above pitch point **b** pitch point and **c** below pitch point

In mixed mode experiments, it is usually observed that the crack extension is not self-similar but occurs at an angle with respect to the original crack. Two criteria exist that predict the direction of crack growth for mixed mode loading. They are

1. The maximum principle stress criterion proposed by Erdogan and Sih [38] and
2. Strain energy density criterion proposed by Sih [39].

The maximum principal stress theory predicts that the crack will grow in the direction perpendicular to the maximum principal stress. The strain energy density criterion states that crack growth takes place in the direction of minimum strain energy density. It has been established that when K_{I} is dominant, the predictions by both these criteria are very close [3] and it is easier to compute the angle of crack growth by using the maximum principal stress criteria. In maximum principal stress criteria, the angle of crack extension is obtained by

$$\left(\tan\frac{\theta_e}{2}\right)_{1,2} = \frac{1}{4}\frac{K_{\mathrm{I}}}{K_{\mathrm{II}}} \pm \frac{1}{4}\sqrt{\left(\frac{K_{\mathrm{I}}}{K_{\mathrm{II}}}\right)^2 + 8} \tag{8.41}$$

Williams and Ewing [40] have reported that improved correlation with test data is obtained when non-singular terms are included in Eq. (8.41).

Table 8.4. Summary of fracture parameters for the problem of a crack emanating from the tensile root fillet of a spur gear for various points of contact

	Above pitch point	Pitch point	Below pitch point
K_{I} (MPa$\sqrt{\mathrm{m}}$)	0.226	0.386	0.275
σ_{ox} (MPa)	–0.260	–0.152	–0.224
K_{II} (MPa$\sqrt{\mathrm{m}}$)	0.002	–0.039	–0.030
Number of mode-I Parameters	5	8	6
Number of mode-II parameters	5	8	6
Convergence error Achieved	0.09	0.09	0.085

The values of the direction of crack propagation obtained using Eq. (8.41) for above pitch point, pitch point and below pitch point contacts respectively are 16.62-, 17.37- and 17.76-deg with respect to the crack axis in the clock wise direction. The crack is artificially extended in the direction of 17-deg with respect to the crack axis in clock wise direction i.e., in the direction opposite to the tooth fillet. The same experiments are repeated for the extended crack. Least squares analysis is carried out and Table 8.5 summarises the results. The crack extension direction for the extended crack respectively for the above pitch point, pitch point and below pitch point contacts are respectively 1.3-, 0.73- and 173-deg [41].

The analysis shows that a crack in the tensile root fillet will follow a curved path before failure. It is interesting to note that for the same torque load (7.12 Nm), the extended crack showed less value of K_{I} and K_{II} but there is significant increase in σ_{ox}. The reduction in K_{I} and K_{II} may be due to artificial increase in the crack length, which was quite long – in the order of 2.5 mm, and it has dipped into the less stressed zone of the gear. Making very small increments to crack extensions could refine the analysis.

Table 8.5. Summary of fracture parameters for the problem of a crack emanating from the tensile root fillet of a spur gear for various points of contact

	Above pitch point		Pitch point		Below pitch point	
	Case 1	Case 2	Case 1	Case 2	Case 1	Case 2
K_{I} (MPa$\sqrt{\mathrm{m}}$)	0.367	0.265	0.418	0.314	0.347	0.295
σ_{ox} (MPa)	0.094	–0.945	0.098	–0.505	–0.070	–0.518
K_{II} (MPa$\sqrt{\mathrm{m}}$)	–0.056	–0.003	–0.067	–0.002	–0.057	0.009
Number of mode-I Parameters	8	8	8	8	7	8
Number of mode-II parameters	8	8	8	8	7	8
Convergence error Achieved	0.04	0.09	0.06	0.08	0.05	0.06

Case 1 – Initial crack with a length of 1.3 mm ; Case 2 – Extended crack

8.10 Closure

In this chapter, the use of digital photoelasticity for evaluating the stress field parameters in high stress gradient parameters has been discussed. The role of statistical methods to improve the analysis of experimental data to obtain meaningful results for finite body problems is illustrated. The use of theoretical reconstruction of fringe patterns with the data points echoed back provides a self-consistent approach for evaluating and presenting experimental results. In the case of evaluating fracture parameters, the theoretical reconstruction of fringe patterns show that the use of a multi-parameter solution is quite useful for solving practical problems involving finite geometries.

Exercises

8.1 For the image processing hardware in your laboratory or you have access to, determine the conversion to real world co-ordinates based on hardware information alone and by direct experimentation for various values of optical magnification. Compare your results. Give reasons for any discrepancies in the results.

8.2 Develop a software code to determine the contact stress parameters in a least squares sense. Incorporate both parameter error minimisation and fringe order error minimisation criteria to stop the iteration. Test your software by collecting 100 data points from Figs. 8.2 and 8.3.

8.3 In problem 8.2, also include the origin of the co-ordinate system to be determined in a least squares sense. Does this approach help to improve the accuracy of the analysis?

8.4 What is the importance of σ_{ox} stress term in fracture analysis? In which problem situations use of a two-parameter analysis is sufficient?

8.5 Develop a software code to determine the parameters K_I and σ_{ox} in a least squares sense.

8.6 The length of the crack shown in Fig 8.11 is 7.86 mm. The material stress fringe value of the model is 12.32 N/mm/fringe and the model thickness is 5.98 mm. Using Irwin's two-parameter method, determine K_I and σ_{ox} by collecting data from fringe orders 3, 4, 5 and 6. Plot the variation of K_I and σ_{ox} as a function of fringe order. Also tabulate the value of r_m/a for each case. Collect 100 data points over the field and determine these parameters in a least squares sense. Compare the results and discuss the importance of least squares analysis.

8.7 Is determination of crack-tip from the experimental results difficult? If so include the co-ordinates of the crack-tip as unknowns in Prob. 8.5 and modify your software to determine these in a least squares sense.

8.8 Develop a software code to determine K_I in a least squares sense by a multi-parameter solution. Why a multi-parameter solution was not thought of in numerical studies?

8.9 A three-dimensional photoelastic model of turbine blade whose thickness varies from point to point was stress frozen. A crack developed in the blade region. Considering the blade as very thin and neglecting the curvature effects, discuss what modifications are to be done in the formulation of the equations to determine the fracture parameters in a least squares sense.

References

1. Morton WB, Close LZ (1922) Notes on Hertz's theory of the contact of elastic bodies. Phil Mag Series 6, 43:320–329

2. Smith JO, Liu CK (1953) Stresses due to tangential and normal loads on an elastic solid with application to some contact stress problems. J Appl Mech 20:157–166
3. Broek D (1986) Elementary engineering fracture mechanics. Kluwer Academic Publishers, Dordrecht, The Netherlands
4. Broek D (1989) The practical use of fracture mechanics. Kluwer Academic Publishers, Dordrecht, The Netherlands
5. Murakami Y (1987) Stress intensity factors handbook Vol I, Vol II, Pergamon Press
6. Tada H, Paris P, Irwin G (1973) The stress analysis of cracks. Handbook. Del research corporation, Hellertown
7. Sirkis JS, Chen YM, Singh H, Cheng AY (1992) Computerised optical fringe pattern analysis in photomechanics: a review. Opt Engng 31(2):304–314
8. Shukla A, Nigam H (1985) A numerical-experimental analysis of the contact stress problem. J Strain Analysis for Engng Des 20(4):241–245
9. Pankhawalla VA, Ramesh K, Dixit PM (1998) Evaluation of contact stress parameters by digital photoelasticity in non-linear mechanics. Kalyanaraman V, Sriram P (eds) Allied publishers, New Delhi, pp 284–291
10. Westergaard HM (1939) Bearing pressure and cracks. J Appl Mech 6:A49–A53
11. Irwin GR (1957) Analysis of stress and strains near the end of a crack traversing a plate. J Appl Mech 24:361–364
12. Wells AA, Post D (1958) The dynamic stress distribution surrounding a running crack – a photoelastic analysis. Proc SESA 16(1):69–92
13. Irwin GR (1958) Discussion of paper by Wells A, Post D. The dynamic stress distribution surrounding a running crack – a photoelastic analysis. Proc SESA, 16(1):93–95
14. Sih GC (1966) On the Westergaard method of crack analysis. Int J Fracture Mech 2:628–631
15. Eftis J, Liebowitz H (1972) On the modified Westergaard equations for certain crack problems. Int J Fracture Mech 8:383–391
16. Eftis J, Subramonian N, Liebowitz H (1977) Crack border stress and displacement equations revisited. Engng Fracture Mech 9:189–210
17. Paris PC, Sih GC (1965) Stress analysis of cracks. ASTM STP 381:30–81
18. Chona R, Irwin GR, Sanford RJ (1983) Influence of specimen size and shape on the singularity dominated zone. ASTM 791:I3–I23
19. Ramesh K (1994) Recent advances on the application of digital computers in experimental evaluation of stress field parameters in fracture mechanics. J Aero Soc India 46(4):224–237
20. Ramesh K, Yadav AK, Pankhawalla VA (1995) Plotting of fringe contours from finite element results. Communications in Numer Meth in Engng 11:839–847
21. Sanford RJ (1979) A critical re-examination of Westergaard method for solving opening mode crack problems. Mech Res Communications 6(5):289–294
22. Ramesh K, Gupta S, Srivastava AK, Kelkar AA (1995) Use of higher order stress field equations for the evaluation of stress field parameters in fracture mechanics. ARDB-STR-TR-95-786-02
23. Williams ML (1957) On the stress distribution at the base of a stationary crack. J Appl Mech 24:109–114
24. Atluri SN, Kobayashi AS (1987) Mechanical response of materials. In: Kobayashi AS (ed) Handbook on experimental mechanics. Prentice Hall, New York pp 1–37
25. Ramesh K, Gupta S, Srivastava AK (1996) Equivalence of multi-parameter stress field equations in fracture mechanics. Int J Fracture Mech 79:R37–R41
26. Bradley WB, Kobayashi AS (1970) An investigation of propagating cracks by dynamic photoelastic analysis. Exp Mech 10(3):106–113

27. Etheridge JM, Dally JW (1977) A critical review of methods for determining stress intensity factors from isochromatic fringes. Exp Mech 17:248–254
28. Ramesh K (1991) Experimental determination of stress intensity factors in isotropic model materials by the technique of photoelasticity. ARDB-STR-TR-91-540–02
29. Sanford RJ, Dally JW (1972) A general method for determining mixed mode stress intensity factors. Engng Fracture Mech 4:357–366
30. Schroedl MA, Smith CW (1975) A study of near and far field effects in photoelastic stress intensity factors. Engng Fracture Mech 7:341–355
31. Rosakis AJ, Ravichandar K (1986) On crack tip stress state: an experimental evaluation of three-dimensional effects. Int J Solids and Structures 22(2):121–134
32. Ramesh K, Gupta S, Kelkar AA (1996) Evaluation of stress field parameters in fracture mechanics by photoelasticity – revisited. Engng Fracture Mech 56:25–45
33. Chauhan DPS (1990) An experimental study of static and dynamic behaviour of cyclically symmetric radial cracks. M.Tech Thesis, Dept of Mech Engg, IIT Kanpur
34. Ramesh K, Kelkar AA, Pankhawalla VA, Dixit PM (1996) Automated photoelastic evaluation of stress intensity factors for cracked gear teeth. In: Lakshmana Rao C (ed) Recent Advances in Solid Mechanics. Allied Publishers, New Delhi, pp 115–110
35. Dally JW, Chen YM (1991) A photoelastic study of friction at multipoint contacts. Exp Mech 31:144–149
36. Inoue K, Kato M, Yamanaka M (1989) Fatigue strength and crack growth of carburised and shot-peened gears. Proc Int Power Transmission and Gearing Conference, ASME:663–668
37. Honda H, Conway JC (1979) An analysis by finite element techniques of the effects of a crack in the gear tooth fillet and its applicability to evaluating strength of the flawed gears. Bulletin of JSME 22(174):1848–1855
38. Erdogan F, Sih GC (1972) On the crack extension in plates under plane loading and transverse shear. J Basic Engng 85:441–446
39. Sih GC (1974) Strain energy density factor applied to mixed mode problems. Int J Fracture Mech 10:305–322
40. Williams JG, Ewing PD (1972) Fracture under complex stress – the angled crack problem. Int J Fracture Mech 8:441–446
41. Viswanadham K (1998) Contact stress and fracture analysis of a spur gear. M. Tech thesis, IIT Kanpur

9 Stress Separation Techniques

9.1 Introduction

Photoelasticity directly provides the information of principal stress difference and the orientation of the principal stress direction at the point of interest. Using these, one can find the normal stress difference and in-plane shear stress by invoking equations in mechanics of solids or Mohr's circle.

Photoelasticity, as a tool for stress analysis, can be useful only when it assists the process of design. For comparative evaluation of various designs, even a qualitative study of isochromatic fringe patterns can provide sufficient information [1]. Shape optimisation can be effected by altering the boundary of the component such that the fringes on the boundary follow the boundary contour rather than cutting it. Such fillets are known as *streamline fillets* [1]. The evaluation of stress concentration factor (SCF) demands only the knowledge of the isochromatic fringe order at selected points in the field (Chap. 2). In the previous chapter, it has been shown how the contact stress parameters and fracture parameters can be determined based on the knowledge of isochromatic fringe patterns alone. Further, in two-dimensional problems, if the principal stresses have opposite signs, then, the isochromatic fringe contour is equivalent to the Tresca yield contour for zones where the yield condition is satisfied. Thus, for a wide variety of design problems, photoelasticity provides direct information.

In two-dimensional problems, if both the principal stresses have the same sign, then, one needs to know their individual magnitudes to apply the Tresca yield criteria [2]. This necessitates the use of auxiliary methods which are known as stress separation techniques. In a restricted sense, they can be used to find the individual principal stresses and in a complete sense, they can provide all the components of the stress tensor. These methods can be either point-by-point or whole field. In early studies for point-by-point stress separation, use of a lateral extensometer was used to provide the lateral strain at the point, which provided the value of sums of principal stresses. This in conjunction with the photoelastic data provided the individual stress components [3]. For whole field stress separation, use of interferometric techniques and electrical analogy methods [3–5] for obtaining sums of principal stresses have been proposed. All these methods require an additional experimental information to be obtained by techniques other than photoelasticity. The introduction of oblique incidence technique by Drucker [6, 7] provided additional data for stress separation using photoelasticity itself for a point-by-point stress separation approach.

With the advent of digital computers, the use of numerical techniques to provide additional data for stress separation became quite feasible [8]. The basic methodology involved the integration of equilibrium equations or compatibility equation or the use of both. In order to account for errors in photoelastic data, Berghaus [9] introduced the use of least squares for stress separation. Attempts have also been made to develop techniques for stress separation based on isochromatic data alone [10–12].

The use of numerical techniques in conjunction with photoelasticity came to be known as hybrid techniques. Various shades of hybrid techniques exist [13–18]. In some, experimental data is given more prominence and in some numerical data is given prominence and in some both are judiciously used.

In this chapter, the various techniques for stress separation are briefly summarised. Techniques that are amenable for digital photoelasticity are discussed in detail.

9.2 Oblique Incidence Method

9.2.1 Secondary Principal Stresses

In the previous chapters, photoelastic equations were developed for normal incidence. Drucker [6] introduced the technique of oblique incidence and in this, the incident ray is sent at an angle to the face of the model. To avoid refraction effects, the model is usually immersed in a fluid whose refractive index is same as that of the model (Fig. 9.1). The interpretation of optical patterns is achieved through the use of secondary stress components. It has been shown that, only those stress components that lie in a plane perpendicular to the incident beam contribute to photoelastic effect. These are termed as secondary stress components. For a plane stress problem, the secondary stress components, viz., σ_x^ξ, σ_y^ξ and τ_{xy}^ξ can be obtained by using stress transformation laws as

$$\begin{bmatrix} \sigma_x^\xi & \tau_{xy}^\xi & \tau_{xz}^\xi \\ \tau_{yx}^\xi & \sigma_y^\xi & \tau_{yz}^\xi \\ \tau_{zx}^\xi & \tau_{zy}^\xi & \sigma_z^\xi \end{bmatrix} = \begin{bmatrix} \cos\xi & 0 & -\sin\xi \\ 0 & 1 & 0 \\ \sin\xi & 0 & \cos\xi \end{bmatrix} \begin{bmatrix} \sigma_x & \tau_{xy} & 0 \\ \tau_{yx} & \sigma_y & 0 \\ 0 & 0 & 0 \end{bmatrix} \begin{bmatrix} \cos\xi & 0 & \sin\xi \\ 0 & 1 & 0 \\ -\sin\xi & 0 & \cos\xi \end{bmatrix} \tag{9.1}$$

where σ_x, σ_y and τ_{xy} are the stress components that constitute the stress tensor for a plane stress problem and ξ is the angle of oblique incidence in the xz-plane. On simplification, the above equation gives

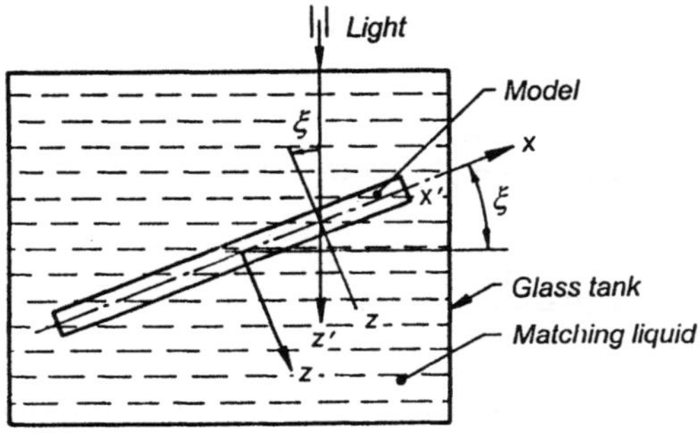

Fig. 9.1. Top view of the model immersed in a liquid bath containing the matching fluid having the same refractive index as the model

$$\begin{bmatrix} \sigma_x^\xi & \tau_{xy}^\xi & \tau_{xz}^\xi \\ \tau_{yx}^\xi & \sigma_y^\xi & \tau_{yz}^\xi \\ \tau_{zx}^\xi & \tau_{zy}^\xi & \sigma_z^\xi \end{bmatrix} = \begin{bmatrix} \sigma_x \cos^2\xi & \tau_{xy}\cos\xi & \frac{\sigma_x}{2}\sin 2\xi \\ \tau_{yx}\cos\xi & \sigma_y & \tau_{xy}\sin\xi \\ \frac{\sigma_x}{2}\sin 2\xi & \tau_{xy}\sin\xi & \sigma_x \sin^2\xi \end{bmatrix} \tag{9.2}$$

The pseudo principal stresses corresponding to the secondary stress components are called secondary principal stresses and are given as

$$\sigma_1^\xi, \sigma_2^\xi = \frac{\sigma_x \cos^2\xi + \sigma_y}{2} \pm \sqrt{\left\{\frac{\sigma_x \cos^2\xi - \sigma_y}{2}\right\}^2 + (\tau_{xy}\cos\xi)^2} \tag{9.3}$$

In oblique incidence, the secondary principal stresses and their directions vary along the light path (Fig. 1.20, Chap. 1). In plane stress problems, if the model is not too thick and if the stress gradient in the plane of the model are not too high, then, for oblique incidence angles less than 45°, the variation of σ_1^ξ, σ_2^ξ and their direction along the light path can be neglected. Noting that the length of the light path is $h/\cos\xi$, the 2-D stress optic law can be extended for oblique incidence and is given as

$$\sigma_1^\xi - \sigma_2^\xi = \frac{N^\xi F_\sigma \cos\xi}{h} \tag{9.4}$$

where F_σ is the material fringe value, h is the model thickness and N^ξ is the fringe order in oblique incidence for the oblique angle ξ.

9.2.2 The Methodology

At the point of interest, the isochromatic and isoclinic fringe orders for both the normal (N, θ) and oblique incidence (N^ξ, θ^ξ) are noted. From Mohr's circle, the shear stress and normal stress difference are related to photoelastic data as

$$\sigma_x - \sigma_y = (\sigma_1 - \sigma_2)\cos 2\theta = \frac{NF_\sigma}{h}\cos 2\theta \tag{9.5}$$

$$\tau_{xy} = \frac{(\sigma_1 - \sigma_2)}{2}\sin 2\theta = \frac{NF_\sigma}{2h}\sin 2\theta \tag{9.6}$$

On similar lines, from Eq. (9.4), the normal stress difference for oblique incidence in terms of σ_x and σ_y can be written as

$$\sigma_x^\xi - \sigma_y^\xi = \sigma_x \cos^2\xi - \sigma_y = \frac{N^\xi F_\sigma \cos\xi}{h}\cos 2\theta^\xi \tag{9.7}$$

Subtracting Eq. (9.7) from Eq. (9.5) one gets

$$\sigma_x = \frac{F_\sigma}{h\sin^2\xi}(N\cos 2\theta - N^\xi \cos 2\theta^\xi \cos\xi) \tag{9.8}$$

Based on oblique incidence, shear stress is

$$\tau_{xy} = \frac{N^\xi F_\sigma}{2h}\sin 2\theta^\xi \tag{9.9}$$

Though in principle, for the determination of in-plane shear stress either Eq. (9.6) or Eq. (9.9) could be used, one will usually find that the results are not always equal due to experimental difficulties. The use of Eq. (9.6) is recommended for general use (*obliq2D.c*, CD). The plane of oblique in

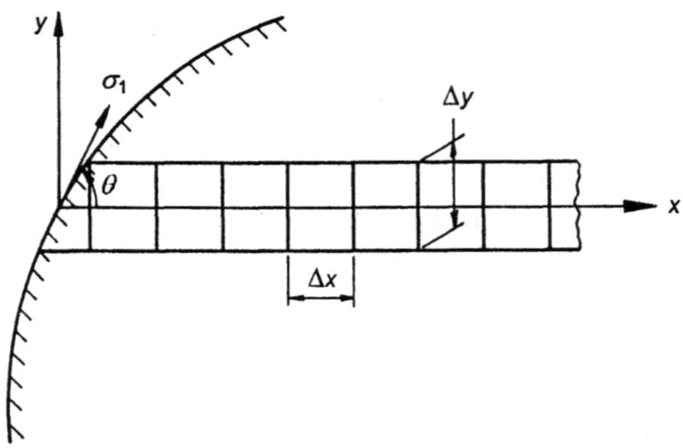

Fig. 9.2. Figure showing the grid needed for employing the shear difference method. The free boundary stress component is also shown

Fig. 9.1 is shown as the *xz*-plane. However, one can choose any plane that is convenient for a problem on hand.

In conventional photoelasticity, the angle of oblique is taken as 15- or 30-deg to provide sufficient change in the fringe data. Further, the technique is basically a point-by-point one and may necessarily involve a compensation technique for data reduction. The angle of oblique can be substantially reduced and the method can be made whole field if digital photoelasticity is used for data reduction, which is discussed in Sect. 9.6.

9.3 Shear Difference Technique

9.3.1 Conventional Method

In the absence of body forces, the equation of equilibrium when applied to the problem of plane stress are

$$\begin{aligned} \frac{\partial \sigma_x}{\partial x} + \frac{\partial \tau_{yx}}{\partial y} = 0 \\ \frac{\partial \sigma_y}{\partial y} + \frac{\partial \tau_{xy}}{\partial x} = 0 \end{aligned} \tag{9.10}$$

In the shear difference method, one of the equations in Eq. (9.10) is integrated to separate the stresses. Thus, one can do stress separation for lines parallel to *x*-axis or *y*-axis. If in a problem, the stresses have to be separated along lines parallel to *x*-axis (Fig. 9.2), the first of Eq. (9.10) is integrated to become

$$(\sigma_x)_j = (\sigma_x)_i - \int_i^j \frac{\partial \tau_{yx}}{\partial y} dx \tag{9.11}$$

Using the finite difference approximation, Eq. (9.11) can be rewritten as

$$(\sigma_x)_j = (\sigma_x)_i - \sum_i^j \frac{\Delta \tau_{yx}}{\Delta y} \Delta x \tag{9.12}$$

where, $(\sigma_x)_i$ is the stress value at the start of the integration procedure. If the starting point is chosen on a free boundary then $(\sigma_x)_i$ is

$$(\sigma_x)_i = \sigma_1 \cos^2\theta = \frac{NF_\sigma}{h} \cos^2\theta \tag{9.13}$$

Once σ_x is calculated for a particular point on the grid, σ_y can be calculated using Eq. (9.5). On similar lines, integration of the equilibrium equations can be done for the y-direction which is left as an exercise to the reader.

The technique has come to be known as shear-difference technique because of the manner in which shear slope is calculated in Eq. (9.12). For performing the integration, one requires the data of both isochromatic and isoclinic fringe orders along the line of interest and for two more lines one above and the other below the line of interest (Fig. 9.2).

One of the basic problems associated with the technique is that the numerical errors accumulate as one proceeds along the line of interest. Nevertheless, the technique has been quite widely used because of its simplicity (*shdiff.c*, CD).

9.3.2 Improvement by Tesar [19]

One of the main sources of error accumulation in the conventional shear difference is the error in calculating the shear slope. It is well known that numerical differentiation is usually less accurate than numerical integration. Further, the calculation of the shear stress involves the use of isoclinic parameter whose measurement accuracy is generally poorer than that of the isochromatic fringe order. Tesar suggested that the shear stress be expressed in terms of the difference of principal stresses as in Eq. (9.6) and be differentiated as a product of two functions. Substituting this into Eq. (9.11) gives

$$(\sigma_x)_j = (\sigma_x)_i - \int_i^j \frac{\partial(\sigma_1 - \sigma_2)}{\partial y}\frac{1}{2}\sin 2\theta\, dx - \int_i^j (\sigma_1 - \sigma_2)\cos 2\theta \frac{\partial \theta}{\partial y} dx \tag{9.14}$$

Equation (9.14) can be approximated using the finite difference expressions as

$$(\sigma_x)_j = (\sigma_x)_i - \sum_i^j \frac{\Delta(\sigma_1 - \sigma_2)}{\Delta y}\frac{1}{2}\sin 2\theta \Delta x - \sum_i^j (\sigma_1 - \sigma_2)\cos 2\theta \frac{\Delta\theta}{\Delta y}\Delta x \tag{9.15}$$

The rest of the procedure is same as that of the conventional method (*shdiff.c*, CD). Vadovic [20] compared the use of the conventional method with that of Tesar and concluded that the later method provides better accuracy. Allison and his co-workers [21, 22] have used Tesar's method to separate stresses for various problems. The usefulness of the method for stress separation in photoelastic coating technique is shown by Trebuna [23].

9.4 Survey of Numerical Methods

Referring to Eq. (1.66), Chap.1, when body forces are constant, the compatibility equation for both plane stress and plane strain cases become

$$\left(\frac{\partial^2}{\partial x^2}+\frac{\partial^2}{\partial y^2}\right)(\sigma_x+\sigma_y)=0 \tag{9.16}$$

In Eq. (9.16), $(\sigma_x + \sigma_y)$ is the first stress invariant for two-dimensional problems. Representing this by a single function ψ, Eq. (9.16) reduces to

$$\left(\frac{\partial^2}{\partial x^2}+\frac{\partial^2}{\partial y^2}\right)\psi=\nabla^2\psi=0 \tag{9.17}$$

The first stress invariant satisfies the Laplace equation. It is harmonic and the evaluation of ψ in the domain is based on the theory of potential functions. The values of ψ on the boundary of the model can be established from photoelastic fringe pattern. For these boundary values, their values inside the region are unique. By combining the solution of Eq. (9.17) with photoelastic data, stress components can be separated.

9.4.1 Integration of Compatibility Condition

For simple problem domains, the solution of Laplace's equation could be obtained by the method of separation of variables. Alternately, one can adopt the electrical analogy method, which is based on the fact that the voltage distribution in a uniformly conducting medium satisfies the Laplace equation. However, for problems of practical interest, generally numerical methods such as the finite difference or the finite element methods are preferred. In this section, the details of the finite difference method are presented.

Finite Difference Approach

In this method a network of squares is drawn over the field of the object. These networks are drawn such that they closely correspond to the shape of the boundary. Where the boundaries are not straight, the network in the object is subdivided so that the squares near the boundary closely represent the shape of the object's boundary (Fig. 9.3).

The solution domain in the xy-plane for a two-dimensional equilibrium problem is shown in Fig. 9.3a with a uniform spacing of Δx and Δy in the x and y directions respectively.

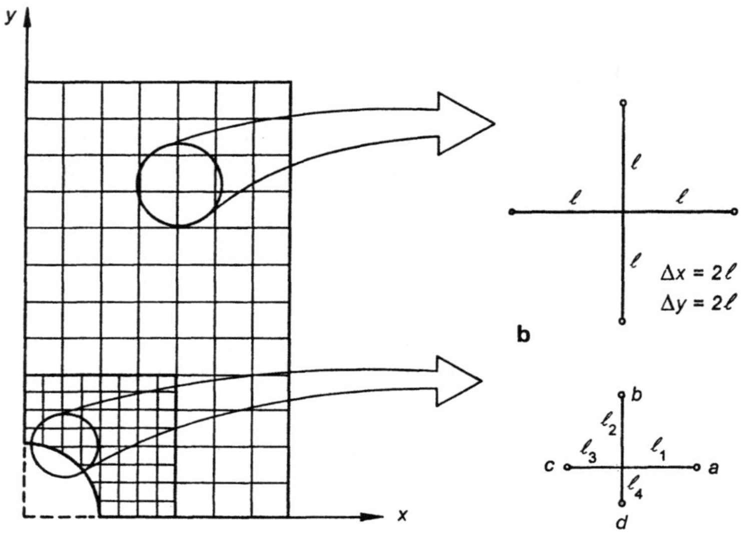

Fig. 9.3. Typical finite difference grid to model the problem domain. Note the change in the fineness of the grid in different zones

Let

$$\psi_{xx} = \frac{\partial^2 \psi}{\partial x^2}, \ \psi_{yy} = \frac{\partial^2 \psi}{\partial y^2} \tag{9.18}$$

Since a numerical approach is used to solve Eq. (9.18), certain approximation is involved in the analysis. In the subsequent discussions, let the exact solution be denoted by $\overline{\psi}(x, y)$ and the approximate solution be denoted by $\psi(x, y)$. Using the grid point (i, j) as the base point, and writing the Taylor's series for $(\overline{\psi})_{i+1,j}$ and $(\overline{\psi})_{i-1,j}$ truncated after the second derivative term, the partial derivative $\overline{\psi}_{xx}$ is obtained as

$$(\overline{\psi}_{xx})_{i,j} = \frac{\overline{\psi}_{i+1,j} - 2\overline{\psi}_{i,j} + \overline{\psi}_{i-1,j}}{\Delta x^2} + r(\Delta x^2) \tag{9.19}$$

where the term $r(\Delta x^2)$ represents the truncation error of the Taylor's series approximation of $(\psi)_{i,j}$. A finite difference approximation of the above equation can be obtained by neglecting the remainder term. Thus

$$(\psi_{xx})_{i,j} = \frac{\psi_{i+1,j} - 2\psi_{i,j} + \psi_{i-1,j}}{\Delta x^2} \tag{9.20}$$

Similarly in the y direction

$$(\psi_{yy})_{i,j} = \frac{\psi_{i,j+1} - 2\psi_{i,j} + \psi_{i,j-1}}{\Delta y^2} \tag{9.21}$$

Substituting Eqs. (9.20) and (9.21) in the Laplace equation gives

$$\psi_{i+1,j} + \alpha^2 \psi_{i,j+1} + \psi_{i-1,j} + \alpha^2 \psi_{i,j-1} - 2(1+\alpha^2)\psi_{i,j} = 0 \tag{9.22}$$

where α is the grid aspect ratio defined as

$$\alpha = \frac{\Delta x}{\Delta y} \tag{9.23}$$

Solving Eq. (9.22) for $\psi_{i,j}$ results

$$\psi_{i,j} = \frac{\psi_{i+1,j} + \alpha^2 \psi_{i,j+1} + \psi_{i-1,j} + \alpha^2 \psi_{i,j-1}}{2(1+\alpha^2)} \tag{9.24}$$

For a point near the boundary, which is not surrounded by equidistant-neighbouring points, a different formula is needed. Referring to Fig. 9.3c for four points located at distances ℓ_1, ℓ_2, ℓ_3 and ℓ_4 from the origin O, the following equation holds true

$$\begin{aligned}&\left(\frac{1}{\ell_1\ell_3} + \frac{1}{\ell_2\ell_4}\right)4\psi_o - \frac{1}{\ell_1(\ell_1+\ell_3)}\psi_a - \frac{1}{\ell_2(\ell_2+\ell_4)}\psi_b \\ &- \frac{1}{\ell_3(\ell_3+\ell_1)}\psi_c - \frac{1}{\ell_4(\ell_4+\ell_2)}\psi_d = 0\end{aligned} \tag{9.25}$$

where, ψ_a, ψ_b, ψ_c and ψ_d are the values of the harmonic function ψ at the four neighbouring points a, b, c and d respectively. For interior points with equally spaced neighbours, Eqs. (9.24) and Eq. (9.25) reduce to Liebmann's formula

$$\psi_{i,j} = \frac{\psi_{i+1,j} + \psi_{i,j+1} + \psi_{i-1,j} + \psi_{i,j-1}}{4} \tag{9.26}$$

The direct problem in the numerical solution of the Laplace equation consists of writing equations of the type [Eq. (9.24)] for each interior node. This yields a system of linear algebraic equations, which are same in number as that of the nodes in the grid. These systems of equations are to be solved based on the boundary data. On a load free boundary, the fringe order N is directly proportional to the sum of the principal stresses as one of the principal stresses is zero. Along the loaded edges

$$\psi = (\sigma_1 + \sigma_2) = (\sigma_1 - \sigma_2) + 2\sigma_2 \tag{9.27}$$

which can be written as

$$\psi_{\text{boundary}} = \frac{NF_\sigma}{h} + 2\sigma_2 \tag{9.28}$$

where σ_2 is given by the condition of the load. Given the boundary values, the values of ψ at the interior points are obtained by solving the system through iterative methods. For this, another set of squares is drawn (of the same dimensions) but displaced, so that their corners lie at the centre of the squares of the first net. The values of ψ are estimated using Eqs. (9.24), (9.25) or (9.26). Using these values of ψ, the values of ψ for the original network of grid are obtained. This is the first improvement. This procedure is carried on iteratively until the convergence is found to be good.

Mahfuz et al. [14] were the first to combine digital photoelasticity and finite difference approximation for stress separation. They however used a very simple hardware, which involved a tedious procedure for data collection in digital photoelasticity. The fringe patterns were projected onto a digitiser board and the data is collected manually using a cursor. The digitising of the boundary was manual and involves the physical traversing of the cursor over the boundary of the digitizer board. Similarly the isochromatics were digitised. Thus, photoelastic data were collected inside the domain by the scanning technique. Fringe value at each nodal point of the finite-difference grid is found by interpolation or extrapolation. Using central difference approximation as discussed earlier, Laplace equation is converted to a set of algebraic equations by the finite difference method. These equations were solved by successive over relaxation technique coupled with the Gauss-Seidel method. Successive over relaxation with optimum accelerating factor has been found to enhance the convergence of the solution [24].

9.4.2 Integration of Stress Difference Equations

For problems involving zero body forces, Segerlind [25] introduced the concept of *stress-difference elasticity*. It is the reformulation of the two-dimensional elasticity equations in terms of the obtainable photoelastic data and the first stress invariant. The advantage of the approach is that a new group of partial differential equations are developed which can be used to check the quality of photoelastic data within the domain.

Letting $\psi = \sigma_x + \sigma_y$ and $D = \sigma_x - \sigma_y$, the normal stress components are

$$\sigma_x = \frac{\psi + D}{2}$$
$$\sigma_y = \frac{\psi - D}{2} \tag{9.29}$$

Substitution of these in Eq. (9.10) gives

$$\frac{\partial \psi}{\partial x} + \frac{\partial D}{\partial x} + 2\frac{\partial \tau_{xy}}{\partial y} = 0 \tag{9.30}$$

$$\frac{\partial \psi}{\partial y} - \frac{\partial D}{\partial y} + 2\frac{\partial \tau_{xy}}{\partial x} = 0 \tag{9.31}$$

By differentiating Eq. (9.30) with respect to x and Eq. (9.31) with respect to y, two equations can be obtained by either adding or subtracting them. Addition gives

$$\frac{\partial^2 D}{\partial x^2} - \frac{\partial^2 D}{\partial y^2} + 4\frac{\partial^2 \tau_{xy}}{\partial x \partial y} = 0 \tag{9.32}$$

Subtraction gives

$$\nabla^2 D = \frac{\partial^2 \psi}{\partial y^2} - \frac{\partial^2 \psi}{\partial x^2} \tag{9.33}$$

By differentiating Eq. (9.30) with respect to y and Eq. (9.31) with respect to x, two more equations can be obtained by either adding or subtracting them. Addition gives

$$\nabla^2 \tau_{xy} = -\frac{\partial^2 \psi}{\partial x \partial y} \tag{9.34}$$

Subtraction gives

$$\frac{\partial^2 D}{\partial x \partial y} - \frac{\partial^2 \tau_{xy}}{\partial x^2} + \frac{\partial^2 \tau_{xy}}{\partial y^2} = 0 \tag{9.35}$$

Using strain-compatibility, Hooke's law and equilibrium equations, several other equations can be obtained which can be found in ref. [25].

Chandrasekhara and Jacob [15] proposed a technique to obtain the complete stress distribution inside a two-dimensional body with minimum experimental data based on stress-difference elasticity. Using the boundary stresses, Laplace equation is initially solved by finite difference technique. Once ψ is known, Eqs. (9.33) and (9.34) are solved to obtain the individual stress components. Mahfuz et al. [14] extended this approach with the use of digitised data. While solving Eq. (9.34), the digitised isoclinic values are used as initial guess to accelerate convergence.

9.4.3 Least Squares Method

The simple concept of the method can be explained by considering the system of equations

$$\begin{aligned} a_{11}x_1 + a_{12}x_2 + \cdots + a_{1n}x_n &= c_1 \\ a_{21}x_1 + a_{22}x_2 + \cdots + a_{2n}x_n &= c_2 \\ \vdots \quad \vdots \quad \vdots \quad \vdots \quad &\vdots \\ a_{m1}x_1 + a_{m2}x_2 + \cdots + a_{mn}x_n &= c_m \end{aligned} \tag{9.36}$$

Here $m \geq n$, a_{ij} are the known coefficients, x_j are the values of the independent variables to be found and c_i are the known values of the dependent variable. If the set of equation is non singular, a unique solution for x_j is obtained when $m = n$, if $m > n$ then an overdetermined solution for x_j is found. The residua is defined as

$$r_i = c_i - \sum_{j=1}^{n} a_{ij}x_j \tag{9.37}$$

Sum of the squares of all the residua is

$$e = \sum_{i=1}^{m} (r_i)^2 = \sum_{i=1}^{m} \left(c_i - \sum_{j=1}^{n} a_{ij}x_j \right)^2 \tag{9.38}$$

If the system of equations [Eqs. (9.36)] is based on experimental information, there is greater confidence in one part of the experiment than in another. These confidence levels may be determined in many problems intuitively or determined statistically. This aspect can be incorporated in calculating the cumulative error using appropriate weighting. The Eq. (9.38) is modified to include the weighting as

$$e = \sum_{i=1}^{m} \left[w_i \left(c_i - \sum_{j=1}^{n} a_{ij} x_j \right) \right]^2 \tag{9.39}$$

The weighting factors have to be chosen carefully. The role of a particular equation can be made to influence the final results by increasing its weighting. On the other hand, weighting of zero removes the equation. For a problem in which equal confidence exists in all equations, the order of magnitude of $w_i a_{ij}$ should be the same in the various equations. This means that w_i's are not unity always but it depends on what equations are used for the overdetermined solution. The solution of Eq. (9.39) is found by

$$\frac{\partial(e)}{\partial x_j} = 0 \quad \text{for} \quad 1 \le j \le n \tag{9.40}$$

From Eq. (9.40) one gets n equations in n unknowns, which can be solved by standard techniques such as Gaussian elimination, Gauss-Seidel etc.

Berghaus [9] used equations of stress equilibrium, boundary conditions and static equilibrium conditions to construct an overdetermined system of equations. The stress equilibrium equation is converted to algebraic equations by finite difference approximation. It is reported that this method gives greater accuracy as compared to the shear difference method. Later Berghaus [26] additionally introduced the compatibility condition in finite-difference form as part of the least squares procedure. The study by him showed that the inclusion of compatibility condition does not improve the accuracy of the results significantly. However, if the static equilibrium equations are not included, then significant error results.

In normal photoelastic analysis, one has to initially find the material stress fringe value of the model material. If not known earlier, the stress-optic relation also could be added to the set of equations and the material stress fringe value can be determined as part of the solution.

Thompson and Negus [12] described a non-linear least squares technique for complete separation of stresses in the stress concentration region near the notches using the asymptotic character of rapidly varying stress field near the root of the notch. The essential feature of the asymptotic analysis technique is the uncoupling by analytical means of the contributions of the notch geometry and of the nominal stresses to the rapidly varying stress field around the root of the notch. This uncoupling is symbolically represented by

$$\sigma_{ij} = F_{ij} \sigma_{ij}^{*} \tag{9.41}$$

where σ_{ij} is the stress component of interest, F_{ij} is an analytically derived rapidly varying function (depends only on the notch geometry) and σ_{ij}^{*} is the slowly varying function of the unknown magnitude (depends on free field stress state). If F_{ij} is known, the polynomial approximation of σ^*_{ij} can be obtained by fitting the right-hand side of the Eq. (9.41) to experimentally obtained values of the σ_{ij} from the limited number of data points within the high gradient zone.

9.4.4 Hybrid Techniques

Berghaus [16] combined the finite element method in displacement formulation with photoelasticity over regions partially bounded by free surfaces and axes of symmetry. Photoelastic data are straightforward to obtain along the free surfaces or lines of symmetry, which are the solution boundaries of the problem. An overdetermined system of equations in displacement is produced when data from these boundary locations are placed into the appropriate displacement equations. For finite element discretization, Berghaus used constant strain triangular elements. For nodes lying on the free boundary considering that the boundary stress is tensile, the photoelastic equation in displacement form is written as

$$(u_{2i} - u_{1i})\cos\phi + (v_{2i} - v_{1i})\sin\phi = \frac{s_i}{E}\left(\frac{N_i F_\sigma}{h}\right) \qquad (9.42)$$

where u and v are the displacements in x and y directions, $\cos\phi = (x_2 - x_1)/s_i$ and $\sin\phi = (y_2 - y_1)/s_i$ and s_i is the length of the free boundary element i. Subscripts '1' and '2' refers to nodes as shown in Fig. 9.4a. If the boundary stress is compressive, which is to be determined by auxiliary means, the R.H.S of Eq. (9.42) should be negative.

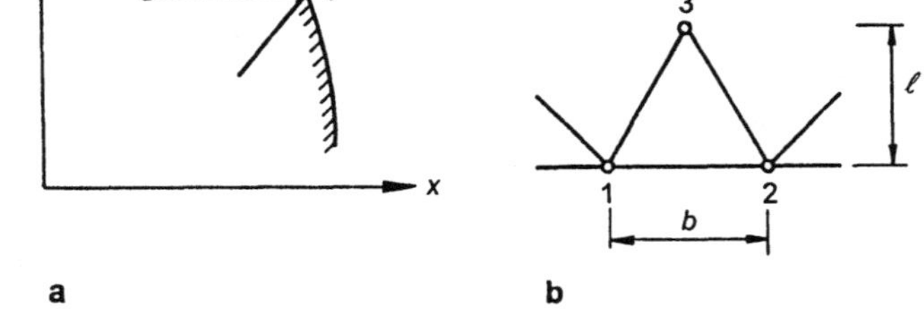

Fig. 9.4. Node numbering of CST elements: **a** curved boundary **b** line of symmetry

For nodes lying on the horizontal line of symmetry, the photoelastic equation in the displacement form is written as

$$(u_{2i} - u_{1i}) - \frac{b_i}{\ell_i} v_{3i} = \frac{b_i(1+\nu)}{E}\left(\frac{N_i F_\sigma}{h}\right) \tag{9.43}$$

Here an isosceles triangle with base b_i and height ℓ_i is taken for the element i. The subscript '1', '2' and '3' refers to the nodes as shown in Fig. 9.4b. Equation (9.43) is written on the premise that $\sigma_x > \sigma_y$. If this condition is not satisfied, the R.H.S of Eq. (9.43) is to be made negative.

In view of excess information on hand, it is not necessary to use applied nodal forces as in conventional FE analysis. However, the overall loading needs to be incorporated using static-equilibrium equations. Static-equilibrium equations relate the total loads applied in the horizontal and/or vertical directions to the loads produced by stresses along appropriate solution lines, in this case, lines of symmetry. One of these needs to be used for least squares analysis of the solution. For force equilibrium in the vertical direction, the static-equilibrium equation in terms of displacement is

$$\left(\frac{1-\nu^2}{Eh}\right)\sum F_y = \sum_{i=1}^{n_i}\left[\frac{v_{3i}}{\ell_i} + \nu\left(\frac{u_{2i} - u_{1i}}{b_i}\right)\right] \tag{9.44}$$

The overdetermined set of equations consists of Eqs. (9.42) and (9.43) along boundary nodes, Eq. (9.44) along the baseline and force-displacement equilibrium equations at all nodes, except those along the region boundary which is opposite the baseline. These equations are solved by the least squares technique.

9.4.5 Methods Using Only Isochromatic Data

Doyle [10] combined equilibrium equation, stress compatibility equation and stress optic relation to separate stresses along lines of symmetry. A non-linear equation of the following form is obtained.

$$\pm\frac{\partial^2\sigma_x}{\partial x^2} - \left(\frac{h}{NF_\sigma}\right)\left(\frac{\partial\sigma_x}{\partial x}\right)^2 = \left(\frac{F_\sigma}{4h}\right)\left[\frac{\partial^2 N}{\partial x^2} - \frac{\partial^2 N}{\partial y^2}\right] \tag{9.45}$$

This equation is written in finite difference form and integrated iteratively by numerical methods. Since the basic equation is non-linear, precautions are needed while integration. In this method N appears in the denominator which causes problem if zero fringe order is present. This can be overcome by selecting appropriate integration segments. In lines of symmetry, the in-plane shear stress is zero and hence after determining σ_x, the stress component σ_y, can be obtained using stress-optic law. Since isoclinics are not measured, it can be applied to transient events such as slowly varying thermal stress problems.

Shibuya et al. [11] proposed a technique to separate stresses along a line using only isochromatic values on the line of interest and two more lines one above and one below it. The technique basically uses the shear difference approach for stress separation. In shear difference technique, one needs to know the shear slope along the line of interest. In conventional analysis, this is directly determined by isochromatic and isoclinic data. To determine this only from isochromatic data, Shibuya et al. [11] used a very involved numerical procedure based on the equations of stress-difference elasticity. A numerical approach based on Eq. (9.35) is used to determine the shear stresses. Initially the value of the difference of the principal stresses is approximated as

$$D_i = (\sigma_x - \sigma_y) = \text{sgn} \times N_i \tag{9.46}$$

where sgn = (+1 or −1) is chosen such that the signs of D_i and N_i at the intersection point of the line of interest and the boundary are coincident with each other. Now considering the interval to be small and the distribution of the shear stresses along the y direction to be linear, the shear stress along the line is determined using shear difference Eq. (9.35) assuming $\partial^2 \tau_{xy} / \partial y^2 = 0$. The shear stresses above and below the line are evaluated by the shear difference equation [Eq. (9.32)]. Shear stresses above and below the line are adjusted such that the boundary conditions for the shear stresses are satisfied. The normal stress difference along the three lines is then determined using

$$D_i = \text{sgn} \times N_i \sqrt{1 - \left(\frac{2\tau_i}{\lambda N_i} \right)^2} \tag{9.47}$$

where $\lambda = \max(1, |2\tau/N|_{\max})$. The term under the square root in Eq. (9.47) gives the value of $\cos 2\theta$. Integrating the equilibrium equation as in the conventional shear difference technique separates the stresses. The method is repeated till the value of D_i has converged.

9.5 Stress Separation by Combined Phase Shifting and FEM

In Chapter 5 it is seen how phase-shifting technique can provide isochromatic data for the whole field. It has also been discussed that the evaluation of isochromatics is more accurate than the evaluation of isoclinic angle. If the basic goal is failure analysis, then it is enough one finds the individual principal stresses. By combining phase shifting and any of the numerical methods for integration of compatibility condition, it is possible to evaluate the individual principal stress components over the complete domain. Among the various techniques discussed in Sect. 9.4, analytical separation method is suitable for simple geometries and loading conditions. Finite difference method when applied to 2-D problem is usually iterative in nature. Further the restriction on the choice of the grid shape makes it difficult to model complicated boundaries. The least squares method requires large amount of data and is also iterative. The solution of Laplace equation by the finite element method overcomes the above difficulties.

9.5.1 Finite Element Formulation

In the finite element method, the domain of interest is initially discretised by geometrically simple shapes called elements. The element can have curved sides to model arbitrary geometry – a distinct advantage than that of the finite difference method. Over the years, a large number of elements have been developed to suit different applications and one has to just select an appropriate element or a combination of these to solve a problem of interest. Using variational or weighted residual methods, discrete equations for an element are formulated. Next, the equations over all elements of the collection are connected by the continuity of the *primary variable* or variables. Finally, the boundary conditions of the problem are imposed and then the connected equations are solved [27].

In weighted residual formulation (or weak formulation), the given differential equation is recast in an equivalent integral form by trading the differentiation between a test function and the dependent variable. The purpose of transferring the differentiation from the dependent variable onto the test function is to equalise the continuity requirements on the dependent variable and the test function. This results in a weaker continuity requirement on the solution of the dependent variable in the variational problem than in the original equation. In the process of transferring the differentiation, one obtains boundary terms, which determine the nature of the boundary conditions in the solution. Two sets of boundary conditions can be identified. One of which is called the essential boundary condition (EBC) and the other is the natural boundary condition (NBC). The boundary condition, which specifies the primary variables, is known as the essential boundary condition. The one that contains the secondary variables of the formulation is called the natural boundary condition.

Let us now formulate the variational form of the compatibility condition. Let Ω^e represents the domain of the eth element and w be the weight function assumed to be once differentiable with respect to x and y. Multiplying the Laplace equation by the weight function and integrating the resultant equation over the domain Ω^e one gets

$$\int_{\Omega^e} w\nabla^2\psi \, dxdy = 0 \tag{9.48}$$

Integrating Eq. (9.48) by parts to distribute the differentiation between ψ and w one gets

$$-\int_{\Omega^e} \nabla w.\nabla \psi \, dxdy + \oint_{\Gamma} \frac{\partial \psi}{\partial \boldsymbol{n}} w ds = 0 \tag{9.49}$$

where, ds is the arc length of an infinitesimal line element along the boundary, $\boldsymbol{n}$ is the unit vector normal to the surface Γ of the domain Ω^e and $\partial/\partial \boldsymbol{n}$ is the normal derivative operator defined as

$$\frac{\partial}{\partial \boldsymbol{n}} = \boldsymbol{n} \bullet \nabla = n_x \frac{\partial}{\partial x} + n_y \frac{\partial}{\partial y} \tag{9.50}$$

Substituting Eq. (9.50) into Eq. (9.49) and expanding Eq. (9.49) gives

$$\int_{\Omega^e} \left(\frac{\partial w}{\partial x} \frac{\partial \psi}{\partial x} + \frac{\partial w}{\partial y} \frac{\partial \psi}{\partial y} \right) dxdy - \oint_{\Gamma} w \left[n_x \frac{\partial \psi}{\partial x} + n_y \frac{\partial \psi}{\partial y} \right] ds = 0 \tag{9.51}$$

Equation (9.51) helps one to identify the natural and essential boundary conditions of the problem. Natural boundary conditions are those that specify the coefficient of the test function and its derivative in the boundary integral. Thus, the natural boundary condition is given by

$$n_x \frac{\partial \psi}{\partial x} + n_y \frac{\partial \psi}{\partial y} = q_n \quad \text{on } \Gamma \tag{9.52}$$

To specify the NBC, for x-axis symmetry, one has $n_x = 0$, $\partial\psi/\partial y = 0$ and hence $q_n = 0$. Similarly for y-axis symmetry, one has $n_y = 0$, $\partial\psi/\partial x = 0$ and hence $q_n = 0$.

The form of the test function in the boundary integral specifies the essential boundary condition. Since, only the test function not its derivatives appear in the boundary integral term of Eq. (9.51), specification of ψ over the boundary is the essential boundary condition (EBC). The EBC is specified depending on the stress applied on the boundary.

The first term of Eq. (9.51), requires that the approximation chosen for ψ should be at least linear in both x and y and satisfy the essential boundary condition. The Galerkin method seeks an approximate solution to Eq. (9.51) in the form of a finite series of ψ. Let

$$\psi(x,y) = \sum_{j=1}^{n} c_j \chi_j^e(x,y) \tag{9.53}$$

where the coefficients c_j are chosen such that $w = \chi_i$ (i = 1, 2, ..., n) and χ_j are the approximation functions. In this case, the Lagrange families of interpolation functions are admissible as approximation functions. Let ψ be approximated over a typical finite element Ω^e by the expression

$$\psi(x,y) \approx \psi^e(x,y) = \sum_{j=1}^{n} \psi_j^e \chi_j^e(x,y) \tag{9.54}$$

where ψ_j^e is the value of ψ^e at the jth node (x_j, y_j) of the element, n is the number of nodes per element, and χ_j^e are the approximation functions satisfying the property

$$\chi_i^e(x_j, y_j) = \delta_{ij} \tag{9.55}$$

where δ_{ij} is Kronecker delta. Substituting Eq. (9.54) into Eq. (9.51), one gets

$$\int_{\Omega^e}\left[\frac{\partial w}{\partial x}\left(\sum_{j=1}^{n}\psi_j^e\frac{\partial \chi_j}{\partial x}\right)+\frac{\partial w}{\partial y}\left(\sum_{j=1}^{n}\psi_j^e\frac{\partial \chi_j}{\partial y}\right)\right]dxdy - \oint_{\Gamma} w q_n ds = 0 \tag{9.56}$$

Replacing $w = \chi_i$, i = 1, ..., n, the ith algebraic equation can be written as

$$\sum_{j=1}^{n}\left\{\int_{\Omega^e}\left[\frac{\partial \chi_i}{\partial x}\frac{\partial \chi_j}{\partial x}+\frac{\partial \chi_i}{\partial y}\frac{\partial \chi_j}{\partial y}\right]dxdy\right\}\psi_j^e - \oint_{\Gamma}\chi_i^e q_n ds = 0 \tag{9.57}$$

or

$$\sum_{j=1}^{n} K_{ij}^e \psi_j^e = Q_i^e \tag{9.58}$$

where

$$K_{ij}^e = \int_{\Omega^e}\left[\frac{\partial \chi_i}{\partial x}\frac{\partial \chi_j}{\partial x}+\frac{\partial \chi_i}{\partial y}\frac{\partial \chi_j}{\partial y}\right]dxdy \tag{9.59}$$

and

$$Q_i^e = \oint_{\Gamma}\chi_i^e q_n ds \tag{9.60}$$

Equation (9.58) can be expressed in matrix form as

$$[K^e]\{\psi^e\} = \{Q^e\} \tag{9.61}$$

The matrix $[K^e]$ in Eq. (9.61) is not a stiffness matrix in the conventional sense and hence can be termed as a pseudo-stiffness matrix. This matrix is to be evaluated numerically using Gauss quadrature with appropriate integration schemes.

It should be noted that the weighted residual form as well as the algebraic form of the finite element model of a given equation does not depend on the type (linear, quadratic, etc.) of the element. Therefore, Eqs. (9.51) and (9.57) are valid for any type of elements. Use of a quadratic element is recommended for stress separation in photoelasticity as curved boundaries can easily be modelled. A quadratic element of the Lagrange family requires an internal node and this does not contribute to the inter-element connectivity. It is desirable to use serendipity elements to avoid the internal nodes present in the Lagrange elements.

The interpolation functions for an eight noded isoparametric element (quadratic serendipity element) are well documented and can be expressed in terms of natural co-ordinates (r, s) (Fig. 9.5) as

$$\begin{aligned} &\text{for corner nodes } (N_1 \text{ to } N_4) && \frac{1}{4}(1+rr_i)(1+ss_i)(rr_i+ss_i-1) \\ &\text{for side nodes } (N_5, N_7), r_i = 0 && \frac{1}{2}(1-r^2)(1+ss_i) \\ &\text{for side nodes } (N_6, N_8), s_i = 0 && \frac{1}{4}(1+rr_i)(1-s^2) \end{aligned} \tag{9.62}$$

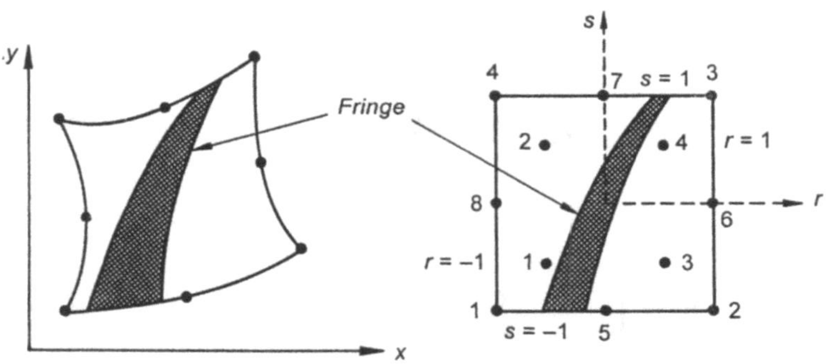

a Global b Local

Fig. 9.5. Eight noded isoparametric element in global and local co-ordinates

9.5.2
Meaningful Discretization of the Domain

Finite element (FE) modelling involves the selection of an appropriate element or a combination of different elements for the problem under consideration. Further, a *meaningful discretization* is very important in evaluating the intended parameters accurately. Ideally, one would like to use a large number of elements to discretise the domain but due to constraints on computer resources, one is normally expected to optimise the discretization. Elements of smaller size may be used in zones of stress concentration, and elsewhere the discretization could be coarser.

In the literature one finds the use of only constant strain triangle (CST) elements [16, 29] in the context of solving Laplace equation for stress separation in photoelasticity. In order to simulate the steep gradient in the stress concentration zones, Seguchi et al. [29] suggested a discretization scheme using CST elements wherein the nodal co-ordinates are taken on fringe contours. This makes the determination of discretization topology complicated. Further the node numbering is difficult.

The developments in automatic mesh generation have helped in solving problems with complicated geometry more easily using FEM. Several software packages are available for automatic discretization of domains using triangular elements. This may be the reason why in initial studies of applying FE for photoelastic analysis, triangular elements were used for domain discretization. Domain discretization using quadrilateral elements can help one to use the most versatile eight noded isoparametric element. In comparison to the CST elements, the curved boundaries and zones of stress concentrations could be modelled with fewer quadratic elements.

The initial developments on automatic mesh generation were concentrated on just to fill the domain with elements. The later developments also paid attention on modifying the discretization scheme based on the accuracy of the final results. The developments in adaptive [30] mesh refinement are a case in point, which emphasised the need for a meaningful discretization of the domain. Another approach to arriving at a *meaningful discretization* is through the use of photoelasticity. The basic guideline is that the discretization scheme for a given problem is meaningful if the fringe patterns observed in a photoelastic experiment are simulated faithfully by finite element modelling. A systematic study has been carried out by Ramesh and Pathak [31, 32] and based on their study the guidelines for discretising the domain using quadrilateral elements for a class of stress concentration problems are shown in Fig. 9.6. Figure 9.6a gives the discretization scheme for a curved boundary. Figures 9.6b and 9.6c show the possible discretization schemes for modelling a concentrated load.

In general engineering problems, no special effort is normally made to model the loading by concentrated forces. However, significant improvement in the results could be achieved by extending the use of theory of elasticity concepts. A concentrated load P acting normally on the edge of a

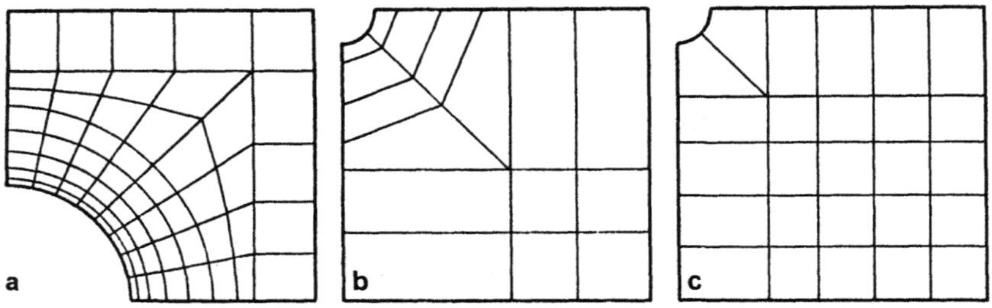

Fig. 9.6. Discretization schemes for various problems. **a** Modelling a curved boundary. Modelling concentrated load: **b** fine mesh **c** coarse mesh

semi-infinite plane (Fig. 9.7a) can be replaced by a radial system of stresses.

$$\sigma_r = -\frac{2P\cos\theta}{\pi r h} \tag{9.63}$$

where r, θ are the polar co-ordinates and h is the thickness of the plate. The resultant of the vertical components of the radial stresses replacing the concentrated load is P and the resultant of the horizontal components acting on each quadrant of the semicircle is P/π (Fig. 9.7b). When these forces are assumed to act on a semi-circle of very small radius, the vertical stress will approach infinity and the horizontal stress will approach zero. Figures 9.7c and 9.7d show the representation of concentrated force as a distributed force over a radius of r and Table 9.1 gives the respective nodal forces for the case of the eight noded isoparametric element.

Table 9.1. Nodal forces in horizontal and vertical direction when concentrated load is replaced by radial load

	1	2	3	4	5
Vertical (P_i)	$0.0590P$	$0.2357P$	$0.0834P$	$0.0976P$	$0.0244P$
Horizontal (Q_i)	$0.0155P$	$0.0622P$	$0.0531P$	$0.1501P$	$0.0375P$

9.5.3 Plotting of Fringe Contours from FE Results

A whole field comparison of numerical results with the experiments is possible by plotting fringe contours from numerical results. The post processing unit of many standard FEM packages only provide facilities for plotting line contours or with arbitrary colours varying over the domain. In contrast to line contours, the fringe contours appear as broad bands.

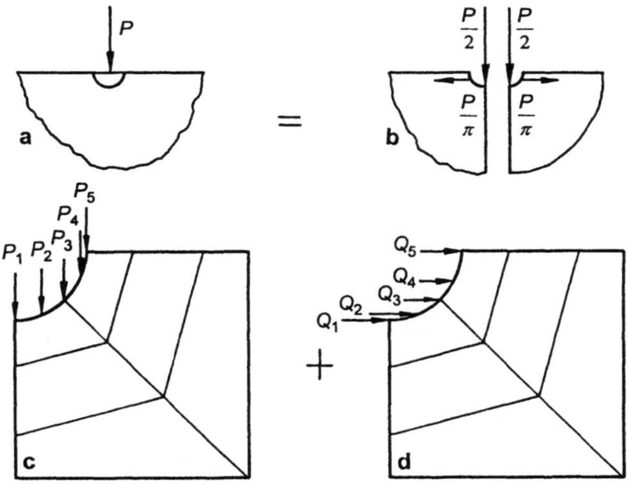

Fig. 9.7. Scheme to replace a concentrated load as distributed load

The thickness of the fringe is indicative of the gradient of the variable. The fringes are very broad when the gradient is small, and vice versa. Plotting of fringe contours using a simple non-iterative approach for various experimental techniques such as photoelasticity, Moiré and holography has been reported by Ramesh et al. [33].

In the context of photoelasticity, one is interested in comparing the stress contours. Finite element calculations directly provide the displacement information, which are continuous across element borders. Unlike displacements, the strains and stresses are related to the first derivatives of displacements with respect to space variables. In order to plot stresses, some precautions are needed to avoid jumps or breaks. The procedure for evaluating the stresses using 8-noded isoparametric quadrilateral element is discussed next.

In every element, the stresses at Gauss points (Fig. 9.5) are calculated using

$$\{\sigma\} = [D]\{\varepsilon\} = [D][B]\{d\} \tag{9.64}$$

where, $[D]$ is the elasticity matrix and $[B]$ is the matrix of derivatives of shape functions. The stresses at the four corner nodes can be calculated using the following equation

$$\{\sigma\}_{\text{nodes}} = [T]\{\sigma\}_{\text{Gauss pts}} \tag{9.65}$$

Stresses at the mid-side nodes can be calculated by averaging the stress values at the two adjacent corner nodes. For an eight-noded quadrilateral, with the Gauss points shown in Fig. 9.5, the transformation matrix $[T]$ using the shape functions takes the following form:

$$[T]=\begin{bmatrix} c_{11} & c_{12} & c_{13} & c_{14} \\ c_{21} & c_{22} & c_{23} & c_{24} \\ c_{31} & c_{32} & c_{33} & c_{34} \\ c_{41} & c_{42} & c_{43} & c_{44} \end{bmatrix}^{-1} \tag{9.66}$$

where

$$\left.\begin{aligned} c_{i1} &= N_{1i} + N_{8i}/2.0 + N_{5i}/2.0 \\ c_{i2} &= N_{4i} + N_{7i}/2.0 + N_{8i}/2.0 \\ c_{i3} &= N_{2i} + N_{6i}/2.0 + N_{5i}/2.0 \\ c_{i4} &= N_{3i} + N_{7i}/2.0 + N_{6i}/2.0 \end{aligned}\right\} \; i = 1,2,3,4 \tag{9.67}$$

N_{ji} is the shape function for node j calculated for the co-ordinates of Gauss point i.

If a node is common to n elements, the stress value σ_i assigned to the node is calculated as

$$\sigma_i = \frac{1}{n}\sum_{j=1}^{n} \sigma_i^j \tag{9.68}$$

For plotting contours corresponding to $(\sigma_1 - \sigma_2)$ or $(\sigma_1 + \sigma_2)$, one has to find σ_x, σ_y and τ_{xy} at the node using the above procedure and use them to calculate the sum or difference of principal stresses. Finally the fringe order at the node is calculated. For plotting $(\sigma_1 - \sigma_2)$ contours, use Eq. (1.15) to find the fringe order, and for plotting isopachics use the following equation.

$$\frac{\nu t}{E}(\sigma_1 + \sigma_2)\cos\alpha = \pm N_p \quad \text{where } N_p = \frac{1}{2}, \frac{3}{2}, \frac{5}{2}, \cdots \tag{9.69}$$

where ν is Poisson's ratio, h is specimen thickness, E is modulus of elasticity and α is the angle of incidence of light. N_p is the order of dark fringes.

Typically, the values of fringe orders for all nodes are calculated and stored in a file for plotting contours. On similar lines one can also plot pseudo fringe contours of σ_1 and σ_2.

Fringe plotting is done by the scanning scheme mentioned in Sect. 1.17.4. A software code named *femfrn.c* is included in the CD. Each element in the domain is scanned in pixel increments in its local co-ordinates. For each of the points scanned, the global co-ordinates and the fringe order (variable *frn* in the program) at the point are calculated using the shape functions as

$$\begin{aligned} x_g &= N_1 x_1 + N_2 x_2 + \cdots + N_8 x_8 \\ y_g &= N_1 y_1 + N_2 y_2 + \cdots + N_8 y_8 \\ frn_g &= N_1 frn_1 + N_2 frn_2 + \cdots + N_8 frn_8 \end{aligned} \tag{9.70}$$

where x_i, y_i are the co-ordinates of the nodes of the element and x_g, y_g are the global co-ordinates of the point, N_i are the shape functions, frn_i are the nodal values of fringe orders. For plotting dark-field fringes, the corresponding point is plotted, when the fringe order satisfies the following condition

$$\text{abs}\,[(frn_g) - \text{round}(frn_g)] \le 0.1 \tag{9.71}$$

where 0.1 is the fringe thickness tolerance. For plotting bright-field fringes, the following condition is to be used.

$$\text{abs}\,[(frn_g + 0.5) - \text{round}(frn_g + 0.5)] \le 0.1 \tag{9.72}$$

9.5.4 Influence of Error in Fringe Data

The epoxies used for making the photoelastic model are sensitive to moisture in the environment and develop spurious fringes over time. This phenomenon is termed as *time-edge* effect. This affects only the boundaries. Thus the solution procedure discussed in Sect. 9.5.2 needs to be evaluated for its suitability to the error in boundary information. Before one does that, it is worthwhile to see the performance of the procedure for ideal data for problems with known analytical solution.

Consider the problem of an infinite plate with a hole subjected to uniaxial tension. The infinite plate is modelled by considering a very small hole of diameter 1.9 mm in a wide strip of 37.46 mm. The thickness of the plate is 6 mm and its length is 153 mm. For a load of 1800 N applied at its ends the contours of sums of principal stresses obtained by FE and the analytical solution for the complete domain are shown in Figs. 9.8a and 9.8b. To obtain the results of Fig. 9.8, the following scheme is adopted. Due to symmetry only one quarter of the plate is modelled. Using the concept of *meaningful discretization*, the stress concentration zone is discretised using radial lines and arcs, which increases in a geometric progression. The remaining regions are filled with rectangular elements. In all, 144 elements with 489 nodes are used. The boundary conditions for solving the Laplace equation using FE is obtained not from photoelastic experiment but from solving the problem domain using the conventional FE analysis. The stress components are then separated by combining the solution from Laplace equation and $(\sigma_1 - \sigma_2)$ values from 2-D FE analysis.

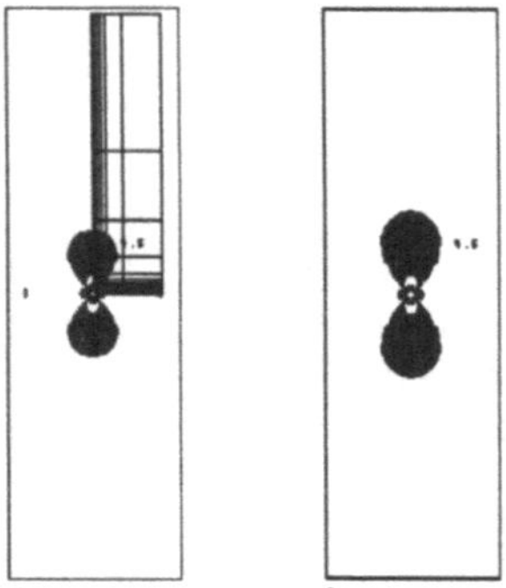

a b

Fig. 9.8. Fringe contours of $(\sigma_1+\sigma_2)$ for the complete domain for the case of a hole (dia = 1.9 mm) in an infinite plate (thickness 6.0 mm, F_σ = 11 N/mm/fringe) under uniaxial tension of 8 MPa: **a** FE (No of elements 144) **b** theoretical

Figure 9.9a shows the σ_1 contours and Fig. 9.9b shows the σ_2 contours obtained analytically. The Figs. 9.9c and 9.9d show the respective contours obtained by combining the solution of Laplace equation by FEM and ideal photoelastic data. The results by FEM and the analytical solution match very well and indicate that the discretization scheme adopted is good.

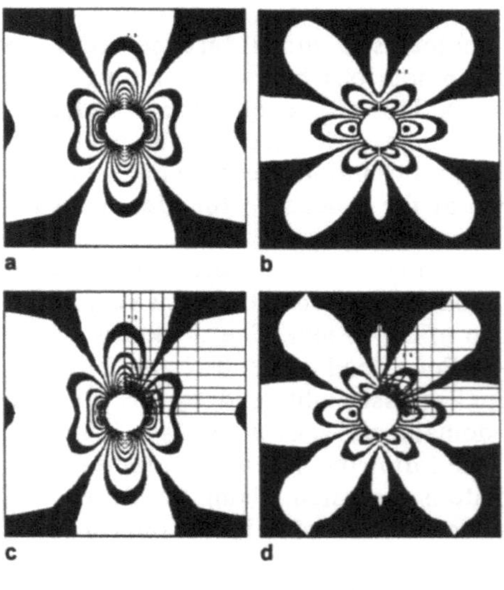

c d

Fig. 9.9. Separated stress contours near the SC zone: **a** σ_1 theoretical **b** σ_2 theoretical **c** σ_1 by summation of the photoelastic data (obtained from 2D FEM) and the solution of Laplace equation **d** σ_2 obtained as in **c**

To study the influence of the error in boundary information on the accuracy of evaluation of sums of principal stresses and subsequently its effect on the separated stress components, an error analysis is carried out numerically [34]. The nodal values of ψ corresponding to different boundaries are introduced with an error of ±5%, ±10%, alternates values of ±10% & ±5%, and a random variation between +15% to −15% at alternate boundary nodes.

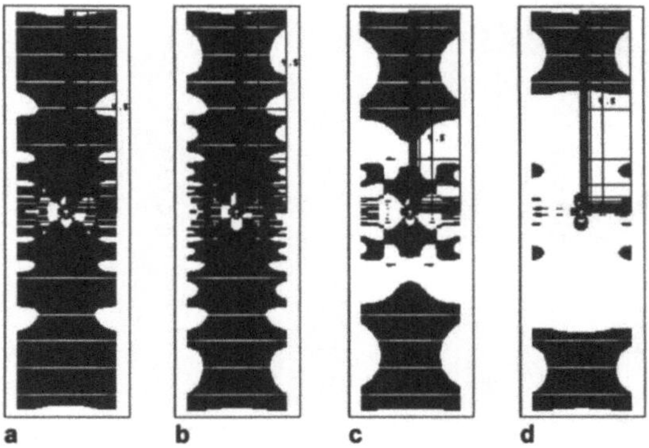

Fig. 9.10. Fringe contours of $(\sigma_1+\sigma_2)$ for the complete domain with different percentage of errors at alternate nodes on the boundary values **a** ±5% **b** ±10% **c** +10, +5, −5 and −10% **d** random error between +15% and −15%

The overall isopachics for the four cases are shown in Fig. 9.10. The isopachics near the hole are shown in Fig. 9.11. It is observed from the figures that the change in the boundary values has noticeable effect on the fringe contours. When random error is introduced on the boundary values, the fringes very near to the hole are not affected. Figure 9.12 show the σ_1 contours for the four error schemes. Figure 9.13 show the σ_2 contours for the four error schemes. From the results, one can observe that for a random error distribution in boundary values, the final results are not greatly influenced. However, a systematic error on the boundary values has a stronger influence in the final results. Hence, in experimentation every effort must be made to avoid the introduction of systematic errors.

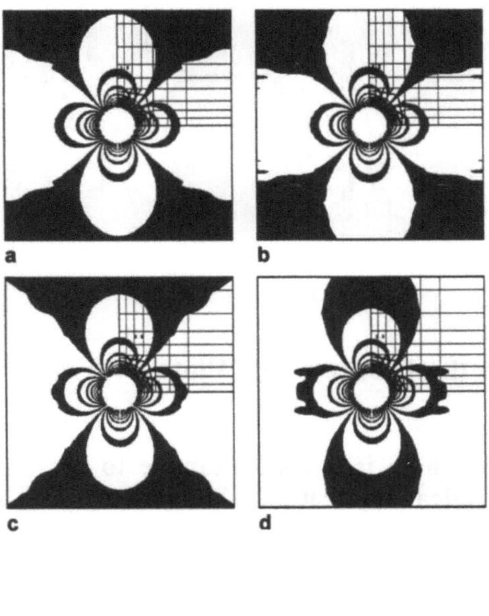

Fig. 9.11. Fringe contours of $(\sigma_1+\sigma_2)$ for the SC zone with different percentage of errors at alternate nodes on the boundary values **a** ±5% **b** ±10% **c** +10, +5, −5 and −10% **d** random error between +15% and −15%

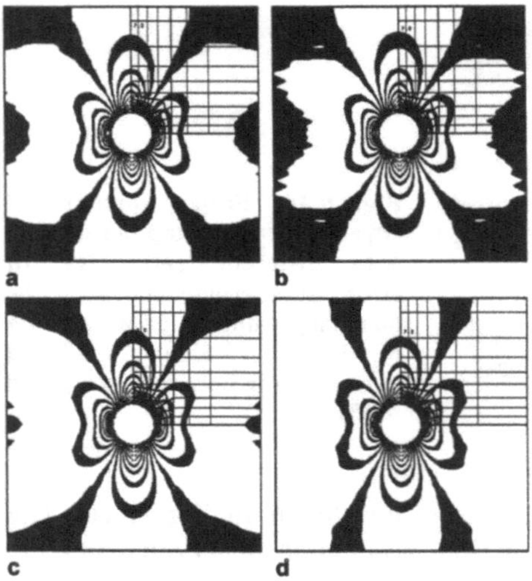

Fig. 9.12. Fringe contours of σ_1 near the SC zone with different percentage of errors at alternate nodes on the boundary values **a** ±5% **b** ±10% **c** +10, +5, −5 and −10% **d** random error between +15% and −15%

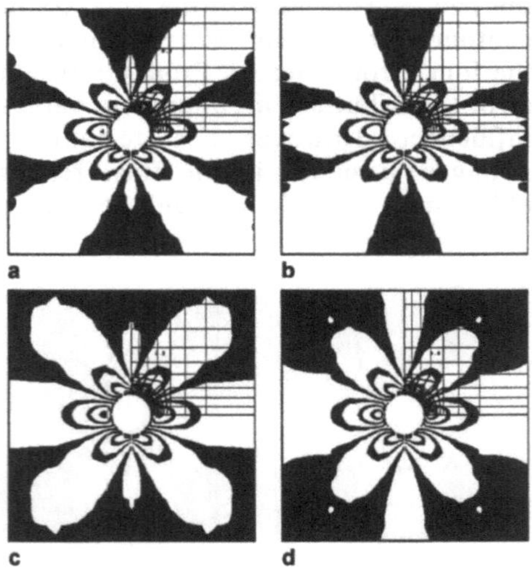

Fig. 9.13. Fringe contours of σ_2 for the SC zone with different percentage of errors at alternate nodes on the boundary values **a** ±5% **b** ±10% **c** +10, +5, −5 and −10% **d** random error between +15% and −15%

9.5.5 Application of the Technique to the Problem of a Plate with a Hole

The problem of a finite plate with a hole is a good example to see the applicability of the technique to practical applications. The boundary of a hole is a region of stress concentration and in Chapter 6, it has been discussed how to extract the fringe data using OET. Use of OET is all the more important as the solution of Laplace equation requires good quality boundary data and its use helps one to avoid the introduction of any systematic error in the analysis. Figure 9.14 gives the three-dimensional representation of the total fringe order for the plate (width 37 mm, hole dia 12.24 mm, length 174 mm and thickness 5.26 mm subjected to a uniaxial tension of 3.38 MPa).

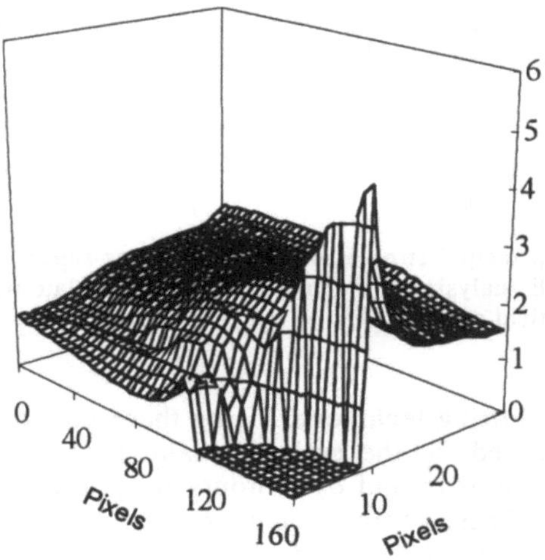

Fig. 9.14. Three-dimensional representation of total fringe order obtained by phase shifting in conjunction with OET for the problem of a finite plate with a hole

For solving the Laplace equation, one quarter of the domain of the problem is discretised using the concept of *meaningful discretization*. The values of the fringe order at the boundary and interior nodes are obtained by

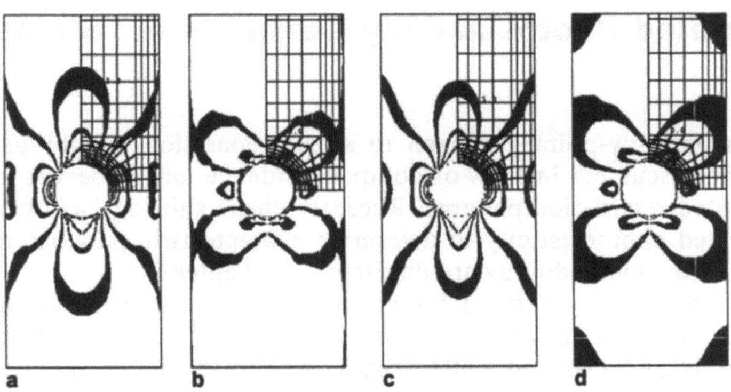

Fig. 9.15. Separated stress contours near the SC zones for a finite plate in uniaxial tension of 3.38 MPa, F_σ = 11.5 N/mm/fringe: **a** σ_1 by summation of the photoelastic data (obtained by PST) and the solution of Laplace equation **b** σ_2 obtained as in 'a' **c** σ_1 by 2-D FEM **b** σ_2 by 2-D FEM

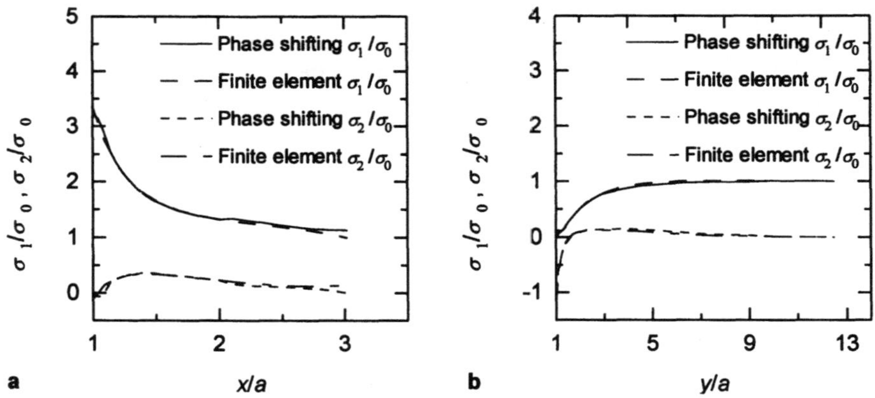

Fig. 9.16. Quantitative comparison of principal stresses obtained by stress separation by combined FEM and PST with 2-D FE analysis for the problem of a finite plate with a hole along **a** horizontal axis and **b** vertical axis

processing the results from phase-shifting technique. Using these boundary values, the Laplace equation is solved for the entire domain. From these results, the whole field representation of σ_1 and σ_2 contours are plotted and are shown in Figs. 9.15a and 9.15b. These contours compare well with those obtained by 2-D FE analysis (Figs. 9.15c and d). For quantitative comparison, σ_1/σ_o and σ_2/σ_o values, where σ_o is far field stress applied, are plotted for the horizontal (Fig. 9.16a) and vertical (Fig. 9.16b) lines of symmetry. These are compared with 2-D FEM results. The comparison is quite good.

9.6 Use of Integrated Photoelasticity Concepts for Stress Separation

In Sect. 9.2.2, a point-by-point approach to stress separation by oblique incidence has been discussed. In view of oblique incidence, one observes in reality the integrated retardation patterns. Recently phase shifting has been applied to integrated photoelasticity to determine characteristic parameters [35]. The details of the methodology are discussed in Chapter 11.

Thus, PST can be directly applied for evaluating the retardation N^{ξ} in oblique incidence. In view of higher accuracy obtainable by PST, small angle of oblique is sufficient to collect data in digital photoelasticity. Smaller the angle of oblique, the lesser is the error introduced in modelling the stress field being uniform over the light path.

The methodology discussed in Sect. 9.2.2 used both isoclinic and isochromatic data for point-by-point stress separation. However, if one has to take the full advantage of PST, the methodology should use only isochromatic data. This requires an iterative approach.

9.6.1 Least Squares Algorithm

The secondary principal stress difference in oblique incidence is given as

$$(\sigma_x \cos^2\xi - \sigma_y)^2 + 4(\tau_{xy} \cos\xi)^2 = \left\{\frac{N^\xi F_\sigma \cos\xi}{h}\right\}^2 \quad (9.73)$$

In Eq. (9.73), one can consider the unknowns as σ_x, σ_y and τ_{xy}. In the least squares techniques discussed in Chap. 8, overdetermined set of equations were obtained by collecting data from a large number of points. In Sect. 9.4, overdetermined set of equations were obtained by using different equations of elasticity, for a point of interest. Here, one can get an overdetermined set by changing the angle of oblique ξ in Eq. (9.73).

From Eq. (9.73) a function g_m is defined for the mth data point as follows

$$g_m = (\sigma_x \cos^2\xi_m - \sigma_y)^2 + 4(\tau_{xy} \cos\xi_m)^2 - \left\{\frac{N_m^\xi F_\sigma \cos\xi_m}{h}\right\}^2 \quad (9.74)$$

The above equation is non-linear in terms of the unknowns σ_x, σ_y and τ_{xy}. If initial estimates are made for σ_x, σ_y and τ_{xy} in Eq. (9.74) it is possible that g_m is not equal to zero since the estimates may not be accurate. Following Sect. 8.4, a non-linear least squares analysis can be carried out for determining the stress components. The usefulness of the technique can be improved further by selecting appropriate weighting functions (Sect. 9.4.3) for each angle of the oblique.

9.6.2 Design of the Loading Frame

For image capturing in oblique incidence, the model should be immersed in a bath of liquid having the same refractive index as that of the model to avoid refraction effects. Figure 9.17 shows a loading frame that can be used to load the model inside the liquid bath. At the bottom of the tank a protractor has been placed with holes at 0, 5, 10, 15, 20, 30 and 45 degrees. Above it, a swivelling platform is placed. The platform has in-built loading plunger, a pointer and pull/push rod which gets fitted in the protractor hole to avoid undue rotation during the image capturing process. Due to the nature of swivelling platform, the model can be rotated in the loaded condition.

Fig. 9.17. Loading attachment to load and rotate a model immersed in a bath of matching liquid

9.6.3 Application to the Problem of Disc Under Diametral Compression

The problem of disc under diametral compression has been the standard problem for verifying any new technique in photoelasticity. The first step in validating the methodology is to see how far the theoretical fringe patterns differ if they are constructed considering the stress being uniform over the light path or varying along the light path as in a 3-D model. For constructing theoretical fringe patterns while considering the model as a 2-D model, the state of stress at the central plane of the disc is considered as the stress over the length of the light path. Since, the state of stress in a circular disc is known, for the length of the light path in oblique incidence, the varying secondary principal stress components also could be computed. Considering the length of the light path as a train of 120 discrete retarders, the Jones matrix for each of them could be written and the product of these will give the net optical effect [Eq. (1.63), Chap. 1] introduced by the model. From this the characteristic parameters corresponding to the point of interest could be theoretically computed [35]. Figure 9.18 shows the fringe patterns for a disc under diametral compression (dia = 55.2 mm, thickness = 5.1 mm, diametral load = 754N, and F_σ = 12.43 N/mm/fringe) for oblique incidence in steps of 5° obtained by considering it as a 2-D problem. Figure 9.19 shows the plots of characteristic retardation by considering the variation of secondary principal stresses and their orientation along the light path [36]. Comparing Figs. 9.18 and 9.19, it is clear that theoretical integrated retardation pattern and the oblique incidence isochromatic fringe pattern match extremely well for smaller angles of oblique incidence.

A numerical validation of the least squares methodology for stress separation has also been carried out [36]. In this, the fringe orders N^ξ for various angles of oblique incidence are calculated from theoretical solution.

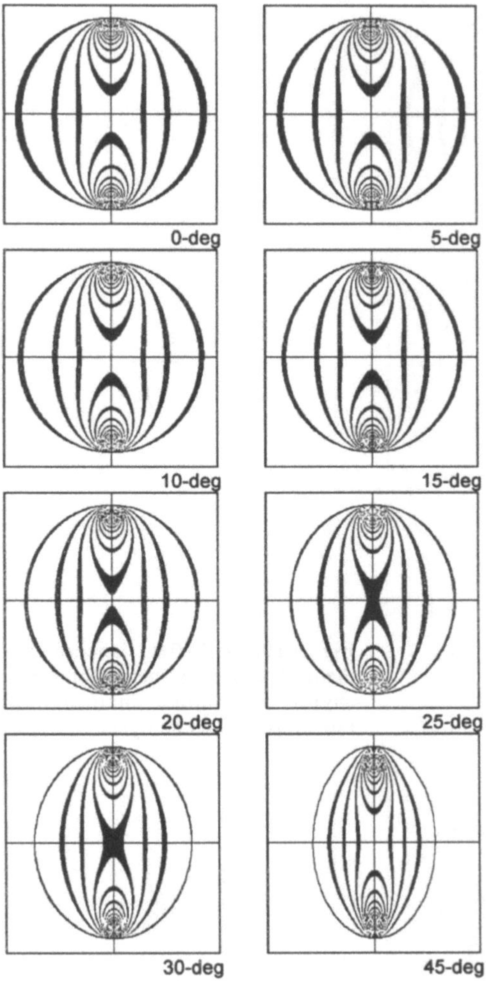

Fig. 9.18. Dark-field isochromatics theoretically constructed for various angles of oblique incidence obtained using 2-D analysis

Based on these values, by invoking the algorithm of Sect. 9.6.1, the individual stress components are calculated. Figure 9.20 shows the pseudo fringe contours of σ_x, σ_y and τ_{xy} obtained theoretically and Fig. 9.21 shows these contours obtained using the least squares algorithm. Leaving the zones of stress concentration, the comparison is quite good. Though the contours compare well with theory, in view of the stress components appearing as powers of two in Eq. (9.74), the signs of the stress components, particularly the sign of the shear stress, has to be determined by auxiliary means. If one is interested in determining only the individual principal stresses, the sign of the shear stress does not pose a problem.

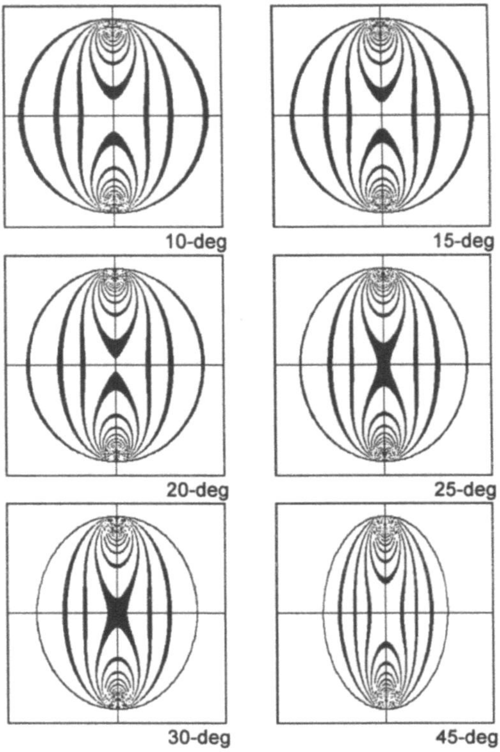

Fig. 9.19. Dark-field isochromatics theoretically constructed for various angles of oblique incidence obtained using 3-D analysis

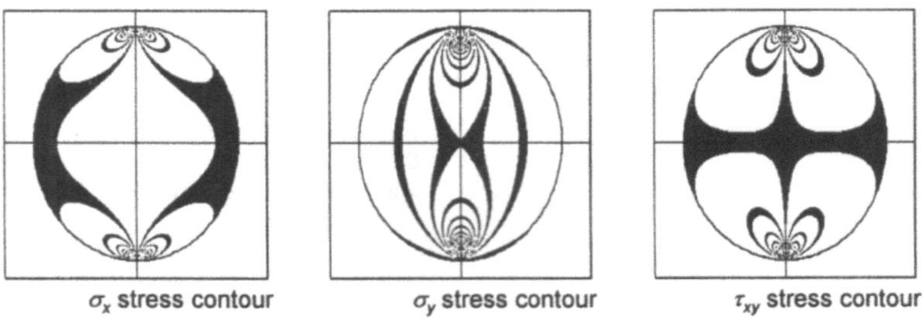

Fig. 9.20. Theoretically obtained pseudo stress fringe contours

For an experimental implementation of the technique using PST, a proper book keeping of the data for a point of interest for various oblique incidences has to be done. Referring to Fig. 9.1, while recording the image one gets the projected image of the fringe patterns in the $x'y'$-plane. The image is divided into pixels on this plane. From this data, one has to work backward to locate the point in the model. A detailed description of the methodology for experimental implementation can be found in ref. [36].

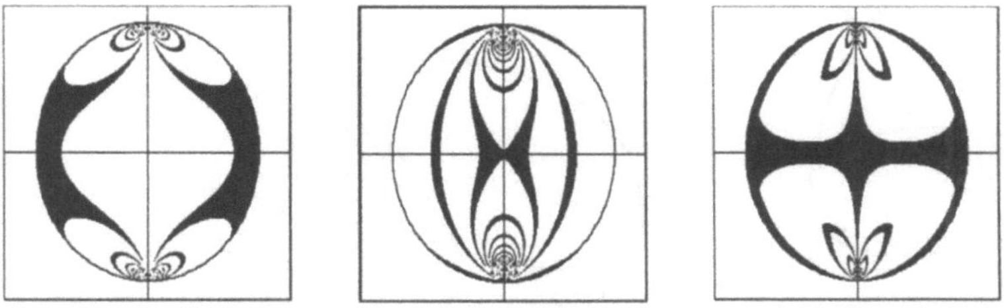

Fig. 9.21. Pseudo stress fringe contours obtained by the new methodology based on theoretically simulated data

9.7 Stress Separation in Three-Dimensional Photoelasticity

In the previous sections, an elaborate discussion has been given on the separation of stresses in two-dimensional problems. In Chapter 1, it has been noted that two-dimensional techniques of photoelasticity could be applied to solve three-dimensional problems through the use of stress freezing and slicing technique. The approach is experimentally intensive. However, the mathematical analysis is simple as the slices can be analysed as two-dimensional models. It is to be noted that at any point in the slice, the stress tensor in general can have all the six stress components present. In view of this even for normal incidence, the concept of secondary stress components have to be invoked (Sect. 9.2.2). As in two-dimensional models, the evaluation of stresses on free boundaries is still relatively simple. However, the separation of stresses inside the body requires great experience and understanding of the stress and strain distribution in three-dimensional bodies. Slicing plan for a three-dimensional model has to be carefully evolved to suit the application. In general, three stages of slicing and recording of the data have to be executed (Fig. 9.22).

Before proceeding for determining all the stress components constituting the stress tensor at a point of interest, one has to ask what is the goal of the overall problem on hand. If the goal is the failure analysis of the three-dimensional model made of isotropic material, then one can still get such data from direct photoelastic analysis. For isotropic materials, one can use either the Von-Mises yield criterion or the Tresca Yield criterion. In Von-Mises criterion, yielding occurs when

$$\left\{\frac{1}{2}\left[(\sigma_1-\sigma_2)^2+(\sigma_2-\sigma_3)^2+(\sigma_3-\sigma_1)^2\right]\right\}^{\frac{1}{2}}=S_y \tag{9.75}$$

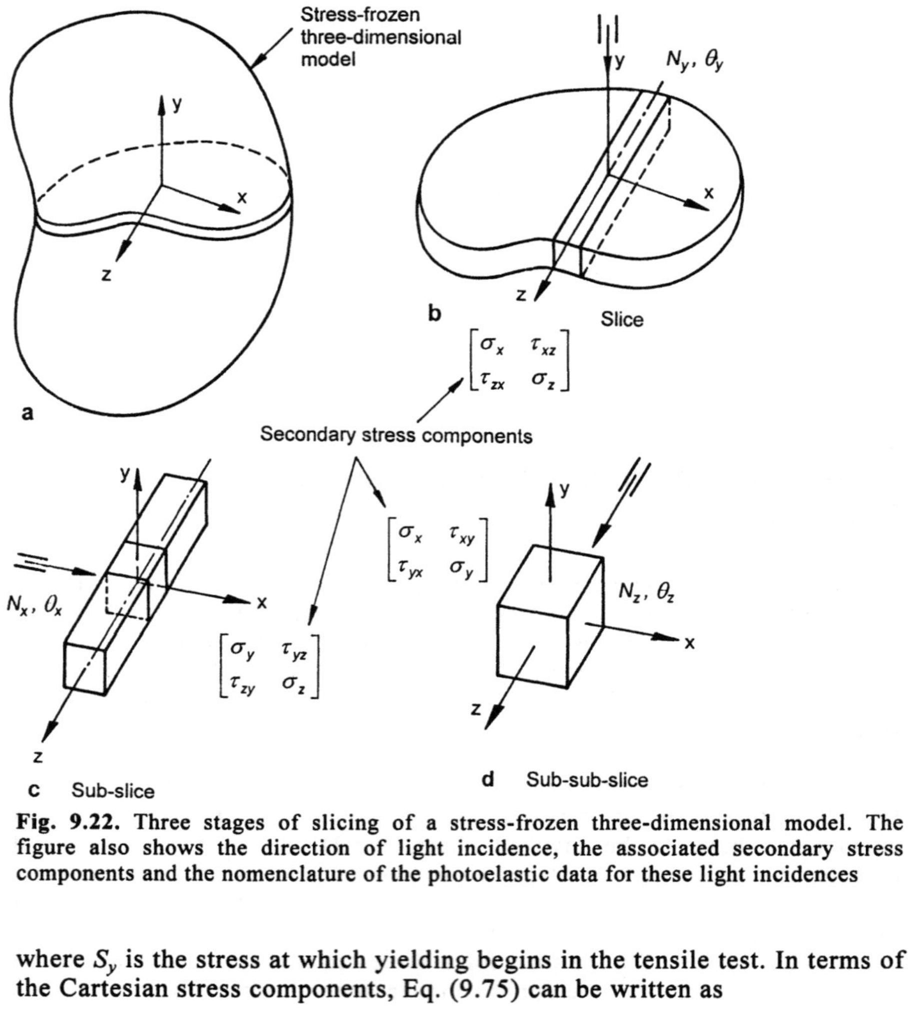

Fig. 9.22. Three stages of slicing of a stress-frozen three-dimensional model. The figure also shows the direction of light incidence, the associated secondary stress components and the nomenclature of the photoelastic data for these light incidences

where S_y is the stress at which yielding begins in the tensile test. In terms of the Cartesian stress components, Eq. (9.75) can be written as

$$\left\{\frac{1}{2}\left[(\sigma_x-\sigma_y)^2+(\sigma_y-\sigma_z)^2+(\sigma_z-\sigma_x)^2\right]+3\tau_{xy}^2+\right.$$
$$\left.3\tau_{yz}^2+3\tau_{zx}^2\right\}^{\frac{1}{2}}=S_y \tag{9.76}$$

Let us consider that in Fig. 9.22d, the sub-subslice is analysed by light incident along x, y, and z direction and let the respective fringe orders be denoted as N_x, N_y and N_z and the isoclinic angles be θ_x, θ_y and θ_z. In terms of photoelastic data, the Eq. (9.76) can be written [37, 38] as

$$\frac{1}{2}\left[\left(\frac{N_z F_\sigma}{h}\right)^2(2+\sin^2 2\theta_z)+\left(\frac{N_x F_\sigma}{h}\right)^2(2+\sin^2 2\theta_x)+\left(\frac{N_y F_\sigma}{h}\right)^2(2+\sin^2 2\theta_y)\right]^{\frac{1}{2}}=S_y \tag{9.77}$$

Thus, the requisite information for checking the Von-Mises yield criterion at a point of interest can be obtained from direct photoelastic data using three normal incidences alone. The technique does not use any oblique incidence but the method is only a point-by-point technique.

The dimensions of the sub-subslice have to be sufficiently small to minimise the inherent error due to the usual assumption that the stress is constant within the sub-subslice. The size of the sub-subslice determines the averaging effect and has the same effect on the accuracy of stress determination as the gauge length of a strain gauge. A wealth of literature is available on strain gauge instrumentation to account for such errors. The smallest practical sub-subslice seems to be 3-mm [39].

For failure analysis if one can use Tresca yield criteria, it is possible to get this over the entire slice by the use of one normal incidence and two oblique incidences. Following the steps mentioned in Sect. 9.2.1, the secondary principal stresses corresponding to a slice cut from a three-dimensional model for an oblique incidence at an angle of ξ in the xz-plane are

$$\begin{bmatrix} \sigma_x\cos^2\xi+\sigma_z\sin^2\xi+\tau_{xz}\sin 2\xi & \tau_{xy}\cos\xi+\tau_{yz}\sin\xi \\ \tau_{xy}\cos\xi+\tau_{yz}\sin\xi & \sigma_y \end{bmatrix} \tag{9.78}$$

Another oblique incidence is needed to get individual shear stress components. It is convenient to make it at an angle $-\xi$ in the xz-plane. The secondary stress components contributing to photoelastic effect for this oblique incidence are obtained by replacing ξ to $-\xi$ in Eq. (9.78). Representing the normal incidence photoelastic data as (N_z, θ_z) and oblique incidence data as (N^{ξ}_z, θ^{ξ}_z) and ($N^{-\xi}_z$, $\theta^{-\xi}_z$) the normal stress differences components can be expressed as

$$(\sigma_x-\sigma_y)=\frac{N_z F_\sigma}{h}\cos 2\theta_z$$
$$(\sigma_z-\sigma_x)=\frac{F_\sigma}{2h\sin^2\xi}(N_z^{\xi}\cos 2\theta_z^{\xi}\cos\xi+N_z^{-\xi}\cos 2\theta_z^{-\xi}\cos\xi-2N_z\cos 2\theta_z) \tag{9.79}$$

The shear stress components are obtained as

$$
\begin{aligned}
\tau_{xy} &= \frac{N_z F_\sigma}{2h}\sin 2\theta_z \\
\tau_{yz} &= \frac{F_\sigma}{4h\tan\xi}\left(N_z^{\xi}\sin 2\theta_z^{\xi} - N_z^{-\xi}\sin 2\theta_z^{-\xi}\right) \\
\tau_{zx} &= \frac{F_\sigma}{4h\sin\xi}\left(N_z^{\xi}\cos 2\theta_z^{\xi} - N_z^{-\xi}\cos 2\theta_z^{-\xi}\right)
\end{aligned}
\tag{9.80}
$$

Using these values, it is possible to get the radius of the Mohr's circle in the planes xy, yz and zx. The maximum of these gives the maximum shear stress at the point of interest.

For stress separation in two-dimensional models, one can choose either the oblique incidence technique or the shear-difference approach among the many techniques available. However, in three-dimensional models one does not get all the stress components by either using only oblique incidence method or the shear difference approach but one has to use both of them for stress-separation. Failure analysis using Tresca yield criteria was possible using a slice cut from the 3-D model. However, for complete determination of stress components, a sub-slice has to be cut from the slice and analysed. For each data point, one has to use normal incidence along y and z-axis and one oblique incidence in the xz-plane. From the normal incidence along z-axis and oblique incidence in xz-plane, the expressions for $(\sigma_x - \sigma_y)$, τ_{xy} and τ_{yz} can be obtained from Eqs. (9.79) and (9.80). From normal incidence along y direction, the expressions for $(\sigma_z - \sigma_x)$ and τ_{zx} can be obtained as

$$
\begin{aligned}
(\sigma_z - \sigma_x) &= \frac{N_y F_\sigma}{h}\cos 2\theta_y \\
\tau_{zx} &= \frac{N_y F_\sigma}{2h}\sin 2\theta_y
\end{aligned}
\tag{9.81}
$$

For stress separation one has to integrate the equilibrium equation in three-dimension [40]. Following Eq. (9.12), it can be written as

$$
(\sigma_x)_j = (\sigma_x)_i - \sum_i^j \frac{\Delta\tau_{yx}}{\Delta y}\Delta x - \sum_i^j \frac{\Delta\tau_{xz}}{\Delta z}\Delta x \tag{9.82}
$$

To integrate the above equation, one has to find the shear stresses for two more lines in the xy-plane and yz-planes as shown in Fig. 9.23. Tesar's modification of expressing the shear stress in terms of principal stress difference and the isoclinic angle as used in two-dimensional problem can be extended here too, which is left as an exercise to the reader.

It is worthwhile to note that while making normal incidence along z direction, the slice is considered as small such that no variation in the

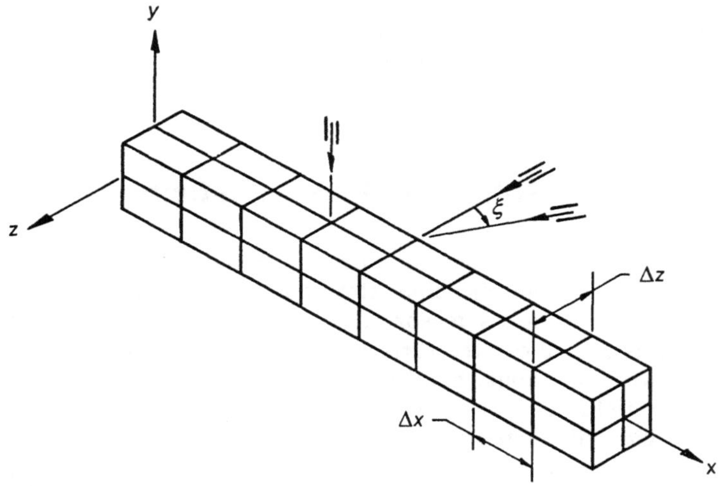

Fig. 9.23. Scheme to collect data for performing shear-difference technique in three-dimension. The directions of two normal and one oblique incidence are indicated

z-direction exists. However, while using normal incidence in y-direction, the slice is considered sufficiently thick to provide a variation of τ_{xz}. The same approximation applies to the determination of τ_{xy} too. This is the inherent contradiction in three-dimensional analysis and one has to live with it for engineering analysis as it provides enormous simplicity to solve problems of practical interest.

Hitherto the techniques have been considered for solving a general three-dimensional problem. Much simplification in the technique is possible if one is interested in solving axi-symmetric problems. In such cases, the use of a meridional slice alone is sufficient for stress separation. Use of one normal and one oblique incidence coupled with integration of the equilibrium equation in polar co-ordinates helps in separating the stresses [3]. Stress freezing and slicing has also been used successfully for stress analysis of torsion problem [41, 42].

Hybrid methods for stress separation in three-dimensional photoelasticity have also been reported. Chandrasekhara and Jacob [43] have discussed a numerically intensive technique for stress separation in three-dimensional problem using only the boundary information from the experiments. Sanford and Beaubien [18] investigated the use of combining 2-D FEM and three-dimensional photoelasticity. They analysed a newly designed petal from a four-part sabot of a long-range artillery shell. The component consisted no more than two components in contact and in which the contact boundary conditions change with increasing load. In view of the complexity of the problem, the methodology of solution neither favoured the approach by 2-D FEM nor 3-D photoelasticity. They observed that 2-D FEM could be used to estimate the loads that need to be applied for stress-freezing operation. Since, the boundary conditions are difficult to idealise without an experimental analysis; they could deduce iteratively the approximate boundary conditions for use in 2-D FEM modelling from 3-D photoelastic analysis of the problem.

9.8 Stress Separation in Reflection Photoelasticity

The stress separation techniques used in reflection photoelasticity confine to the determination of in-plane principal strains/stress components rather than the complete strain or stress tensor at the point of interest. Since, photoelastic coating is pasted on free surface of the components, the stress field is essentially two-dimensional but the strain field is three-dimensional. In view of the stress field being two-dimensional, Tresca yield criteria can be easily applied if the individual principal stresses are known.

For flat objects, oblique incidence method as used in transmission photoelasticity is possible. It is to be noted that the basic information obtained in photoelastic coating is strains. In view of the three-dimensional nature of the strains in the coating, the equations of oblique incidence are much more complex. The individual principal strains at a point of interest can be written in a general form as

$$\begin{aligned} \varepsilon_1^s = \varepsilon_1^c = \frac{F_\varepsilon}{2h_c}[AN^\xi - BN] \\ \varepsilon_2^s = \varepsilon_2^c = \frac{F_\varepsilon}{2h_c}[AN^\xi - CN] \end{aligned} \tag{9.83}$$

where the coefficients A, B and C are functions of Poisson's ratio of the coating material and the angle of oblique. If the physical properties of the coating material are not known, then the coefficients A, B and C have to be found by calibration using a specimen with known values of ε_1 and ε_2. Standard attachments for performing oblique incidence are available [44]. While using standard attachments and commercially obtained coating materials, the manufacturer also supplies the coefficients A, B and C of Eq. (9.83). The technique is basically a point-by-point one. Following the methodology discussed in Sect. 9.6, the technique could be extended for a whole field analysis. For curved objects, oblique incidence method cannot be applied.

Many other techniques exist, where the coating is modified to provide either separation of stresses locally or separation of stresses globally. Local separation of stress is possible by the *hole method* proposed by Durelli and Rajaiah [45]. A whole field stress separation is possible by using a *strip coating*, which consists of closely spaced parallel strips. The strips respond primarily to their long axes and they are largely unresponsive to transverse and shear strains. Several ways of using strip coatings for stress separation can be found in ref. [46]. In the *hole method*, the coating is perforated with small circular holes of a diameter of the order of the thickness of the coating and at a distance from each other, sufficiently large, to avoid interference. The introduction of such holes creates free boundaries in the coating and at these boundaries, the stress is uniaxial and can be determined completely in a direct manner. Empirical relations are derived to find the principal stresses at the centre of the hole from free boundary stresses. In a refinement of the technique, it is suggested to acquire data at a radius of 1.4 times the radius of the hole, and use this to improve the previous estimate of principal stresses based on the free boundary stresses. The technique is reported to provide better accuracy than the strip coating method. In practice the use of both *strip coating* and *hole method* are experimentally quite intensive.

Use of a photostress separator gauge is now available [47]. The technique is equally applicable for flat and curved objects. In this approach, one has to just paste the strain gauge at the point of interest on the birefringent-coated model and the strain gauge signal is proportional to the sum of the principal strains. This is because, the gauge has two sensing elements mutually perpendicular to each other connected in series (Fig. 9.24). Combining this with photoelastic data, principal strains can be obtained. While pasting the strain gauge no special care needs to be taken for angular alignment of the gauge as the construction of the gauge accounts for this. The gauge also comes with pre-soldered lead wires for ease of mounting.

Fig. 9.24. Stress separator gauge. The two grid elements are mutually perpendicular to each other (Courtesy of Measurements Group, Inc., Raleigh, NC, USA)

In view of the strain gauge being pasted on the birefringent plastic, it locally reinforces the coating and there will be drift in the reading due to poor thermal conductivity of the coating. The residual reinforcement error is accounted for by using an effective gauge factor and the use of specialised instrumentation minimises the thermal effects due to poor conductivity. In Chap. 2, it has been discussed how correction factors need to be employed for interpreting the coating results when it is applied on members subjected to bending, torsion etc. On similar lines, additional correction factors have been developed for photostress separator gauge too and the correction factors are available in the literature [48].

In many applications involving engineering assessment of the design, photoelastic coating is used mainly to validate the numerical technique such as finite elements for stress field evaluation or design certification. Though for flat objects, stress separation is simpler, for curved objects, the evaluation of stress field becomes quite involved. In such instances, the validation of the numerical techniques becomes much easier if the results of the numerical technique are plotted in the form of fringe contours. Using the colour code given in Chap. 7, one can plot fringe contours in colour for both flat and curved objects. A quick assessment of the numerical technique is possible by comparing the fringe patterns of the experiments with the fringe contours plotted.

9.9 Closure

The various methods for stress separation in photoelasticity are discussed in this chapter. The way to utilise directly the photoelastic data for engineering analysis is highlighted. The role of digital photoelasticity for improving stress separation techniques have been brought out.

Exercises

9.1 In what class of problems can photoelastic data be directly used?

9.2 What are secondary principal stresses. How this concept has simplified data reduction in 2-D and 3-D photoelasticity?

9.3 How can you reduce the accumulation of errors in shear difference technique?

9.4 What is the contribution of stress-difference elasticity in photoelastic analysis?

9.5 Develop a software code to solve Laplace equation by FEM.

9.6 Based on Tables 7.2 and 7.3, develop a software code to plot fringe orders from FE calculations in colour. How can this help in solving industrial problems?

9.7 A three-dimensional model of a crane hook has to be analysed by 3-D photoelasticity. The hook has a trapezoidal cross section. Mention the slicing plan you need to solve this problem.

9.8 Compare the various methods of stress separation in reflection photoelasticity.

9.9 How digital photoelasticity has contributed to problems regarding stress separation? Which areas of digital photoelasticity need to be strengthened for it's use in stress-separation?

References

1. Heywood RB (1952) Designing by photoelasticity. Chapman and Hall Ltd
2. Crandall SH, Dahl NC, Lardener TJ (1978) An introduction to the mechanics of solids. McGraw-Hill, New York
3. Srinath LS, Raghavan MR, Lingaiah K, Gargesha G, Pant B, Ramachandra K (1984) Experimental stress analysis. Tata McGraw-Hill, New Delhi
4. Dally JW, Riley WF (1991) Experimental stress analysis. edn (3) McGraw-Hill, New York
5. Haake SJ, Patterson EA (1992) The determination of principal stresses from photoelastic data. Strain 28:153–158
6. Drucker DC (1943) Photoelastic separation of principal stresses by oblique incidence. J Appl Mech :A156-A160
7. Frocht MM (1944) Discussion of ref 6. J Appl Mech :A125–A126
8. Ramesh K, Pathak PM (1998) Numerical techniques for stress separation in photoelasticity – a review. ARDB technical report. ARDB-SP-TR-919-01. India
9. Berghaus DG (1973) Overdetermined photoelastic solutions using least squares. Exp Mech 13(3):97–104
10. Doyle JF (1986) Photoelastic stress separation along lines of symmetry. Exp Mech 26:297–300
11. Shibuya T, Koizumi T, Takakuda K, Maeno T (1986) New separation method of photoelastic principal stresses using a microcomputer. Bulletin of JSME 29 (248):400–407
12. Thompson JC, Negus KJ (1984) Developments in least squares asymptotic analysis of isochromatic data from stress concentration regions in plane problems. Strain 20:133–134
13. Chandrashekhara K, Jacob KA (1976) A method for separation of stresses in two and three-dimensional photoelasticity. J Indian Institute of Science 58(8) :331–344
14. Mahfuz H, Wong TL, Case RO (1990) Hybrid stress analysis by digitized photoelastic data and numerical methods. Exp Mech 30:190–194
15. Chandrashekhara K, Jacob KA (1977) An experimental-numerical hybrid technique for two-dimensional stress analysis. Strain 13:25–31
16. Berghaus DG (1991) Combining photoelasticity and finite element method for stress analysis using least squares. Exp Mech 31:36–41
17. Chambles D, Suhling J, Swinson WF, Turner JL (1986) A new hybrid photoelastic – finite element technique for stress analysis. Proc 1986 SEM spring conference on Experimental Mechanics :991–998
18. Sanford RJ, Beaubien LA (1977) Stress analysis of a complex part: photoelasticity vs. finite elements. Exp Mech 17:441–447
19. Tesar V (1935) la photoelasticimetrie et ses applications dans la construction aeronautique. La Science Aerienne II:372–394
20. Vadovic F (1965) Contribution to the analysis of errors in photoelasticity. Proc SESA, XII (2):413–416
21. Allison IM Blakemore RH (1968) The analysis of photoelastic data for stress separations. In: Recent advances in stress analysis. Royal Aeronautical Society, London :4.7–4.12
22. Allison IM, Nurse P (1978) Optimisation of photoelastic stress separation procedures. VDI-Berichte (313):41–44
23. Trebuna F (1990) Some problems of accelerating the measurements and evaluating the stress fields by the photostress method. Exp Tech 14: 36–40
24. Mahfuz H, Wong TL, Case RO (1990) Separation of principal stresses by SOR techniques over arbitrary boundaries. Exp Mech 30:319–327

25. Segerlind LJ (1971) Stress-difference elasticity and its application to photomechanics. Exp Mech 11:441–445
26. Berghaus DG (1989) Adding the Laplace Equation to Least Squares Photoelastic stress solutions. Exp Tech 13:18–21
27. Reddy JN (1993) An Introduction to the Finite Element Method. edn(2) McGraw-Hill, New York
28. Zienkiewicz OC, Taylor RL (1989) The finite element method. McGraw-Hill, New York
29. Seguchi Y, Tomita Y, Watanabe M (1979) Computer aided fringe pattern analyzer – a case of photoelastic fringe. Exp Mech 19(10):362–370
30. Zienkiewicz OC, Zhu JZ (1991) Adaptivity and mesh generation. Int J Numer Methods in Engng 32:783–810
31. Ramesh K, Pathak PM Role of photoelasticity in evolving discretisation schemes for FE analysis. Exp Tech 23:
32. Pathak PM, Ramesh K (1999) Validation of finite element modelling through photoelastic fringe contours. Communications in Numerical Methods in Engng 15:229–238
33. Ramesh K, Yadav AK, Pankhawalla VA (1995) Plotting of fringe contours from finite element results. Communications in Numerical Methods in Engng 11:839–847
34. Mangal SK, Pathak PM, Ramesh K (1999) Use of finite elements for stress separation in digital photoelasticity. J Aero Soc of India, (in press)
35. Mangal SK, Ramesh K (1998) Automation of data acquisition in integrated photoelasticity. Proc of Int Conf on Optics and Optoelectronics, Instruments Research and Development Establishment, Dehra Dun, India pp 1292–1297
36. Ramesh K, Mangal SK Whole field stress separation by oblique incidence using phase shifting technique. Unpublished work
37. Brock JS (1958) The determination of effective stress and maximum shear stress by means of small cubes taken from photoelastic models. Proc SESA, XVI:1–8
38. Chaudhari UM (1985) Quick assessment of damage potential using photoelastic data. Exp Tech 9:26–27
39. Burger CP (1993) Photoelasticity. In: Kobayashi AS (ed) Handbook of experimental mechanics. SEM and VCH publishers, New York, pp165–266
40. Frocht MM, Guernsey R (1955) Further work on the general three-dimensional photoelastic problem. J Appl Mech 22:183–189
41. Frocht MM (1965) Photoelasticity, vol I . John Wiley and Sons, New York
42. Frocht MM (1967) Photoelasticity, vol II. John Wiley and Sons, New York
43. Chandrasekhara K, Jacob KA (1977) An experimental-numerical hybrid technique for three-dimensional stress analysis. Int J for Numer Methods in Engng 11: 1845–1863
44. Introduction to stress analysis by the photostress method, Measurements Group Technical Note TN-702, Measurements Group Inc, Raleigh, North Carolina
45. Durelli AJ, Rajaiah K (1980) Determination of strains in photoelastic coatings. Exp Mech 20:57–64
46. Zandman F, Redner S, Dally JW (1977) Photoelastic Coatings. SESA Monograph No.3, Society for Experimental Stress Analysis, USA
47. Nickola WE (1986) A summation strain gauge-alternative to oblique incidence in photoelastic coatings. Proc of 1986 SEM spring conference on Experimental Mechanics
48. Principal stress separation in photoelastic measurements, Measurements Group Technical Note TN-708, Measurements Group Inc., Raleigh, North Carolina

10 Fusion of Digital Photoelasticity, Rapid Prototyping and Rapid Tooling Technologies

10.1 Introduction

The recent advances in rapid prototyping (RP) [1–5] have made it possible to produce prototypes of very complicated parts directly from three-dimensional computer aided design (CAD) models without using part specific tooling. Application of these techniques for producing prototypes leads to considerable reduction in the total cycle time. There are a number of RP techniques developed so far, such as stereolithography (STL), fused deposition modelling (FDM), solid ground curing (SGC), selective laser sintering (SLS) etc. This technology has emerged in response to the need for reducing the lead-time to produce physical prototypes. These models could be directly used for checking form, fit and function, getting management approval for detailed design and development, conducting market research, submitting for price negotiations and as models for wind tunnel testing.

A design process is complete only if the prototypes are validated for their structural response under operating loads. Thus, stress analysis of RP models is important and the technique chosen must be such that the testing time is quite short. Among the various experimental stress analysis tools, the whole field techniques are ideal to make measurements over large model areas quickly.

Among the various RP technologies available, stereolithography (STL) has found initial acceptance by photoelasticians [6] because of its ability to provide models that are birefringent and transparent, which facilitated transmission photoelastic analysis. In principle, techniques of photoelastic coating can also be extended for analysing the models made by RP. In such a case, a careful selection of appropriate photoelastic coating is necessary for the analysis of RP models as these are made of plastic [7]. A systematic study on the applicability of various coating materials on RP models made by FDM is discussed in this chapter.

Rapid prototyping is a material additive manufacturing process and the models are generated layer by layer. In view of the layered manufacturing approach adopted, the models usually have residual stresses and are porous to various degrees depending on the type of RP process selected and its process parameters. The inherent limitations of the RP process in this regard are highlighted in this chapter. This has made the early initiatives for stress analysis inadequate.

Fabrication of functional prototypes becomes necessary for conducting meaningful stress analysis. This could be done in a short time by rapid tooling (RT) technologies using RP models as patterns. In this chapter, the use of MCP vacuum casting machine of HEK GmbH (Germany) for RT is discussed. The key issue here is the identification of a suitable resin for making models, which could facilitate a three-dimensional photoelastic analysis.

10.2 Difficulties in Conventional Three-Dimensional Photoelasticity

In conventional three-dimensional photoelastic stress analysis, the model is initially stress frozen and then mechanically sliced for evaluating the stress field. The slices removed from the model are analysed based on concepts developed in two-dimensional photoelastic analysis. This has considerably simplified the mathematical analysis and has provided acceptable solution for engineering analysis. However, the technique is experimentally intensive.

With the advent of digital photoelasticity, the analysis of these slices has become very simple. However, the model making and its subsequent slicing are still cumbersome. The preparation of three-dimensional photoelastic model requires skill. It is here, the use of RP for model making is attractive.

A single model may not be sufficient to complete the design evaluation. For each loading condition, one needs an additional model for stress freezing. Even for a particular loading condition, the evaluation of loads to be applied at the stress-freezing temperature can be quite tricky. If the loading is more, the model may break or may have unacceptable large deformation after stress freezing. If the loading is small, then the slices may not have sufficient number of fringes for analysis. From this point of view, the combination of RP and RT would be ideal for photoelastic analysis.

10.3 Rapid Prototyping in Model Making

One of the limitations of the RP machines is that only small sized parts can be manufactured, usually within an envelop of approximately 250 mm × 250 mm × 250 mm. The size limitation is partly due to the complexity of the RP mechanism as well as the long time taken for manufacturing even small parts. This does not pose a limitation for the use of RP models for photoelastic analysis as appropriate scaling could be arrived at for model studies.

Some manufacturers solved the size problem by dividing the model into separate parts and later gluing them together.

There are three essential components of a RP system.

1. Software for solid modelling.
2. Software to convert the solid models into instructions for the RP machine and
3. The RP machine itself.

10.3.1 Software Issues in RP

For getting an RP model, initially, the model to be made has to be represented digitally as a solid model. A solid model can be defined as a geometric representation of a bounded volume. Closed boundaries alone do not describe solid objects. The boundary data must also carry the orientation of the solid areas. A requirement of solid modelling CAD systems is the generation of an unambiguous description of the geometry being modelled. Surface normal information pertaining to the object's boundary is used to indicate the orientation of the object's mass. The connectivity between the various faces and surfaces of the geometry is achieved through an appropriate topological structure.

Most existing CAD systems employ a hybrid array of mathematical concepts toward solid modelling. They are constructive solid geometry (CSG), boundary representation (B-rep) and faceted approximation. CSG modellers use simple geometric shapes such as cubes, spheres and cones as basic building blocks. In B-rep, non-uniform rational B-splines (NURBS) is used to precisely define the geometry that allows for easy manipulation of two-dimensional, three-dimensional and surface entities. Faceted approximation is a technique of dividing the solid object into planar surfaces. The planes may be divided into triangles or quad facets. A two-dimensional example is approximating a circle with straight-line segments (Fig. 10.1). Some CAD systems widely in use are CATIA of Dassault Systems, I-DEAS of SDRC and Pro/ENGINEER of Parametric Technologies.

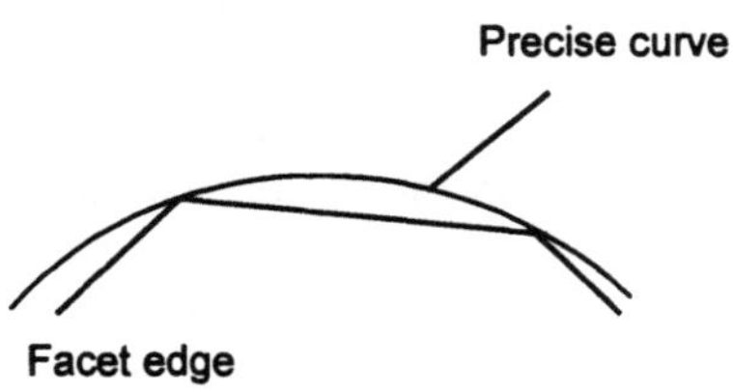

Fig. 10.1. Faceted approximation of a curve

For generating a RP model, the basic definition of the part may be obtained from surface modellers too. Surface modellers are often used to describe aerodynamic and aesthetic shapes. A major functional difference between solid-based CAD systems and surface modellers is the absence of topological data connecting the surfaces. In surface modelling, the user develops two-dimensional wireframe profiles that are revolved, extruded, swept or blended together to form the desired surface. The surface entities can be simple, such as planes and tabulated cylinders. Most advanced systems utilise NURBS-defined entities. Such definitions bestow optimum control over the surface shape, while allowing editing to be easily accomplished. To create a design for RP using surface modellers, the user must 'sew' the adjacent surfaces to enclose the volume. This needs additional work by the users. Nevertheless, this is preferred, as surface modellers may be ideal to define complex surfaces where many solid-based systems will have trouble. Most advanced CAD/CAM systems (CATIA, I-DEAS, Pro/ENGINEER etc.) strive for the best of both worlds; combining solid and surface generation into one homogeneous system.

Most CAD systems use proprietary formats for storing the solid model data. For exchanging the data from solid modellers to other CAD/CAM software, one has to convert the data into standards such as IGES (initial graphics exchange specifications), PDES (product data exchange) etc. Although, early RP systems proclaimed compatibility to IGES, the user has to fix several problems in data conversion separately.

IGES is an example of a *shape-based format* in which all the geometric entities and their attributes are described in a very comprehensive manner. However, the data is more of a drawing than that of an object. On the other hand, RP machines require the solid model to be described in terms of a set of contiguous triangular facets, similar to a finite element mesh representation, completely enveloping the surface of the part [3]. This format has now been standardised and is referred to as STL format. Advanced CAD systems now support the STL interface.

The STL interface is based on a polyhedral representation found in many solid modellers. Precise surfaces (NURBS-based or primitive CSG objects) are approximated to planes and straight lines into a tessellated (triangulated) facet format. The parameters such as chord height and angle control have to be defined properly while transferring the part file to STL file. The level of accuracy on non-planar surface is controlled by the number of facets used to represent that surface. Each facet is defined by four data items, three of which represent the co-ordinates of the vertex of the triangle and the fourth represents the facet normal. Facet normal information and the vertex order are used to distinguish the model's outside from inside. The other rule of facet creation is the vertex-to-vertex rule. This rule says that the vertex of one triangle cannot be in the middle of the side of an adjoining triangle [3].

For layered manufacturing, the solid model has to be sliced into a number of 2-D layers. This is achieved by using an algorithm, which computes the sequential points of intersection of a horizontal plane with the faces and edges of the solid model. The cutting plane is positioned at different elevations, based on the layer thickness or number of layers specified by the user, to compute the various cross-sections of the model.

Choice of part orientation is important as the orientation within the processing chamber influences the build time, part resolution and surface finish. Sloping surfaces that proceed along the build direction will have a distinct 'stair-step' appearance (Fig. 10.2). The height of each step is the layer thickness used in that portion of the part. Stair-step effect could be avoided if the curved surface is oriented in the horizontal plane. In situations where, part orientation could not be changed, then the stair-steps have to be hand finished to improve their surface quality. In liquid-based RP systems, the part-orientation has to be so selected that no trapped volume of uncured resin is left behind.

Fig. 10.2. Stair-stepping effect

Large scaling for parts whose applications are visualisation aides should be done inside the CAD system. Shrinkage factors for casting applications can also be incorporated during the CAD modelling phase, although these can be accomplished in the RP software.

10.3.2 Stereolithography Process

Stereolithography was the first system commercially available, which creates parts by 'laser-curing' of successive layers of liquid resin. The system was launched by 3D systems who pioneered the field of RP technology. In this process, a beam of ultraviolet (UV) light is focused on the top layer of liquid photo-curable resin contained in a vat (Fig. 10.3). Earlier systems used He-Cd lasers (40 mW, 325 nm) which are now replaced by high-powered Argon-ion lasers (600 mw, 351 nm) [5]. The beam is positioned and moved in horizontal x and y-directions using a set of mirrors to crosshatch the area contained within the boundary of a particular layer of the model. Once the layer is formed, the platform attached to the model is lowered and a fresh layer of liquid resin covers the previously formed layer. The laser beam generates the next layer on top of the previous one. The process is repeated until all the layers are formed. Complicated parts may need appropriate supports to be designed as part of the CAD model. Further, while generating the RP model, the part orientation has to be appropriately selected so that undue entrapment of liquid resin is avoided.

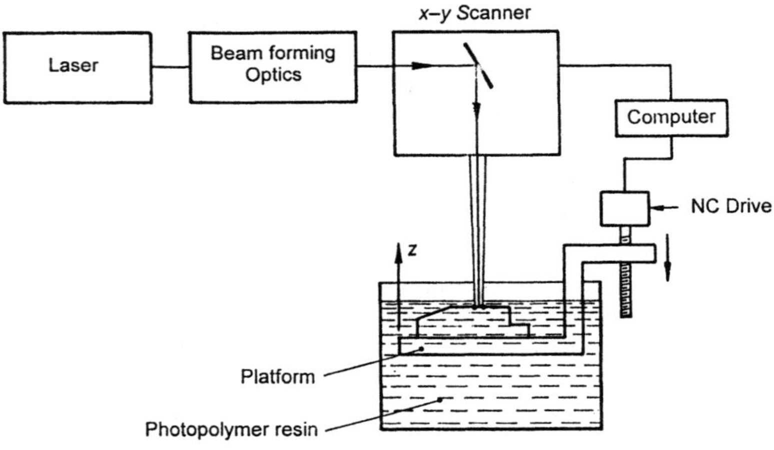

Fig. 10.3. Schematic of a stereolithography system

For each layer, the borders are drawn first and then the interior is scanned by any one of the several scanning styles available with the hardware. Tri-Hatch was the scanning scheme developed initially, which uses a scanned line parallel to the x-axis combined with lines at 60° and 120° to the x-axis. This produces an internal structure of equilateral triangles. Vertical triangular columns are thereby created which contain resin that has not been solidified. The model distortion is attributed to the presence of uncured resin. This trapped resin has to be cured during post-processing and hence the RP model becomes elastically inhomogeneous.

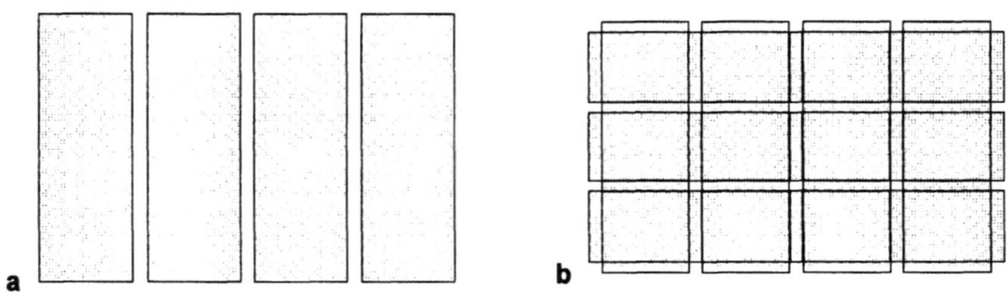

Fig. 10.4. a Top surface of the layer after y-scan **b** Weave like appearance of the surface of the layer after xy scan

In order to minimise the presence of trapped resin, a new procedure called WEAVE™ was then developed. In this, the scanning is parallel to the x and y-axis only. Initially, scanning of the layer is done along the y-direction and it is then scanned along the x-direction. The top surface of the layer after each such scanning is shown in Fig. 10.4. The cloth like appearance of the surface on each cured cross-section accounts for the origin of the name WEAVE™. Figure 10.5 shows the cross-sectional view of the WEAVE™ pattern. Over the depth of the layer, the curing of the resin is not rectangular. For the case of a Gaussian laser and a resin obeying the Beer-Lambert law [3], over the depth, the shape of the cured resin will be parabolic. This leads to pockets of uncured resin within the model. The line spacing is one fifth of what was used in Tri-Hatch. This leaves very little uncured resin in between the scanned lines.

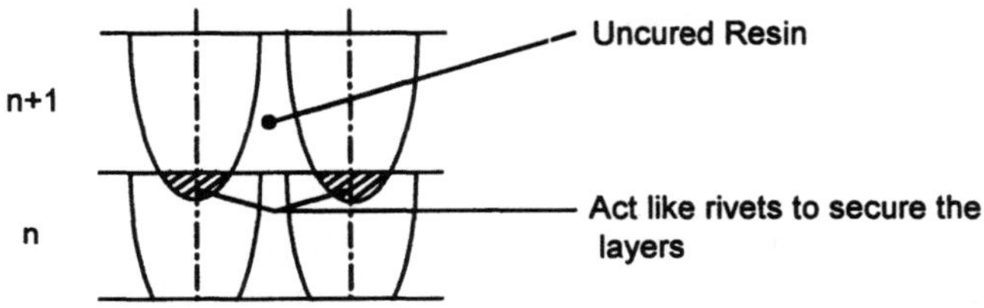

Fig. 10.5. Cross-sectional view of the cured layers obtained by WEAVE™

Later research showed that more than the presence of entrapped uncured resin, several other parameters such the type of scanning scheme, the scanning procedure followed between different layers and the connectivity of the internal scanning with the borders of the layer contribute to model distortion. This lead to the development of STAR-WEAVE™ in which pockets of uncured resin was distributed in an ordered sequence. In STAR, the ST stands for STaggered hatch, the 'A' stands for alternate sequencing of layers and 'R' stands for retracted hatch. In alternate sequencing, the scanning sequence of the layers is reversed and in retracted hatch, the scan patterns are attached to the border alternately (Fig. 10.6). Note the gaps in the scanning procedures, which lead to trapped uncured resin, which needs post curing. This alters the structural nature of the final component. Though the model looks like a solid from the exterior, the model made by RP is not as solid as one would expect it from a casting or forging process.

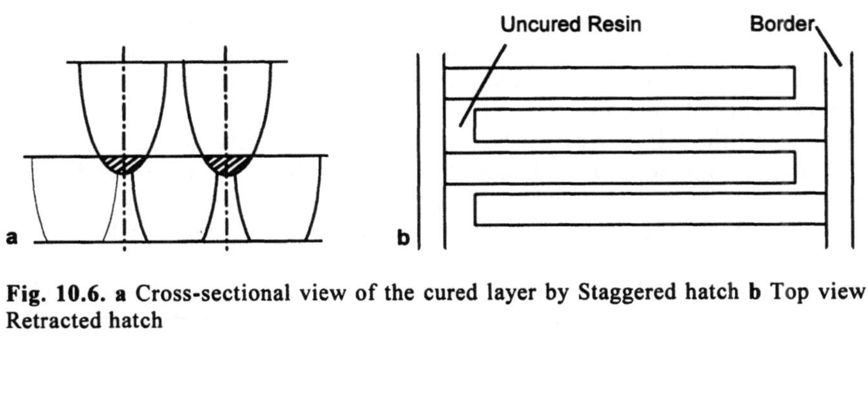

Fig. 10.6. **a** Cross-sectional view of the cured layer by Staggered hatch **b** Top view of Retracted hatch

Final accuracy of the part built by STL depends on various parameters. Several diagnostic tests and models have been developed by 3D systems to calibrate the STL system and to assess the accuracy achievable. The use of WINDOWPANE™ technique is used to obtain the appropriate process parameters for a given stereolithography resin. The use of CHRISTMAS-TREE™ diagnostic part is suggested to obtain the parameters such as line-width compensation (LWC) and shrinkage compensation factor (SCF). The choice of internal hatch scheme also has a significant influence on the part accuracy. It is reported that the accuracy achievable increases as one goes from Tri-Hatch to WEAVE™ to STAR-WEAVE™. For a standard test part, the accuracy achievable is about 0.127 mm of the CAD value [3].

10.3.3 Solid Ground Curing

This is similar to stereolithography. Both create the model layer by layer by curing a liquid photocurable resin. The difference is that in SGC, the layer is simultaneously cured rather than a scanning procedure adopted by stereolithography. This eliminates shrinkage and warping considerations. Simultaneous curing is achieved by first creating a mask on a glass plate and exposing it uniformly by an UV flash lamp. UV radiation passing through the mask solidifies the exposed areas of the resin. The remaining uncured resin, while still a liquid, is then removed and replaced by melted wax. A chilled plate is pressed down on the wax to solidify it. Finally, the slice is milled to the desired thickness forming a flat substrate on which the next layer is built. The milling device also allows the machine to erase flawed slices.

Thus, in SGC a multi-step process generates each layer of the part. Two cycles of operation continue simultaneously in the process. One of them prepares the mask for every layer and the other takes care of building the part layer by layer. After all the layers have been formed, the result is a part embedded within a solid block of wax. This is then melted and dissolved away to get the fully cured model.

The freedom from designing and removing supports, facilitates building many parts simultaneously because they can be suspended anywhere within the building chamber without special provisions. One-piece fabrication of entire functioning assemblies is possible in SGC. Since, resin is applied and removed on each layer, different resins could theoretically be used on different layers of the part.

Among all the RP processes, SGC has promise for producing parts with high accuracy and with low distortion. However, there is scope for further improvements. The ionography process, used to create the masks, deposits electric charge on the glass plate in raster lines with a pitch of 11.8 line per mm. This process introduces stairsteps on horizontal edges whose orientation is not perfectly parallel or perpendicular to the raster line. This is in addition to the usual 'stair stepping' effect in the vertical direction.

The horizontal edge stair stepping effect could be reduced if the raster scanning is made finer, say, 40 lines/mm.

Though, the basic SGC process is designed to simultaneously cure a layer of the model, practical limitations do exist in achieving this. For example, the air-knife wiper does not remove every drop of unexposed liquid resin. This residual resin is to be cured during a second UV exposure.

The SGC technology is well suited for building very large parts. Cubital introduced the Solider system based on SGC. SGC is significantly a larger machine and fairly noisy due to the fly cutter and vacuum system.

10.3.4 Fused Deposition Modelling

It is well known that the cost of a stereolithography system is quite high and cost-effective technologies were later developed for RP. One such approach is Fused Deposition Modelling (FDM). Stratasys Inc., (Minneapolis, MN) introduced FDM.

In this process, a model is built layer-by-layer using a thin filament of thermoplastic material such as ABS 400 polymer. The model is built in a closed cabinet maintained at 70°C. The model to be formed rests on the foam foundation provided on Z-stage platen and the z movement is given to the Z-stage platen. A unit called the FDM head, deposits the plastic material. The FDM head is mounted on a carriage which moves in the x and y direction. It has two nozzles; one for the model material (ABS plastic) and the other for the support material. The liquefier in the head melts the materials at 270°C for the model material and 265°C for the support material.

For each layer, the FDM head first deposits a perimeter road, which follows the shape of a slice curve. After the perimeter road, the head follows fill roads, which fill the solid areas inside the part. The selection of attributes such as road width, road height (slice thickness) and fill patterns (Fig. 10.7) is dependent on the part accuracy and part building time. When a layer is deposited completely, the platform is lowered by the amount of slice thickness and the process is repeated.

a b

Fig. 10.7. Various scanning schemes used in FDM: **a** annular hatch **b** zigzag hatch

The FDM features the breakaway support system (BASS), allowing the designer to create models with greater speed and precision. The supports are designed to prop-up the overhanging portions of the part during modelling. The supports detach easily, making the finished product look better with minimal post-modelling finishing.

The thickness of each slice of the part affects the surface texture and the accuracy in the *z*-axis. FDM gives slice thickness of 0.178, 0.254 and 0.356 mm. More the slice thickness, lesser will be the running time, but that will also degrade the surface finish.

The FDM process is inherently slow during layer deposition. However, no post curing is required for the model. RP models made from FDM are, in general, opaque. Among all the RP processes, the FDM technique is economical as it has relatively lower initial, as well as running cost. The systems simple operation, inert materials and lack of fumes/ash make the FDM process environment friendly and quite compatible with an office environment.

The usual materials used in FDM are such, that the models made are not directly amenable for transmission photoelastic stress analysis. However, if RP based on FDM is coupled with RT – one can make models that are photoelastically sensitive. One can also use photoelastic coatings directly on the RP model.

10.4 Direct Analysis of RP Models by Photoelastic Coatings

The motivation for direct analysis of RP models stems from the fact that the design evaluation could be made faster. The list of the parts built on FDM1650 machine for experimental purposes is given in Table 10.1. The default value for road width of 0.508 mm is used for making the prototypes. The applicability of photoelastic coatings and its limitations are discussed in this section.

Table 10.1. Parts built on FDM

No.	Model	Dimensions (mm)	Slice Thickness (mm)	Build Time (hrs)
1.	Plate with a hole	230 × 37 × 4	0.356	2.6
2.	Circular disc	ϕ60 × 6	0.254	2.2
3.	Tensile test specimen	ϕ20 × 120	0.356	5
		ϕ20 × 120	0.254	9

10.4.1 Experimental Results

Use of photoelastic coatings for metallic prototypes is well developed and has also been applied successfully for analysing fibre-reinforced plastics. The use of photoelastic coatings for RP model is new and several technological issues have to be looked at. The RP models are thin and also porous. In view of the low modulus of the model material, one should ideally use a coating that is meant for a low-modulus material. However, the low-modulus material has a low value of K, which may result in poor optical response. To increase optical response, one may have to increase the loading, which may not be desirable always. Use of coatings meant for high modulus materials may lead to undue reinforcement of the specimen.

Figure 10.8 shows the dark-field isochromatics observed for a disc (dia 59.35 mm, thickness 5.88 mm) made of FDM process with a slice thickness of 0.254 mm coated with the PS1 (Measurements Group, USA) coating material. The dark-field isochromatics shown correspond to the green image plane of the colour image recorded by a Handycam [7].

Fig. 10.8. Green channel of the dark-field isochromatics observed for the RP disc coated with the PS1 coating

In Chapter 2 it was mentioned that the coating could reinforce the model. The following properties are needed to calculate the reinforcing factor R_f. They are, E_s = 2482 MPa, h_s = 5.88 mm, ν_s = 0.35, E_c = 2500 MPa, D = 59.35 mm, ν_c = 0.38, h_c = 3.073 mm, K = 0.15, $\lambda = 574 \times 10^{-06}$ mm. The value of R_f, for the example problem chosen is 1.497. Table 10.2 compares the evaluation of difference in principal stress at the centre of the disc for various loads by experiment and theory. The results are more closer to theory, if R_f is used in the calculation.

Another example problem viz., the problem of a plate with a hole is selected to see the effect of various coating materials on the RP model [7]. The models were made with a slice thickness of 0.356 mm. Figure 10.9a shows the dark-field fringe patterns for a plate coated with PS1 material

Table 10.2. Principal stress difference at the centre of a disc with PS1 coating

Load (N)	Fringe order	$(\sigma_1 - \sigma_2)^s_{expt}$ without R_f (MPa)	$(\sigma_1 - \sigma_2)^s_{expt}$ with R_f (MPa)	$(\sigma_1 - \sigma_2)^s_{the}$ (MPa)	% diff without R_f	% diff with R_f
509.66	2.200	2.518	3.770	3.722	32.3	1.3
628.30	2.762	3.162	4.733	4.588	31.1	3.2
837.80	3.717	4.255	6.370	6.117	30.4	4.1
1047.20	4.514	5.167	7.735	7.647	32.4	1.2
1256.70	5.323	6.093	9.122	9.176	33.6	0.6

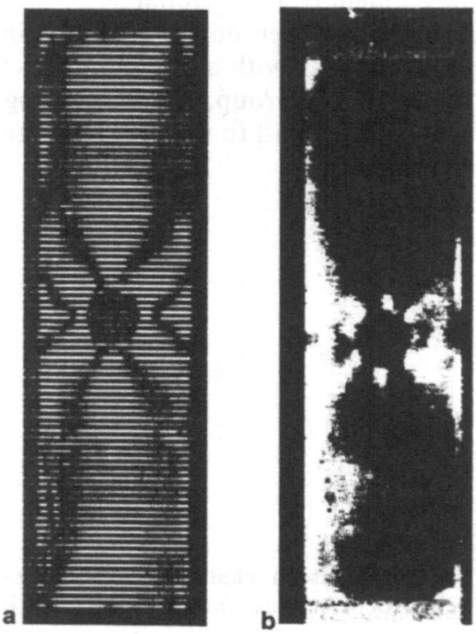

Fig. 10.9. Green channel of the dark-field isochromatics observed for the RP model coated with: **a** PS1 coating (304.9 N) and **b** PS3 coating (1088.9 N)

subjected to a load of 304.9 N. Figure 10.9b shows the fringe pattern for the plate with PS3 coating for a tensile load of 1088.9 N. As the strain coefficient of PS3 is much lower (about 7 times) than that of PS1, one has to apply a very high load to get a reasonable number of fringes [7].

10.4.2
Analysis of the Results

10.4.2.1
Evaluation of Young's Modulus by Tensile Test

The FDM process introduces pores in the models and hence their actual elastic properties could be different from the bulk properties of the plastic used. To get properties in actually manufactured conditions, it is desirable that standard tensile specimens are made by the RP process for different slice thicknesses of 0.254 and 0.356 mm and tested. The Table 10.3 shows that the Young's modulus is drastically reduced when a coarser slice thickness is used.

Table 10.3. Properties of RP models obtained from the tensile test

Slice thickness (mm)	E^s (MPa)	ν^s
Bulk material	2482	0.350
0.254	768.7	0.296
0.356	582.9	0.189

10.4.2.2
Study on the Seepage of the Adhesive

In the plate of slice thickness 0.356 mm, the seepage has been observed throughout the thickness [7]. Though the seepage for a model made of 0.254 mm is not visible to the naked eye; the seepage is visible when viewed through an optical microscope. It is shown in ref. [7] that the adhesive is embedded between the roads of the RP model. The adhesive has seeped up to a certain depth through these gaps in a zigzag pattern.

10.4.2.3
Numerical Simulation of Fringe Patterns

The results of the tensile tests (Table 10.3) have highlighted the porosity of the RP model. The adhesive used is aluminium filled epoxy preparation. Its seepage has contributed to the reinforcement of the porous RP model. In view of this, while performing the photoelastic coating experiment, no significant loss of rigidity was observed [7]. In view of the complexity of the problem, only a numerical simulation can help to understand better the reinforcement phenomenon.

Figure 10.10a shows the fringe pattern for a circular disc (with PS1 coating) under a diametral load of 509.66 N, obtained by FEM using the bulk material properties supplied by the manufacturer. The fringe patterns matches fairly well with those obtained by photoelastic experiment (Fig. 10.8). If the model had behaved as in the tensile test, then one should observe a large number of fringes (Fig. 10.10b).

For the case of a plate with a hole, Fig. 10.11a shows the fringe patterns reconstructed using the bulk material properties and Fig. 10.11b shows the same with properties obtained in the tensile test. By trial and error, the

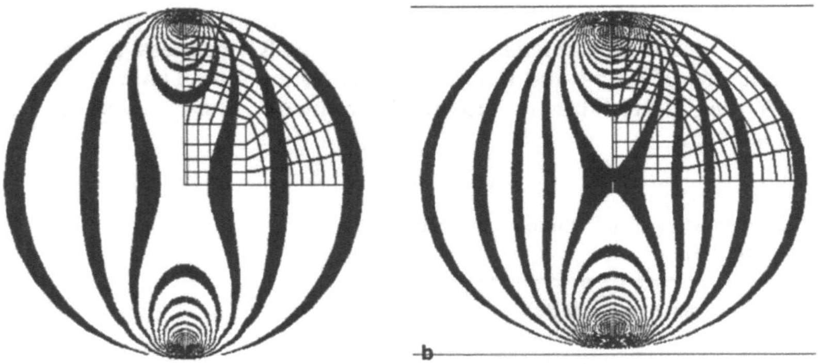

a b

Fig. 10.10. Dark-field isochromatics evaluated by FEM: **a** using bulk material properties **b** using the properties obtained from direct tensile test

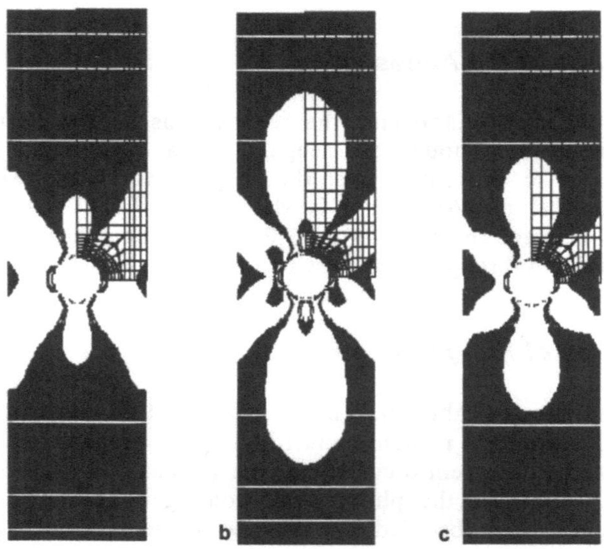

a b c

Fig. 10.11. Numerically evaluated dark-field isochromatics: **a** using bulk material properties **b** using tensile test properties **c** using those elastic properties that give fringes similar to the experiment

specimen Young's modulus is changed (ν is kept constant as 0.35) until the fringe contours are similar to those obtained from the experiment. Young's modulus of 3300 MPa was found to be suitable [7]. The respective fringe contours are shown in Fig. 10.11c and they compare well with the experiment (Fig. 10.9a).

10.4.3 Recommendations

Comparison of Figs. 10.8 and 10.10a shows that the porosity effect due to RP process and the reinforcement effect due to the seepage of the adhesive have balanced themselves for a slice thickness of 0.254 mm. The amount of seepage for the model of 0.356 mm slice thickness is quite high and the seepage of the adhesive is through the thickness. In view of this, the properties and the behaviour of the coated models are quite different. The coated RP specimen has to be modelled as a metal impregnated composite for comparing the model results to actual prototypes. This may be quite cumbersome in real life situations.

To get an engineering assessment of the problem, it is recommended that the RP model by FDM process be made of a slice thickness of 0.254 mm.

10.5 Direct Use of RP Models for Transmission Photoelastic Analysis

One of the basic requirements for RP models to be used for a direct transmission photoelastic analysis is that the model should be transparent. Further, the model material should exhibit the property of birefringence when stressed. Most plastics exhibit some level of birefringence. If the choice of the model material is such that if it can be stress frozen, then a 3-D stress analysis of the RP model could be performed by the conventional slicing technique.

Transparent models could be made by the stereolithography process. The use of CIBA-GEIGY 5170 resin, an epoxy material, for preparing photoelastically sensitive RP models has been reported in the literature [6]. The resin has favourable optical and elastic properties for the stress freezing process. Figure 10.12 shows the fringe patterns observed for a slice cut from a model of an automobile steering knuckle [6]. To improve the transparency of the cut slice, a film of mineral oil is coated. The Fig. 10.12 shows that the transparency of the slice is good but the optical clarity of the fringe patterns is not good. Depending on the weave patterns (Tri-Hatch, WEAVE™, STAR-WEAVE™), used to construct the STL model, the formation of the photoelastic fringes may appear disjointed. This makes the measurement process slightly more difficult when compared to the analysis of fringe patterns obtained from conventional models. The material stress fringe value at stress freezing temperature for this model material is 1.05 N/mm/fr [6]. For conventional photoelastic materials, this is about 0.35 to 0.53 N/mm/fr. Thus, to get a similar optical response, the RP models have to be loaded twice than that of the conventional model materials. This may pose problems for analysing thin-walled models. Another available stereolithography resin that exhibits photoelastic properties for stress

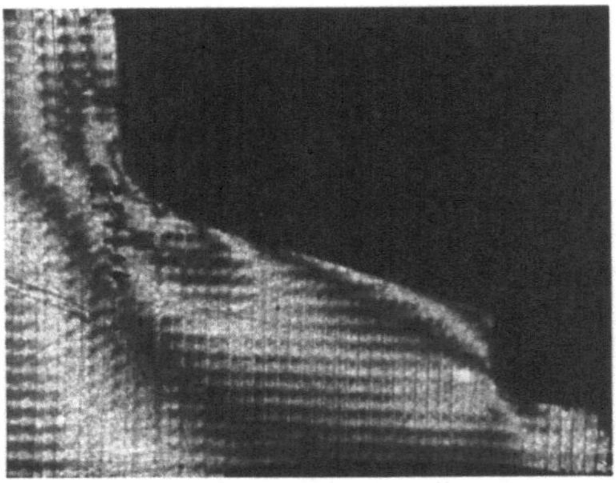

Fig. 10.12. Dark-field isochromatics observed for a slice cut from a STL model of an automobile steering knuckle (Courtesy of Measurements Group, Inc., Rayleigh, NC, USA)

freezing is SOMOS™ 6100 epoxy resin made by DuPoint. The value of material stress fringe value for this resin is found to be 0.83 N/mm/fr, which is slightly better than the previous one [6].

The optical clarity of the fringes may be improved by making models by SGC. This is because, in SGC, each layer is cured simultaneously and no problem of scanning the area within the slice exists. Figure 10.13 shows the fringe patterns for the problem of a disc under diametral compression at room temperature live load conditions. The model is made by using the G-5601 resin. The fringe patterns clearly indicate that the transparency of the model is quite poor. Nevertheless, the fringe clarity is better than the STL model. One does observe scattered dots in the field of view. This may be due to the uncured resin that was left back by the air knife during the wiping process. The material stress fringe value of the G-5601 resin at room temperature condition is 15.2 N/mm/fr, which is quite high. Since, STL and SGC use UV radiation for curing, in principle, the materials used in STL could be used in SGC as well. This may result in improving the transparency of the SGC models and by properly controlling the SGC process, the optical clarity of the fringes could be made similar to those observed in conventional model materials.

Fig. 10.13. Bright-field isochromatics observed for a SGC model of a circular disc under live load conditions

Though the use of RP models directly for stress freezing shortens the overall time for a design evaluation of the RP model, one should keep in mind that the model is quite expensive. The loading of the model has to be properly chosen to avoid its breakage or poor birefringent response. From this point of view, combining RP and RT is quite attractive, which is discussed next.

10.6 Rapid Tooling for Model Making

Among the various rapid tooling (RT) systems, the MCP vacuum casting machine (HEK GmbH, Germany) creates a silicone mould and using this models can be cast from a variety of polymers.

10.6.1 Basic Steps in Rapid Tooling

In RT, a mould is made of silicone rubber using the master model. One of the outstanding properties of the silicone rubber is its release action towards many materials. Further, it has high moulding accuracy, less shrinkage and heat stability. Surface variations up to 1 μm in size can be reproduced. The sequence of steps needed to make a mould by RT is discussed next.

For preparing the silicone mould, the model has to be suspended in a casting frame and provisions are to be made for the runner and riser by attaching the required diameter rods. In order to get the mould into two halves, one has to decide the parting line and mark it on the model by using a dark coloured adhesive tape so that it is visible through the semi-transparent mould. The selection of the parting line has to be done based on experience. Silicone rubber is to be mixed with the catalyst in the ratio 100:10 by weight and stirred for five minutes. Once the catalyst is added, vulcanisation starts and within seventy minutes the whole process of making the mould has to be completed. The silicone rubber is to be then poured into a container approximately five times the volume of the rubber taken (to allow for the increase in volume) and is then de-gassed for twenty minutes in the vacuum chamber. After this, the rubber is poured slowly and steadily all around the model in the casting frame. The mould is then placed in a vacuum chamber to do secondary de-gassing followed by curing at 40°C for eight hours. After the mould is cured, the casting frame and the gating rods are removed, and using a sharp scalpel, the mould is cut along the parting line. The mould is now ready for pouring the resin.

Among the resins compatible with the RT apparatus, the resin SG 95 of HEK GmbH (Germany) is found to be more transparent and also has a higher optical response [8]. This comes in two parts SG 95A (hardener) and SG 95B (resin). The shrinkage of the resin is 0.2% and its heat deflection temperature is 72°C. The silicone mould is applied with appropriate releasing agents and then pre-heated to 70°C. The two components of the resin, A and B, are measured by weight in the ratio 100:150. The mould and the resin are kept in the appropriate chambers of the vacuum casting machine. Resin B is to be stirred for twenty minutes in vacuum. After this, within 10 seconds, resin A is to be poured into the beaker containing resin B and 40 seconds are allowed for these to mix. The mixture is then poured into the mould. The mould assembly is removed from the vacuum chamber and is then placed in the oven at 70°C and allowed to cure for 1 hour. As the whole process of pouring the resin takes place under vacuum, there would be no air inclusions in the end component.

10.6.2 Digital Photoelastic Characterisation of the Process

To verify the suitability of the resin for stress freezing, a circular disc of diameter 60 mm and thickness of 5.7 mm is cast from the mould prepared by the RT process. The disc is loaded by a dead weight load of 98.1 N and allowed to go through a stress-freezing cycle wherein the soaking temperature is kept at 65°C. Figures 10.14a and 10.14b show the dark and bright-field isochromatics of the stress-frozen disc [9]. Two holes and a slot are carefully cut on the disc. The respective dark and bright-field images are shown in Figs. 10.14c and 10.14d. The fringe contours show no distortion and thus the resin SG 95 is amenable for stress freezing. Using a DIP procedure [10], the material stress fringe value is calculated and found to be 1.03 N/mm/fr. The room temperature material stress-fringe value of

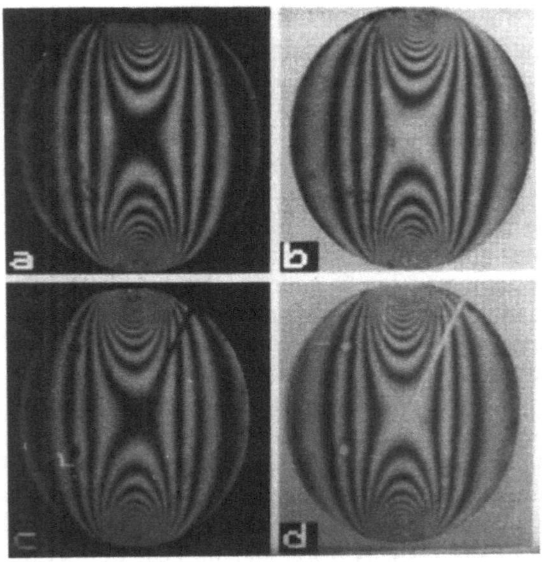

Fig. 10.14. Isochromatics observed for a stress-frozen disc made by RT using SG 95 resin. Note that the fringes are not distorted after introducing two holes and a slot

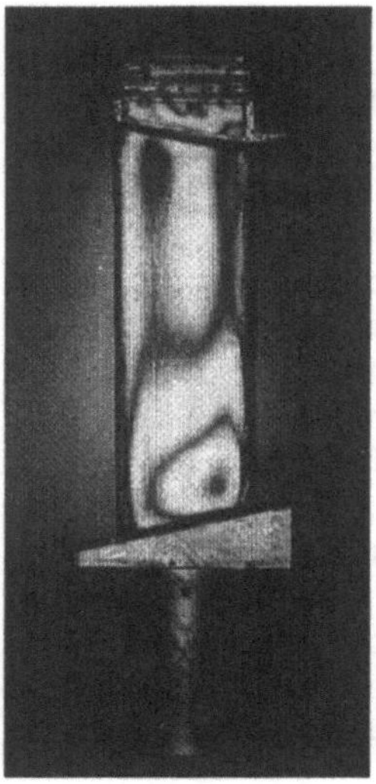

Fig. 10.15. Residual stresses observed in a turbine blade model due to a poor design of the mould

SG 95 is 7.5 N/mm/fr. A comparison of the properties of SG 95 with the conventional and STL resin is presented in Table 10.4.

The model being used for making the mould must be quite smooth. This is necessary for the cast model to be transparent. It is well known that the models made of RP process will have rough surface due to stair stepping effect. A finishing operation is thus needed before RP models could be used for making the mould. Further, a good design of the mould in terms of runners and raisers is necessary for getting a model free of residual stresses. If not, significant residual stresses could be present as shown in Fig. 10.15.

Table 10.4. Comparison of model materials for three-dimensional photoelastic stress analysis

Properties	Epoxy resin [6]	SG 95 Rapid Tooling Resin [9]	CIBA 5170 STL Resin [6]
Stress Freezing Temperature	104°C to 135°C	65°C	65°C
F_σ (N/mm/fringe)	0.35 to 0.53	1.03	1.05

10.7 Closure

In this chapter, the role of RP in providing models for 3-D photoelastic analysis is discussed. The model made by a RP process invariably has pores or elastic inhomogeneity in it. This will have a direct influence on its structural response. The reasons for the introduction of pores or elastic inhomogeneity by various RP processes have been highlighted.

The initial initiatives on using RP models directly for reflection and transmission photoelastic analysis are summarised. Though, these were pioneering efforts, in view of the model porosity, the comparison between the actual prototype and the model study may become quite involved and difficult. Conducting stress analysis by other experimental techniques like strain gauges, Moiré etc. will also have the same difficulty in extending the results of model studies to actual prototypes. In this context, the use of SGC process for RP model making may be better than other RP processes.

Development of functional prototypes by combining RP and RT is the ideal approach for evaluating the RP design for its structural response. The use of MCP vacuum casting machine for preparing a silicone mould from RP patterns is discussed in this chapter. Though, the use of SG 95 is recommended for model making, once the silicone mould is ready, even conventional 3-D photoelastic model materials could be used for making the model.

The use of photoelasticity has an edge over other experimental techniques, since, by choosing an appropriate resin in RT, the model could be used directly for a 3-D photoelastic analysis. The basic CAD data needed to create a RP model could also be directly used for FE analysis of the model. Many CAD packages such as I-DEAS, NASTRAN accepts input data of the component for FE analysis as a solid model. It is expected that the combination of digital photoelasticity, RP and FEM will dominate the design process in the years to come.

Exercises

10.1 What is STL format? Why is this required in making RP model?
10.2 Is an RP model made by stereolithography elastically homogeneous? Explain your answer with neat sketches.
10.3 Compare the RP processes stereolithography, solid ground curing and fused deposition modeling.
10.4 How do the pores in RP model affect the photoelastic coating analysis?
10.5 Compare direct transmission photoelastic study of RP model with RT model.

References

1. Kai CC, Fai LK (1996) Rapid prototyping – principles and applications. Wiley Eastern, New York
2. Wood L (1993) Rapid automated prototyping – an introduction. Industrial Press Inc, New York

3. Jacobs PF (1992) Rapid prototyping and manufacturing – fundamentals of stereolithography. Society of Manufacturing Engineers, McGraw-Hill Inc, New York
4. Ghosh A (1999) Rapid prototyping – a brief introduction. Affiliated East-West Press Ltd, New Delhi
5. Kruth JP, Leu MC, Nakagawa T (1998) Progress in additive manufacturing and rapid prototyping. Annals of the CIRP 47(2):525–540
6. Stress analysis of rapid prototype stereolithography models (1996) Reflections, The PhotoStress® Newsletter, Measurements Group Inc, USA
7. Ramesh K, Mehta RG, Dhande SG (1998) Photoelastic Analysis of Rapid Prototype Models. Proc of the Int Conference on Theoretical, Applied, Computational and Experimental Mechanics, IIT Kharagpur
8. Krishna Rao NCHSV (1999) Three-dimensional photoelastic analysis of rapid prototype models. MTech thesis, Department of Mechanical Engineering, IIT Kanpur
9. Ramesh K, Anilkumar M, Dhande SG (1999) Fusion of digital photoelasticity, rapid prototyping and rapid tooling technologies. Exp Tech 23(2):36–38
10. Ramesh K, Ganesan VR, Mullick SK (1991) Digital image processing of photoelastic fringes – a new approach. Exp Tech 15(5):41–46

11 Recent Developments and Future Trends

11.1 Introduction

The application of digital techniques for data acquisition in 2-D transmission photoelasticity and reflection photoelasticity is well developed and these were discussed in the earlier chapters. The extension of digital techniques for data acquisition to *integrated photoelasticity* and *scattered light photoelasticity* have just made a beginning. In integrated photoelasticity, one has to determine three parameters experimentally and these are known as characteristic parameters. In scattered light photoelasticity, the illumination levels are quite low due to light scattering. This necessitates the use of high-resolution CCD cameras. The ultimate aim of integrated photoelasticity or scattered light photoelasticity is to evaluate the stress field interior to the model. The current developments on tensorial tomography are discussed in this chapter.

There have been continuous developments in DIP hardware and software over the years. These developments have a direct bearing on the future of digital photoelasticity. The various developments on CCD camera modes are summarised. Portability of DIP analyses software is being recognised as one of the crucial issues for the widespread use of the techniques developed in digital photoelasticity. An attempt towards developing a device independent software for photoelasticity is also discussed in this chapter.

Photoelasticity has significantly contributed to the understanding of stress wave propagation, crack propagation and stress wave – crack interaction studies. It is appropriate that digital techniques are extended for data acquisition in such dynamic studies. In this chapter the challenges involved in data recording for dynamic problems is discussed followed by the current developments in digital data acquisition. Finally, the application of digital techniques for data acquisition in fibre composites is discussed.

11.2 Evaluation of Characteristic Parameters

A general three-dimensional photoelastic model with varying secondary principal stress differences and their directions along a given light path can be replaced by an optically equivalent model consisting of a linear retarder and a pure rotator. The parameters defining these are θ, the axis of the retarder, 2Δ, the retardation introduced by the retarder, and γ, the rotary power of the rotator. These are termed as characteristic parameters for a given length of the light path. The characteristic parameters are unique, and they depend only on the nature of stress distribution along the light path and the wavelength of the light.

The optically equivalent model reveals that if linearly polarised light is incident on an actual model at an azimuth θ or $\theta+\pi/2$, it will emerge still linearly polarised but rotated through an angle γ so its azimuth at the exit will be $\theta+\gamma$. These two directions at the entrance (θ, $\theta+\pi/2$) are called the primary characteristic directions and the two directions at the exit ($\theta+\gamma$, $\theta+\gamma+\pi/2$) are called the secondary characteristic directions. The primary and secondary characteristic directions do not, in general, coincide with the secondary principal stress axes of the stressed photoelastic model either at the entrance or at the exit.

Various methods exist for determining the characteristic parameters experimentally [1]. These methods can be broadly classified into direct and indirect methods. In the direct methods, the experimental procedure is such that, at the final stage, the orientation of the optical elements directly determines the characteristic directions. In this regard, several investigators have proposed methods involving iterative steps [2] and the use of auxiliary devices (half-wave plates) [3], rotating polarizers and rotating analyzers [4]. In the indirect methods, experimentally either the azimuth or the ellipticity or both of the exit light for various incident light conditions are measured. Characteristic parameters are then determined using the transformation relationships existing between the incident light conditions (i.e., the incident light ellipses) and the emerging light conditions (i.e., ellipses). Automation of characteristic parameters will go a long way in simplifying the experimental effort.

11.2.1 Srinath and Keshavan's Method

This is a direct method for determining the characteristic parameters. In this method, the model is placed between a polarizer and an analyzer combination. These two units are free to rotate independently; i.e., they are not kept crossed and are not coupled. To determine the characteristic directions, the polarizer is rotated until the transmitted light intensity at the point of interest becomes a minimum. The analyzer is next rotated to reduce this intensity to a further minimum. Next, the polarizer is again rotated to reduce the intensity to a further minimum. These operations are continued until the intensity becomes the least minimum (which is, theoretically, zero). When this occurs, the polarizer will be along one of the primary characteristic directions and the analyzer will be along one of the secondary characteristic directions. It can be shown [2] that the convergence process is quite fast and for each cycle of iteration, the offset angle of the optical elements are reduced by a factor of $\cos^2 2\Delta$. Once the characteristic directions are known, the characteristic retardation is determined by a procedure similar to the Tardy method of compensation used in 2-D photoelasticity.

11.2.2
Whole Field Determination of Characteristic Parameters by Phase Shifting

11.2.2.1
Development of Relevant Equations

For evaluating characteristic parameters by phase shifting technique, one has to use both circular and plane polariscope arrangements [5]. If in Fig. 5.2 (Chap. 5), the 2-D model is replaced by a retarder (θ, 2Δ) and a rotator (γ) representing a 3-D model, the components of light vector along the analyzer axis and perpendicular to the analyzer axis for the circular polariscope arrangement are obtained as

$$\begin{Bmatrix} E_\beta \\ E_{\beta+\pi/2} \end{Bmatrix} = \frac{1}{2}\begin{bmatrix} \cos\beta & \sin\beta \\ -\sin\beta & \cos\beta \end{bmatrix}\begin{bmatrix} 1-\mathrm{i}\cos 2\eta & -\mathrm{i}\sin 2\eta \\ -\mathrm{i}\sin 2\eta & 1+\mathrm{i}\cos 2\eta \end{bmatrix}$$
$$\times\begin{bmatrix} \cos\Delta\cos\gamma-\mathrm{i}\sin\Delta\cos(2\theta+\gamma) & -\cos\Delta\sin\gamma-\mathrm{i}\sin\Delta\sin(2\theta+\gamma) \\ \cos\Delta\sin\gamma-\mathrm{i}\sin\Delta\sin(2\theta+\gamma) & \cos\Delta\cos\gamma+\mathrm{i}\sin\Delta\cos(2\theta+\gamma) \end{bmatrix} \quad (11.1)$$
$$\times\begin{bmatrix} 1 & \mathrm{i} \\ \mathrm{i} & 1 \end{bmatrix}\begin{Bmatrix} 0 \\ 1 \end{Bmatrix} k e^{\mathrm{i}\omega t}$$

where $ke^{\mathrm{i}\omega t}$ is the incident light vector, E_β and $E_{\beta+\pi/2}$ are the components of light vector along the analyzer axis and perpendicular to the analyzer axis respectively.

The intensity of light transmitted for arbitrary positions of the second quarter-wave plate and the analyzer is obtained as

$$I_i = I_\mathrm{b} + \frac{I_\mathrm{a}}{2} + \frac{I_\mathrm{a}}{2}[\sin 2(\beta_i-\eta_i)\cos 2\Delta - \sin 2(\theta-\eta_i+\gamma)\cos 2(\beta_i-\eta_i)\sin 2\Delta] \quad (11.2)$$

where I_a accounts for the amplitude of the light vector and the proportionality constant while I_b accounts for the background/stray light.

It is interesting to note that in Eq. (11.2) if γ is zero, it reduces to the equation that is used in two-dimensional photoelasticity [Eq. (5.27), Chap. 5]. Six-step phase shifting method used in 2-D photoelasticity could be extended to determine the characteristic retardation and secondary characteristic direction ($\theta+\gamma$). The relevant optical arrangements and the intensity equations are summarised in Table 11.1. From the intensity equations in Table 11.1, the secondary characteristic direction is obtained as

Table 11.1. Table showing the six unique intensity equations for the different orientations of second quarter-wave plate and analyzer

ξ	η	β	Intensity equation
$3\pi/4$	0	$\pi/4$	$I_1 = I_b + \frac{I_a}{2}(1+\cos 2\Delta)$
$3\pi/4$	0	$3\pi/4$	$I_2 = I_b + \frac{I_a}{2}(1-\cos 2\Delta)$
$3\pi/4$	0	0	$I_3 = I_b + \frac{I_a}{2}[1-\sin 2(\theta+\gamma)\sin 2\Delta]$
$3\pi/4$	$\pi/4$	$\pi/4$	$I_4 = I_b + \frac{I_a}{2}[1+\cos 2(\theta+\gamma)\sin 2\Delta]$
$3\pi/4$	$\pi/2$	$\pi/2$	$I_5 = I_b + \frac{I_a}{2}[1+\sin 2(\theta+\gamma)\sin 2\Delta]$
$3\pi/4$	$3\pi/4$	$3\pi/4$	$I_6 = I_b + \frac{I_a}{2}[1-\cos 2(\theta+\gamma)\sin 2\Delta]$

$$\tan 2(\theta+\gamma) = \frac{I_5 - I_3}{I_4 - I_6} \tag{11.3}$$

and the characteristic retardation is then obtained as follows

$$\tan 2\Delta = \frac{I_5 - I_3}{(I_1 - I_2)\sin 2(\theta+\gamma)} = \frac{I_4 - I_6}{(I_1 - I_2)\cos 2(\theta+\gamma)} \tag{11.4}$$

In Eq. (11.4), in the regions where $\cos 2(\theta+\gamma)$ or $\sin 2(\theta+\gamma)$ are small, intensities $(I_4 - I_6)$ and $(I_5 - I_3)$ will also be small. In view of digitisation, the evaluation of 2Δ will involve the ratio of two small discrete numbers and will be in considerable error in such regions.

To avoid this, a new expression for 2Δ is obtained as follows

$$\tan 2\Delta = \frac{(I_5 - I_3)\sin 2(\theta+\gamma) + (I_4 - I_6)\cos 2(\theta+\gamma)}{(I_1 - I_2)} \tag{11.5}$$

The numerator of Eq. (11.5) always guarantees a high modulation over the field and the values of 2Δ calculated are expected to be more accurate. Equation (11.5) is similar to the one derived by Quiroga and González-Cano [6] for two-dimensional photoelasticity.

For evaluating θ and γ independently, one requires additional equations and these are to be obtained by viewing the 3-D model in a plane polariscope. Replacing the 2-D model in Fig. 5.1 (Chap. 5) by a 3-D model consisting of a retarder and a rotator, the components of light vector along the analyzer axis and polarizer axis are obtained by Jones calculus as follows

$$\begin{Bmatrix} E_\beta \\ E_{\beta+\pi/2} \end{Bmatrix} = \begin{bmatrix} \cos\beta & \sin\beta \\ -\sin\beta & \cos\beta \end{bmatrix} \times \begin{bmatrix} \cos\Delta\cos\gamma - \mathrm{i}\sin\Delta\cos(2\theta+\gamma) & -\cos\Delta\sin\gamma - \mathrm{i}\sin\Delta\sin(2\theta+\gamma) \\ \cos\Delta\sin\gamma - \mathrm{i}\sin\Delta\sin(2\theta+\gamma) & \cos\Delta\cos\gamma + \mathrm{i}\sin\Delta\cos(2\theta+\gamma) \end{bmatrix} \times \begin{Bmatrix} \cos\alpha \\ \sin\alpha \end{Bmatrix} k e^{\mathrm{i}\omega t} \tag{11.6}$$

On simplification, the intensity of light transmitted can be expressed as

$$I_i = I'_\mathrm{b} + I'_\mathrm{a}[\cos^2\Delta\cos^2(\beta_i - \alpha_i - \gamma) + \sin^2\Delta\cos^2(\beta_i + \alpha_i - 2\theta - \gamma)] \tag{11.7}$$

where I'_a accounts for the amplitude of the light vector and the proportionality constant while I'_b accounts for the background/stray light for plane polariscope arrangement. In these quantities, a prime is added to indicate that in a general situation these could be different from what is used in a circular polariscope arrangement. However, while deriving the expressions for characteristic parameters, this distinction is not followed. Thus, if an expression involves intensities from both plane and circular polariscope it is to be understood that the expression is only approximate.

If one takes all possible combinations of α and β in steps of $\pi/4$, one finds that only eight optical arrangements are independent and unique which are shown in Table 11.2. The respective intensity equations are summarised in Table 11.3. The starting number of the intensity equation sequence is labelled as 7 for ease of book keeping in experimentation.

Table 11.2. Table showing that only eight intensity equations turn out to be independent for all the possible combination of α and β in step of $\pi/4$

β \ α	0	$\pi/4$	$\pi/2$	$3\pi/4$	π
0	I_{13}	I_{14}	I_7	I_{10}	I_{13}
$\pi/4$	I_{11}	I_9	I_8	I_{12}	I_{11}
$\pi/2$	I_7	I_{10}	I_{13}	I_{14}	I_7
$3\pi/4$	I_8	I_{12}	I_{11}	I_9	I_8
π	I_{13}	I_{14}	I_7	I_{10}	I_{13}

Table 11.3. Table showing the eight unique intensity equations for different orientations of the polarizer and analyzer.

α	β	Intensity equation
0	$\pi/2$	$I_7 = I'_b + I'_a[\cos^2\Delta\sin^2\gamma + \sin^2\Delta\sin^2(2\theta+\gamma)]$
0	$3\pi/4$	$I_8 = I'_b + I'_a/2[1 - \cos^2\Delta\sin 2\gamma - \sin^2\Delta\sin(4\theta+2\gamma)]$
$\pi/4$	$\pi/4$	$I_9 = I'_b + I'_a[\cos^2\Delta\cos^2\gamma + \sin^2\Delta\sin^2(2\theta+\gamma)]$
$\pi/4$	$\pi/2$	$I_{10} = I'_b + I'_a/2[1 + \cos^2\Delta\sin 2\gamma - \sin^2\Delta\sin(4\theta+2\gamma)]$
0	$\pi/4$	$I_{11} = I'_b + I'_a/2[1 + \cos^2\Delta\sin 2\gamma + \sin^2\Delta\sin(4\theta+2\gamma)]$
$\pi/4$	$3\pi/4$	$I_{12} = I'_b + I'_a[\cos^2\Delta\sin^2\gamma + \sin^2\Delta\cos^2(2\theta+\gamma)]$
0	0	$I_{13} = I'_b + I'_a[\cos^2\Delta\cos^2\gamma + \sin^2\Delta\cos^2(2\theta+\gamma)]$
$\pi/4$	0	$I_{14} = I'_b + I'_a/2[1 - \cos^2\Delta\sin 2\gamma + \sin^2\Delta\sin(4\theta+2\gamma)]$

Using the intensity equations in Table 11.3, the characteristic rotation is obtained as follows

$$\tan 2\gamma = \frac{I_{10} - I_8}{I_9 - I_7} = \frac{I_{10} - I_8}{I_{13} - I_{12}} = \frac{I_{11} - I_{14}}{I_9 - I_7} = \frac{I_{11} - I_{14}}{I_{13} - I_{12}} \tag{11.8}$$

By substituting γ in Eq. (11.3), θ can be found. Thus, a minimum of ten optical arrangements, six in circular and four in plane polariscope, are needed to evaluate the three experimental parameters. From the set of equations in Table 11.3, it is also possible to evaluate the value of $(4\theta+2\gamma)$, which is related to the intensities as

$$\tan(4\theta + 2\gamma) = \frac{I_{11} - I_{10}}{I_{12} - I_7} = \frac{I_{11} - I_{10}}{I_{13} - I_9} = \frac{I_{14} - I_8}{I_{12} - I_7} = \frac{I_{14} - I_8}{I_{13} - I_9} \tag{11.9}$$

The above equation can be used to check the value of θ obtained using the circular polariscope arrangement. Comparing Eqs. (11.8) and (11.9), to include the evaluation of $(4\theta+2\gamma)$, one has to capture two extra images viz., I_{11} and I_{12}. Thus a total of 12 intensities from I_1 to I_{12}, six in circular and six in plane polariscope, are needed to apply phase-shifting methodology for integrated photoelasticity. Using Eqs. (11.8), (11.9) and (11.3) the expression for $\tan 4\theta$ and $\tan 2\theta$ are obtained as follows

$$\tan 4\theta = \frac{(I_{11} - I_{10})(I_9 - I_7) - (I_{10} - I_8)(I_{12} - I_7)}{(I_{12} - I_7)(I_9 - I_7) + (I_{11} - I_{10})(I_{10} - I_8)} \tag{11.10}$$

$$\tan 2\theta = \frac{(I_5 - I_3)(I_9 - I_7) - (I_{10} - I_8)(I_4 - I_6)}{(I_4 - I_6)(I_9 - I_7) + (I_5 - I_3)(I_{10} - I_8)} \tag{11.11}$$

11.2.2.2 Experimental Evaluation of Characteristic Parameters

A circular disc under diametral compression, (disc diameter 55.24 mm, thickness 5.10 mm, diametral load 753.95 N and material stress fringe value 12.62 N/mm/fringe) is viewed in oblique incidence of 30°. This simulates a situation of variation of secondary principal stresses and their orientation along the light path. Figure 11.1 shows the twelve images for the problem under consideration for the positions indicated in Tables 11.2 and 11.3. Figure 11.2a shows the phase map when the secondary characteristic direction $(\theta+\gamma)$ is evaluated by circular polariscope arrangement [Eq. (11.3)]. The figure shows that in the region of stress concentration, the phase map is not good. Referring to Eq. (11.5), the value of characteristic retardation is affected by the accuracy of evaluation of $(\theta+\gamma)$. It is shown in ref. [5] that $(\theta+\gamma)$ can be improved by using Eqs. (11.8) and (11.10), which essentially use plane polariscope arrangement. The result of phase map calculated using the improved estimation of $(\theta+\gamma)$ is shown in Fig. 11.2b. Here one can see a prominent improvement in the quality of phase map in the stress concentration zones. Figure 11.2c shows the simulated dark-field using the improved estimation of secondary characteristic direction. Figure 11.2c compares well with the experimental dark-field integrated retardation pattern (Fig. 11.1b).

Figure 11.3a shows the plots of 0° contour of primary characteristic direction evaluated by Eq. (11.10). In Fig. 11.3a, spikes can be seen around integer fringe orders, which represent the undefined zones for the parameter being evaluated. These spikes are considered as noise. To remove these, the plots of θ contour is to be skipped in the neighbourhood of integer fringe order. Figure 11.3b shows the θ contour with noise removed. A tolerance of ±0.1 fringe order would be suitable to effect this. Figure 11.4a shows the $\gamma = 0°$ contour without noise removal and Fig. 11.4b shows with noise removal.

11.2.2.3 Whole Field Theoretical Evaluation of Characteristic Parameters

A whole field comparison of characteristic parameters obtained by phase shifting with theoretical solution is desirable. Since, the stress field for the problem of a circular disc under diametral compression is known, it is possible to evaluate the characteristic parameters theoretically. For theoretical determination, one has to model the continuous variation of stress field as a set of discrete retarders. For a length of 8 mm light path, about 120 retarders would be adequate. Each retarder gives retardation at

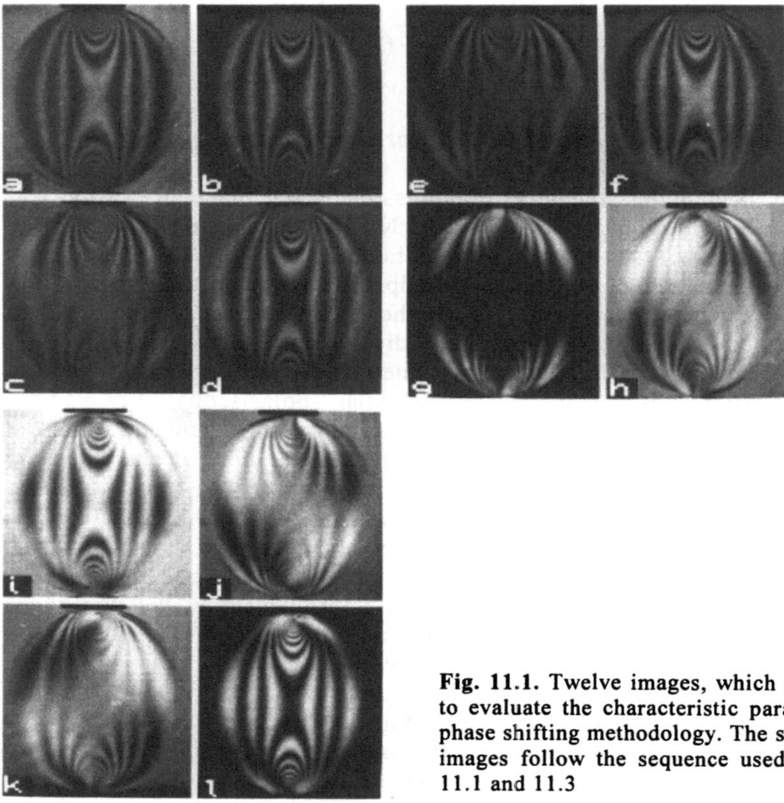

Fig. 11.1. Twelve images, which are needed to evaluate the characteristic parameters by phase shifting methodology. The sequence of images follow the sequence used in Tables 11.1 and 11.3

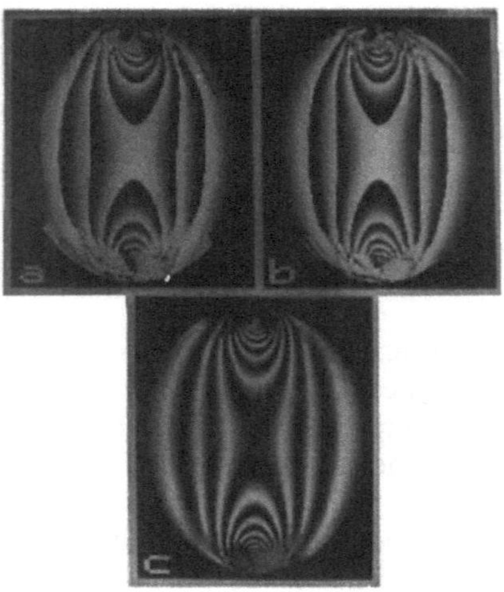

Fig. 11.2. Characteristic retardation obtained by phase shifting technique: **a** phase map when $(\theta+\gamma)$ is obtained by Eq. (11.3) **b** phase map when $(\theta+\gamma)$ is obtained by Eqs. (11.8) and (11.9) **c** simulated dark-field based on 'b'

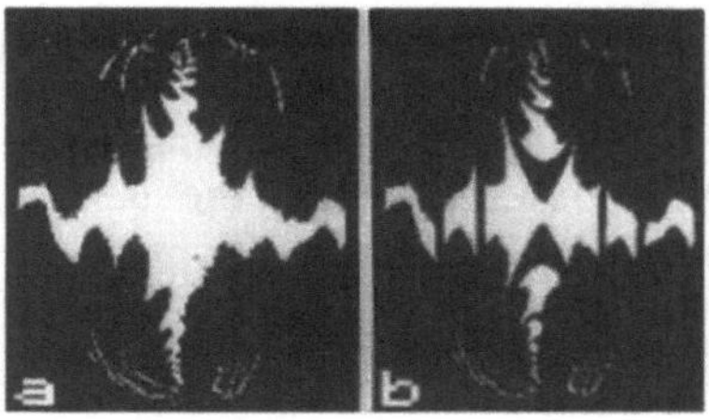

Fig. 11.3. Zero degree primary characteristic direction contour obtained by Eq. (11.10): **a** without noise removal **b** after noise removal

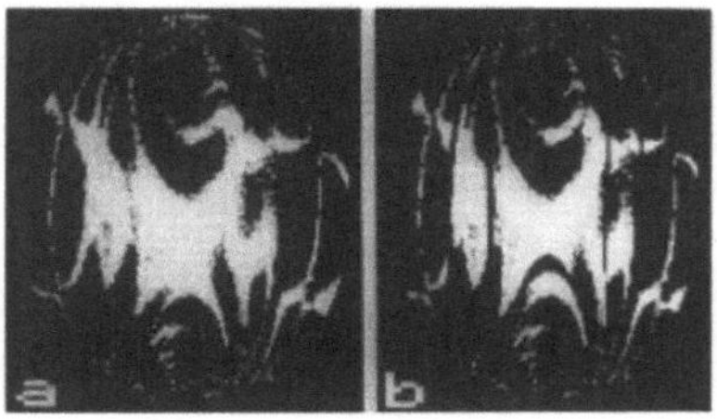

Fig. 11.4. Zero degree characteristic rotation contour: **a** without noise removal **b** after noise removal

varying azimuths. The appropriate secondary principal stress difference and their orientation are to be calculated from the theoretical solution to form the Jones Matrix of each retarder. These matrices are to be then appropriately multiplied to get the final matrix. For the complete train of discrete retarders, the Jones Matrix can be written as

$$[M]=\begin{bmatrix} P-\mathrm{i}Q & -R-\mathrm{i}S \\ R-\mathrm{i}S & P+\mathrm{i}Q \end{bmatrix} \tag{11.12}$$

where,

$$\begin{aligned} P &= \cos\varDelta\cos\gamma \\ Q &= \sin\varDelta\cos(2\theta+\gamma) \\ R &= \cos\varDelta\sin\gamma \\ S &= \sin\varDelta\sin(2\theta+\gamma) \end{aligned} \tag{11.13}$$

From the values of P, Q, R and S the theoretical characteristic parameters can be obtained. Sufficient care has to be taken to account for the multiplicity of the solutions involved in solving the inverse trigonometric equations.

The above mentioned procedure is to be repeated in pixel increments to plot the whole field contours. It is to be noted, that unlike the two-dimensional theoretical simulation of isoclinics, the theoretical simulation of contours of θ and γ will not be continuous. This is because in Eq. (11.13), γ is not defined when $\varDelta$ takes the values of $\pi/2$, $3\pi/2$... etc. Since γ is not defined for fringe orders of 0.5, 1.5 ..., θ is also not defined in these zones and the contours would be discontinuous. For the problem under consideration, the whole field representation of 2Δ, and the contours of $\theta = 0^\circ$ and $\gamma = 0^\circ$ are shown in Fig. (11.5). Along the vertical axis, leaving the central zone, γ is not zero (Fig. 11.5c) and hence, the $\theta = 0^\circ$ contour is not symmetric with respect to y-axis (Fig. 11.5b).

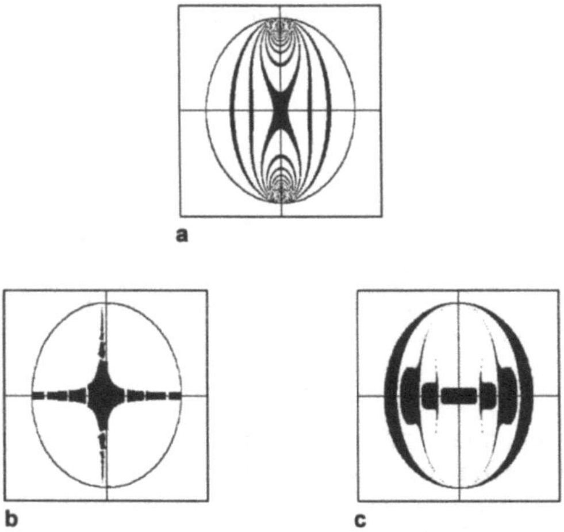

Fig. 11.5. Whole field representation of theoretical characteristic parameters for the problem of disc under diametral compression: **a** 2Δ **b** $\theta = 0^\circ$ and **c** $\gamma = 0^\circ$ contours

Comparison of Figs. 11.2c and 11.5a shows that the characteristic retardation matches well with the theory. However, the comparison of Figs. 11.3b and 11.5b shows that the evaluation of primary characteristic direction is only satisfactory. A similar situation exists in the evaluation of characteristic rotation too (Figs. 11.4b and 11.5c).

To assess the quantitative comparison of the evaluation of the parameters, 2-D graphs of these parameters are obtained for a line at y/R = 0.126. These are shown in Fig. 11.6. In these graphs, the theoretical plots of θ and γ are smoothed out by using a fringe tolerance of ±0.1. The graphs show very close comparison of characteristic retardation for both these lines. The evaluation of θ and γ are, in general, not very good. Future research should pay attention on improving this.

11.3 Tensorial Tomography

The problem of evaluating 3-D elastic stress analysis is a problem of tensor field tomography. Stress is a tensor of rank two; thus one is concerned with

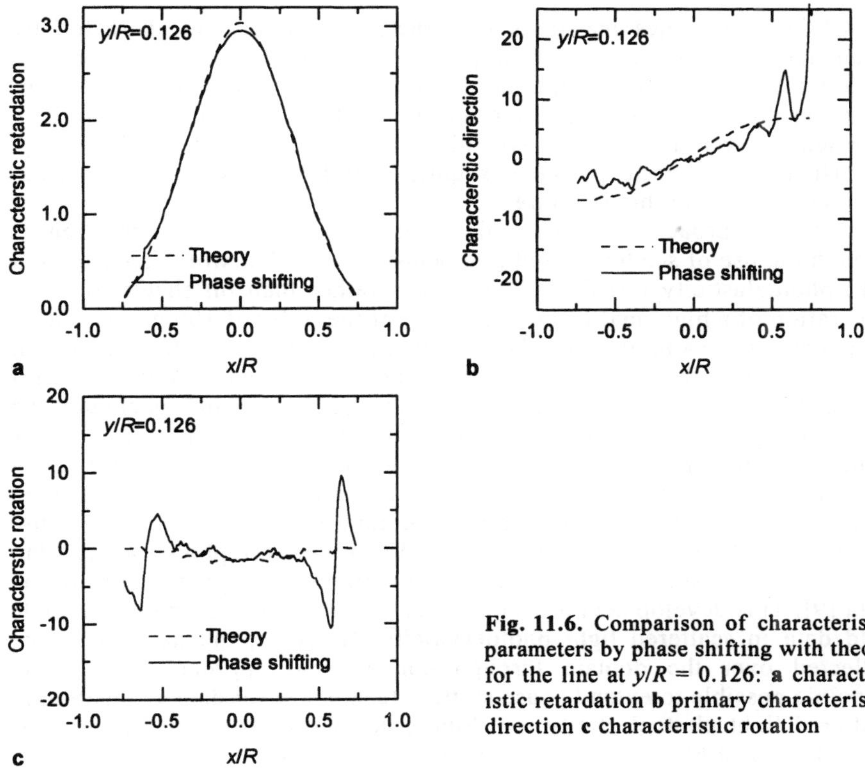

Fig. 11.6. Comparison of characteristic parameters by phase shifting with theory for the line at $y/R = 0.126$: **a** characteristic retardation **b** primary characteristic direction **c** characteristic rotation

determining six unknown distributions which are mutually dependent. The classical Radon inverse transform that is the theoretical basis of a scalar field tomography, cannot be applied here in a direct way [7].

In scalar field tomography, the distribution of the parameter of interest is reconstructed from integrated measurements of that parameter taken at multiple angles. The task of reconstruction in tensor field tomography is much more difficult and it is still at the early stage of exploration. In scalar field tomography, the reconstruction of the parameter for 2-D cross sections of the field is successfully done. This is not possible for the tensor field as the tensor value depends not only on the co-ordinates of the point but also on the ray direction used for the integrated measurement. Only a hybrid approach involving the use of theoretical, experimental and numerical study can help to solve such a problem. Such an approach is widely used in integrated photoelasticity [8].

In integrated photoelasticity, an approximate analytical solution of the stress field is assumed and Jones matrix for the ray of interest is analytically evaluated. The situation is similar to the one discussed in Sect. 11.2.2.3. Instead of using the closed form solution, the secondary principal stress difference and their orientation have to be calculated on the basis of an assumed stress field over the light path. From experimental analysis, for the ray of interest, Jones matrix is constructed based on the characteristic parameters. The parameters defining the approximate analytical solution are numerically obtained by comparing the Jones matrices. The recent developments on the use of phase shifting techniques for evaluating characteristic parameters over the field have given a boost to the methodology. Nevertheless, the basic problem is quite complex and only simple cases have been solved in the literature.

Another approach to acquire data for tomographic reconstruction is through the use of scattered light photoelasticity. Earlier works on scattered light photoelasticity relied on light beams rather than on sheets of light. Generation of a high quality sheet of light, of the order of 100 mm in width, for conducting scattered light photoelastic recording is reported in ref. [9]. In this, the light from a 2 mm dia laser source is allowed to fall on a front-coated mirror traversing slowly along a straight line, which generates the sheet of light. A computer controlled stepper motor and a lead screw achieve the traversing of the mirror (at the rate of 0.5 mm/minute). Even though a line of light illuminates the model at any instant of time, a camera placed with its shutter open throughout the duration of traversal will capture the fringe patterns along the entire path of traverse on a photographic film. Typically four traverses were used to obtain high quality photographs of the field [9]. This development has showed an easy way for recording the whole field data in scattered light photoelasticity. However, the data has to be collected over the domain through conventional approaches. It is in principle possible to extend phase shifting techniques for data acquisition in scattered light photoelasticity too. Since, the phenomenon of scattering is used either as a polarizer or as an analyzer in scattered light photoelasticity, the phase shifting technique envisaged can have only an output or an input quarter-wave plate, as the case may be. Automated data acquisition techniques in this direction can be expected to appear in the future.

Dupré and Lagarde [10, 11] introduced an interesting approach, which simplifies the mathematical analysis of the scattered light patterns. In this, two sheets of light are used to delimit a slice optically from a 3-D model (Fig. 11.7). The sheets of light were obtained by using appropriate optics involving cylindrical lenses unlike the scanning mechanism discussed previously.

The model is viewed perpendicular to the incident light direction as in scattered light photoelasticity. The intensity of scattered light is recorded by a CCD camera for each sheet of light separately (I_1 & I_2) and finally with both the sheets together (I). By introducing the speckle pattern properties, it is shown that the correlation factor of the two scattered beams is similar to the intensity equation obtained in a plane polariscope for the investigation of a slice in a classical frozen-stress technique. The distance between the two sheets of light determines the thickness of the delimited slice. The total scattered light intensity I is

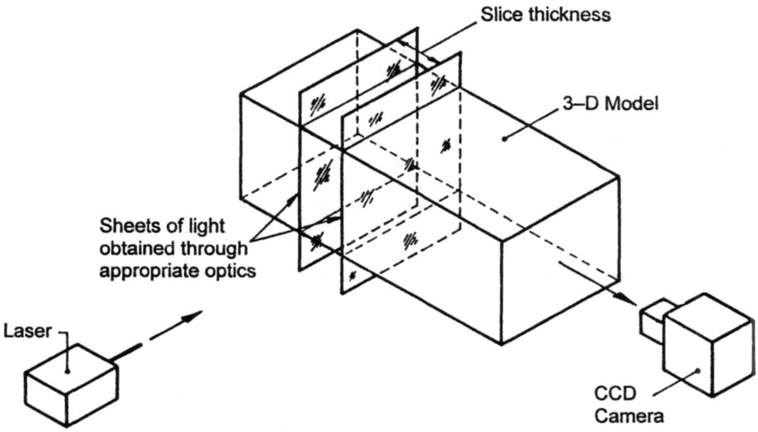

Fig. 11.7. A slice in a three-dimensional model delimited by two sheets of light for conducting a digital scattered light photoelastic analysis

$$I = I_1 + I_2 + 2\sqrt{I_{1s} I_{2s}}\, C_R \cos(\psi_1 + \psi_2 + \chi) \tag{11.14}$$

where I_1, I_2 are the intensities due to the two plane laser beams, ψ_1, ψ_2 are the random phases of the speckle fields, χ is a function of the optical characteristics of the slice and C_R is the correlation factor of the two speckle fields given by

$$C_R^2 = 1 - \sin^2 2\theta \sin^2 \frac{\delta}{2} \tag{11.15}$$

which is same as the intensity of light transmitted in a plane polariscope bright-field.

From the record of intensities I, I_1 and I_2 one has to calculate C_R^2. A whole field plot of it will reveal the isochromatic and isoclinic fringes. To simplify the approach, it is assumed that the intensities of the two incident beams are equal and also the average of scattered light to be equivalent. Thus

$$k = \frac{\langle I_{1s} \rangle}{\langle I_1 \rangle} = \frac{\langle I_{2s} \rangle}{\langle I_2 \rangle} \tag{11.16}$$

where k is a factor related to the scattered properties of the material. The value of C_R^2 is to be obtained from

$$k^2C_R^2 = \frac{\sigma^{*2}}{2\langle I_1\rangle\langle I_2\rangle} \tag{11.17}$$

where σ^{*2} is the variance of I^* defined by

$$I^* = I - I_1 - I_2 = 2\sqrt{I_{1s}I_{2s}}C_R\cos(\psi_1 + \psi_2 + \chi) \tag{11.18}$$

For a calculation area of $2N \times 2N$, the square of the variance is obtained by

$$\begin{aligned}\sigma^{*2}(m,n) &= \left\langle [I^*(m,n) - \langle I^*(m,n)\rangle]^2 \right\rangle \\ &= \frac{\left\{\sum_{o=-N}^{o=+N}\sum_{p=-N}^{p=+N}[I^*(m+o,n+p) - \langle I^*(m,n)\rangle]^2\right\}}{(2N+1)^2}\end{aligned} \tag{11.19}$$

where,

$$\langle I^*(m,n)\rangle = \frac{\left\{\sum_{o=-N}^{o=+N}\sum_{p=-N}^{p=+N} I^*(m+o,n+p)\right\}}{(2N+1)^2} \tag{11.20}$$

The image obtained by plotting $k^2C_R^2$will not be clear to reveal isochromatic and isoclinic fringes. A low pass filter has to be used to improve this [10].

The intensity of the speckle field is very low and special care has to be taken for selecting the CCD camera. It is reported that the use of commercial cameras with spatial resolution of 512×512 with 256 quantization levels is not adequate for this purpose. Use of a high-resolution camera with a spatial resolution of 1000×1000 and quantization levels of 1024 is recommended [10]. Even with such a high-resolution camera, the appearance of isochromatic and isoclinic fringes is very poor. Future research should concentrate on improving this. Here again the use of phase shifting technique for automatic data acquisition could be explored.

Since the optically isolated slice could be analysed as a 2-D model, the mathematical complexities of the general tomographic analysis are greatly reduced. Nevertheless, the experimental technique has to be further refined for obtaining a meaningful data reduction.

11.4 Developments in DIP Hardware

Continuous developments in DIP hardware have been reported over the years. The development includes improvement of frame grabbers and development of high resolution and special purpose cameras. With developments in PCI architecture, frame grabbers that are compatible to this have appeared in the market. These are Matrox Meteor, Pulsar, Genesis etc.

Since, image acquisition is an important component of digital photoelasticity; the new developments in camera technology have a direct bearing on the future of digital photoelasticity. High-resolution cameras of the order of 4k × 4k are available now. Also 3CCD RGB cameras having separate sensors for each image planes of R, G and B have also come into the market. With further developments in hardware, these could become affordable.

In Chap. 3, interline transfer CCD cameras were described. Several other modes of scanning of CCD arrays have been developed to suit various applications. Progressive scan and time-delay and integration (TDI) mode of CCD operation are some of these.

Progressive scan implies non-interlaced or sequential line-by-line scanning of the image. The primary advantage of progressive scan is that the entire image is captured at one instant of time. For capturing moving objects, cameras based on progressive scan could be used with ease. In interlaced systems, since the two fields are collected sequentially in time, it may so happen that if a strobe light is used to stop motion, the image will appear only in the field that was active during the light pulse. This will lead to sharp reduction on the vertical resolution of the image.

Two different architectures are available in progressive scan cameras. These are frame transfer CCD (FT CCD) and full-frame interline transfer CCD. In FT CCD, the sensor array is vertically twice the normal sensor array size and the lower half is masked. The masked portion is used as a storage area. Full-frame interline transfer CCD is the recent development and is a derivative of the interline transfer CCD (Fig. 3.26, Chap. 3). The difference is that each active pixel has its own storage site. This type of construction offers the advantage of progressive scan coupled with fast charge transfer.

Time-delay and integration (TDI) is a particular mode of operation of the CCD sensor developed to record dimly illuminated scenes that have a constant relative velocity with respect to the sensor. For example, to record moving objects in a conveyor belt, the use of a conventional interline transfer CCD camera may lead to image blur. To reduce image blur, one may have to reduce the shutter timing, which leads to the requirement of a high power light source for illumination. The illumination requirement to record the same scene using a TDI sensor would be much lower. The image will appear sharp if the TDI scan speed is exactly matched with the moving scene. The principle of TDI operation is discussed next.

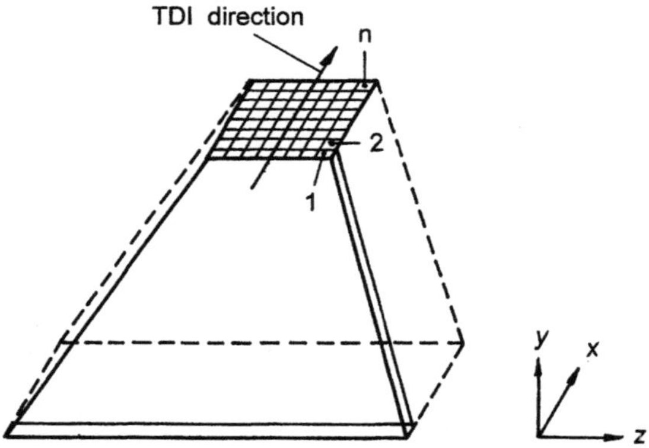

Fig. 11.8. Principle of operation of a TDI sensor. For simplicity, the optics is not shown

For recording scenes that have a constant relative velocity, a simple linear array of sensor elements could be used. The satisfactory performance depends on the proper illumination of the scene. If the light levels are low then one may not get a clear image. To overcome this, the scene has to be imaged several times and the charges added before image interpretation. This is precisely achieved in a TDI sensor [12, 13]. Instead of a linear array, a rectangular array is used for recording the scene. The number of columns of the rectangular array represents the number of TDI stages. The columns are electrically delayed in the x-direction at the same speed as the scene is scanned across the image plane (Fig. 11.8). The output from all the elements in a column is added to provide a signal that is n-times larger than that from a single linear array. For a typical TDI camera commercially available (EDC-1000, Electrim Corporation, USA), the scanning speed of the camera can be varied with functional keys from 1 to 247 000 lines/min.

Performance of a TDI architecture is compared with the addition of many line scan sensors for image recording [14]. The study showed that the noise levels are much lower in a TDI architecture. This is because; the charges are summed up before passing through the output stage, which is the main source of noise in image recording.

The signal (charge) increases linearly with the number of detectors in TDI. The well capacity of the CCD limits the maximum number of TDI elements that can be used. Large area TDI array having 1024 × 128-TDI elements have been developed for reconnaissance applications [15].

11.5 Developments in DIP Software

Developments in software are equally important on the efficient use of the available resources. Basically, DIP involves a large volume of data handling and processing. Frame grabbers usually come with certain basic software libraries, which utilises the hardware features of these efficiently. The softwares reported so far use these hardware libraries and are invariably hardware dependent and thus not portable. The software package recently introduced by Matrox, viz., Matrox Imaging Library (MIL) is an effort to improve the portability of the image processing software. It is a generic image processing package and in view of its generality it is highly priced (in fact three times the cost of a frame grabber!). Many general purpose DIP packages, including MIL, do not provide a module to collect information which is directly useful for photoelastic analysis. It is necessary that dedicated software modules have to be developed for digital photoelasticity, which would make it cost effective.

11.5.1 Development of a Device Independent Software

In general, the images recorded by photoelastic technique have high contrast and require minimal image enhancement for further processing. In techniques such as three fringe photoelasticity (TFP), phase shifting and Fourier transform analysis, the DIP hardware is used only to record the intensity data over the fringe field conveniently. Extraction of fringe data is essentially done by processing the intensity data. This requires only the knowledge of accessing the pixels forming the digitised image and reading the grey level shades corresponding to it.

11.5.1.1 Selection of Software Features

To develop a device independent software, one has to select an appropriate image storage format, so that future expansions are easily effected. Digitised images can be stored in various formats [16] such as BMP, TIFF, TARGA, FLAT, GIF etc. Among these, the TARGA format represents true colours easily. Further, accessing separately the R-Plane, G-plane, and B-plane of the colour image stored is simple.

A large class of problems such as evaluation of stress intensity factor (SIF), contact stress parameters and material stress fringe value etc. requires only the isochromatic fringe order and the corresponding positional co-ordinates. In all these problems, one has to first identify the origin and then measure the distances taking into account the aspect ratio of the pixel for the particular hardware chosen.

11.5.1.2 FRN_DAT Software

The FRN_DAT software enclosed in the accompanying CD can load any image stored in TARGA format. The program code can easily display images from monochrome to true colour. The software has the capacity to show the different channels of a colour image as a grey scale image. The software has the following options:

Mode	Function
a	Aspect ratio determination
m	Magnification determination
c	Origin selection
f	Assigning of fringe orders
d	Interactive selection of data points

The various *mode options* can be selected by pressing the appropriate letter key. Cursor movement can be effected by a mouse. For determining the aspect ratio one has to specify the real world length between two selected points on the image horizontally and then vertically.

The assignment of fringe orders is an important task and could be interactive as in fringe thinning methodologies. For fringe order assignment, a buffer is invoked by pressing the mode 'f' and one has to assign the fringe order interactively. The buffer is unaltered until 'f' is pressed once again. This helps in collecting fringe orders by just clicking the points using the mouse.

11.5.1.3 An Application

The problem of evaluation of SIF from isochromatic data is taken as an example to illustrate the use of FRN_DAT software. For this purpose a SEN (single edge notched) specimen made of aluminium (width = 40 mm and thickness = 6 mm) is coated with a polycarbonate sheet (PS-1, Measurements Group, USA) of thickness 3.07 mm. The K factor of the coating is 0.15. A thin slit of width 0.2 mm and a length of 12 mm, modelling the crack, is cut through the specimen and the coating. The specimen is loaded in an MTS machine to a load of 12,000 N. The isochromatics in colour are recorded using a reflection polariscope equipped with a Sony Handycam. The digitised image is stored in TARGA format. Figure 11.9a shows the dark-field isochromatics (green channel) observed at the tip of a crack and the data points collected by the FRN_DAT software. Though the loading is primarily Mode-I, a small amount of influence of shear is also seen in the fringes. The aspect ratio (y/x) of the image measured by FRN_DAT software is 1.097 and the magnification of the image in the x-direction is 4.95 pixels/mm [17]. The mode I and mode II SIF are determined by the overdeterministic least squares technique discussed in Chap. 8. Figure 11.9b shows the crack tip stress field reconstructed with the data points echoed back. The figure clearly brings out the fact that the analysis is self-consistent. In Fig. 11.9b, several fringes closer to the crack-tip are seen. These are lost due to insufficient optical magnification and digitisation in Fig. 11.9a. One can observe a common feature between Figs. 11.9a and 11.9b that the top loops are larger than the bottom loops of the corresponding fringe order.

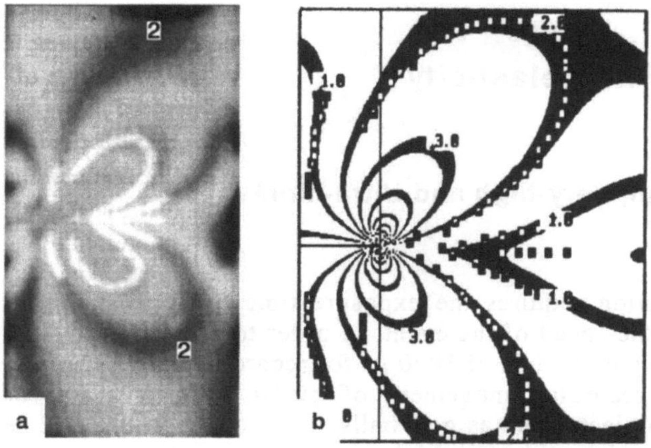

Fig. 11.9. a Isochromatics observed in the vicinity of the crack-tip in a SEN specimen. The figure also shows the data points collected by FRN_DAT software b Theoretically reconstructed fringe patterns with the data points echoed back

Thus, the use of a multi-parameter solution has correctly captured the nature of the stress field.

11.5.2 Future Possibility

The software discussed in the previous section can provide the positional co-ordinates easily and accepts interactive assignment of fringe orders. Thus, it has indicated the possibility of developing device independent simplified software modules exclusively for digital photoelasticity. Further improvements in image handling methodology should be the focus of future research to incorporate techniques such as phase shifting, polarization stepping, Fourier transform methods etc.

These days Windows based systems are quite common. Moving from Dos based systems to Windows based system, has removed some of the restrictions in developing the software for image analysis. The traditional 64K program segment of Dos based system is no longer there now. This can greatly simplify the software structure. The fringe unwrapping could be much easily accomplished by declaring a large array rather than by using files as indicated in Chap. 6.

The introduction of World Wide Web has introduced a new dimension to software methodologies. The software has to work on many types of computers, operating systems and CPUs. This has been achieved through the use of Java as a programming language. For basic digital image processing, several Java *applets* are available. An *applet* is an application designed to be transmitted over the Internet and executed by a Java-compatible Web browser. In future, *applets* specifically designed for digital photoelasticity may become available.

11.6 Digital Dynamic Photoelasticity

11.6.1 Classification of High, Very-high and Ultra-high-speed Photography

Photography of fast action requires the exposure time to be as small as possible depending on the speed of the event. In order to prevent blur [18], one would require an exposure time of 1000 μs for recording slowly moving subjects, 100 μs for recording movement of bird's wings, 10 μs for recording fast moving objects such as golf balls, 1 μs for recording bullets in flight and stress waves in solids. The high-speed photographic committee of Society of Motion Picture and Television Engineers (SMPTE) defined various types of high-speed photography in 1948 [19]. If the exposure time is between 10 μs to 100 μs it is classified as high-speed photography. If it is between 1 μs to 10 μs it is classified as very high-speed photography and if it is less than 1 μs it is classified as ultra high-speed photography.

For impact studies, the speed of the stress waves determines the order of exposure time required for dynamic photography. Commonly used photoelastic materials are CR-39, Homolite 100, and Epoxy. In these materials, the longitudinal wave speed lies in the range of 1800 m/s to 2150 m/s and the Rayleigh wave speed is in the range of 960 m/s to 1130 m/s. In order to adequately record these high velocity fringes, a camera is required which operates in the range between 30,000 and 8,00,000 frames/sec. A camera designed for impact studies can very well be used for recording stress field around a propagating crack, as it moves at a much smaller velocity and for plate glass the terminal velocity of crack is found to be 47% of shear wave velocity [16]. Some of the early studies on dynamic photoelasticity were done on polyurethane rubber. This is because, the stress wave speed is only 200 m/s and a high exposure time of 20 μs is admissible for freezing the dynamic action.

In high-speed photography, the framing rates are expressed in pictures per second and it does not mean that all the pictures would be recorded. If pictures are made with a microsecond interval, the sequential rate is 10^6 frames/sec (fps). If one second worth of 16-mm film were used, the length of the film required would exceed 2100 m. In practice, for such cameras, the sequence is limited to about 20 pictures.

11.6.2 Classical Methods for High-speed Photography

Several systems have been reported for high-speed photography. The various systems can be classified into six main groups [20]. These are,

1. High-speed intermittent cameras.
2. High-speed rotating prism cameras.
3. Rotating-mirror cameras.
4. Short duration spark, flashes, and laser pulses.
5. Kerr cell shutters.
6. Cameras using electron optics.

In intermittent Cameras, the film is held stationary at the time of exposure but is then moved rapidly to be ready for the next exposure. These cameras are limited by the precision needed in the moving mechanical parts and by the strength of the film and designed up to 500 fps. Conventional movie cameras that operate from 16 fps to 24 fps are of this type.

In rotating prism cameras, the problem of moving film intermittently is overcome by continuously moving the film from one spool to another and a rotating prism moves the image at the film speed. Rotating prism cameras can achieve framing rates of 10^4 fps to 4×10^4 fps

In rotating mirror cameras (Cordin, USA), the film is held stationary and the event is first imaged on a rotating mirror, which then re-images the light through framing lenses and slits onto the stationary film. The mirror is rotated at a very high speed of 10 to 20 thousand revolutions per second. This creates tremendous centrifugal forces in the mirror and the distortion introduced is considerable. Technological aspects for designing this type of camera are dealt with by Dubovik [21]. Framing rates of the order of 2×10^6 fps are possible with this type of camera. Rotating mirror cameras are widely adopted for dynamic photomechanics studies. For obvious reasons the maintenance of a rotating mirror camera is cumbersome. Further, in rotating mirror cameras, the exposure time is a function of the operating speed. At higher framing rates such as 10,00,000 fps exposure time is 0.33 μs which is acceptable. However, at low framing rates of the order of 10,000 fps, the exposure time is 3.3μs, which is too long to cause blur of the picture.

Technically the simplest and historically the first method to freeze high-speed motion is to use a spark. The duration of the spark controls the exposure time. The idea was first used by Fox Talbot in 1851 [22] using the spark discharge of a Leiden jar. In 1890 Rayleigh used this approach to photograph air jets, Mach projectiles etc. Fox Talbot's sparks probably had durations of 10 μs. It is worthwhile to note that the ordinary electronic flash has a decay time of 10 to 80 milliseconds. With the development of Xenon flash lamps [18] obtaining sparks of 10 μs with any level of energy requirement is made simpler. However, obtaining sparks of sub-microsecond duration, while also increasing their energy, is much more difficult. In Xenon flash lamps due to the after glow of the gas filled-in, decay time less than 10 μs is not possible. On the other hand an open-air spark gap does not have the problem of after glow and it provides a spark of sub-microsecond duration. Image separation in high-speed photography by means of film transport or rotating mirrors becomes more difficult with increasing picture frequency. Cranz and Schardin [23] developed a camera in 1929 in which

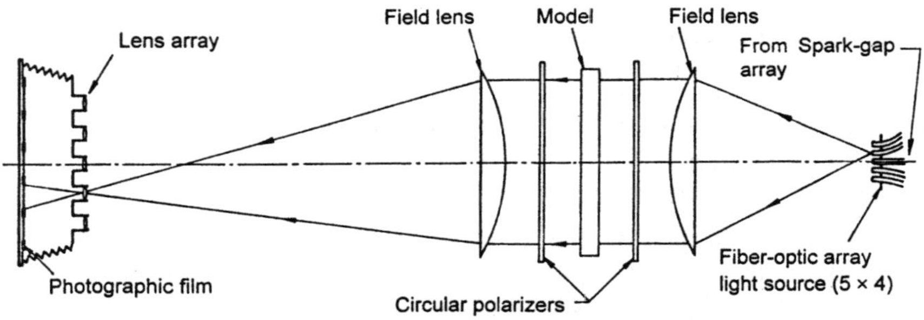

Fig. 11.10. Optical diagram showing the image separation principle used in a Cranz-Schardin Camera

the image separation was done by purely optical means. The camera has multiple units each with its own energy source and spark gap, which are essentially open-air spark gaps. The spark gaps are fired in a time-controlled sequence to provide both the light and the shutter action required to stop image movement during a dynamic event. The sequence frequency is bounded only by electronic limitations. The original electrical firing circuit is a simple RLC circuit known as Mach's circuit. Figure 11.10 shows the optical diagram of the camera and illustrates the image separation principle. From the optics point of view, the light from the spark gaps must be available in a compact form. This is achieved by using fibre optic cables that collect the light from the spark gaps and provide a compact array of light source at the front panel of the camera. The system shown in Fig. 11.10 has an array of 5 × 4 fibre optic cables and a corresponding array of camera lenses of 5 × 4. For recording the dynamic event, camera portion of the recording system is kept at open shutter with the room kept under total darkness. Trigger signal from the model initiates the firing of the spark gaps. Depending on the inter-frame time selected, the spark gaps fire in a time ordered sequence. The inter distance of fibre optic array, camera lens array and the focal length of the field lenses are so chosen that the light from a fibre optic maps on to a corresponding camera lens. With appropriate selection of the lenses and the size of the film, it is possible to get an image size of the order of ϕ50 mm. This helps in providing clarity of information recorded. In rotating mirror cameras, usually a 16-mm film is used. At normal framing rates, the image occupies one full frame of a 16-mm film. At higher framing rates, the film is further divided to facilitate recording more frames, which further reduces the image size.

Since its development in 1929, the camera has been in continuous development till today. Several groups have built their own versions of this camera. Several groups retained the basic Mach's circuit and suitable synchronisation circuits for specific application and appropriate optics have been designed to their need. Christie [24] in England was the first one to use a Cranz-Schardin type camera for dynamic photoelasticity in 1955. Wells and Post [25] introduced this camera in the United States in 1957.

Dally and Sanford have made several improvements on the Cranz-Schardin camera. They reported in 1982 the development of a multiple ruby laser system [26] in which the spark gaps of a Cranz-Schardin camera was replaced by Q-switched ruby lasers. A microprocessor was incorporated into the control circuits to dynamically program, in real time, the firing of the flash lamps and the Q-switching of the laser to obtain extremely precise synchronisation with the dynamic event. Very short exposure time of the order of 0.035 μs was achieved by them. On the other hand an open air spark gap can provide an exposure time of 0.5 μs only, which is however sufficient for most dynamic photomechanics studies.

Winkler [27] in 1987 reported some of the recent developments on the Cranz-Schardin camera. Using semi-conductor technology and a triggerable spark gap of Stenzel [28] he developed a camera in which the sparks can be fired independently from others, which gives the advantage of using pulse generators for timing purpose. This eliminates the need for recording the time separately for each experiment.

Kerr cell shutters are used to record self-luminous events. The Kerr cell shutter opens while the spark emits light. Using a Kerr cell the exposure time could lie in the range of 9 ns to 0.2 μs [19].

Cameras using electron optics were developed in the later part of 1980's. These cameras are called image converter cameras (ICC). In this, light from the scene is changed into electrons, which are suitably amplified; shuttered and deflected and finally converted back to photons to record the image. The ICC incorporates a shuttered image converter tube (CRT-type) with two pairs of mutually perpendicular deflecting plates. For recording a sequence of dynamic events, there has to be a relative motion of the film with the dynamic event. This relative motion is achieved in ICC by proper shutter/deflection control timing of the deflection plates. Up to 16 temporally shifted images formed in two rows can be enabled on the screen of the commercially available Imacon camera (Hadland Photonics Ltd.). The images are recorded on Polaroid or high-speed negative film. In general, the resolution of the images recorded are not as good as those obtainable using a Cranz-Schardin camera. ICC can work on a wide range of framing rates from 1×10^4 to 2×10^7 fps by selecting appropriate plug-in modules. For a given plug-in module, the exposure time and inter-frame time period are fixed.

11.6.3 Digital Dynamic Recording

Early use of the CCD camera for high-speed imaging was done in conjunction with the image converter camera (ICC). One of the first applications of video imaging for high-speed photography was to provide a CCD readout to an image converter camera. Use of FT CCD is reported for this purpose. It is seen in Chap. 3 that the framing rates of the CCD cameras remain constant (50 fps for CCIR or 60 fps for EIA). The CCD camera is just used to transfer the image to a PC and no actual dynamic recording is done.

Efforts to develop CCD cameras with higher framing rates have been attempted. One example is the Kodak EXTAPRO 100 imager, which uses a 192 × 239 pixel array to capture full-screen images at a rate of 1000 per second. Higher rates are achieved by displaying up to six consecutive frames with a reduced vertical size, stacked on a single screen. The traditional storage medium for the acquired images is high velocity videotape, which allows a recording time of several minutes. Kodak has introduced solid state digital recording for recording 1.2 s of the event (EKTAPRO EM). Being a solid state recording system, the system can be asynchronously triggered and thus provides enough flexibility for recording a dynamic event.

Hadland Photonics has developed a high-speed video to record 60 full frame colour images or 180 full frame B&W images per second. The dynamic motion is frozen by Xenon flash lamps having a flash duration of 20 μs. Such systems are suitable for applications such as monitoring the machines in packaging industries, sports science etc. In view of the large exposure time, the system is not suitable for dynamic photoelasticity.

In general, high framing rate CCD sensors are quite expensive and innovative use of commercial grade consumer CCD cameras have been reported by investigators. Hiller and Kowalewski [30] have reported a technique to use FT CCD camera for high-speed imaging. They used a NXA 1011 CCD imager made by Volvo and demonstrated two possibilities of using it for dynamic recording.

In FT CCD, during half period at 50Hz (20 ms), the charges generated by the incident photons are integrated and then the frame as a whole is transferred to the masked storage area rapidly. The frame transfer time is of the order of 475 μs for the sensor used by them. They suggested a multiexposure either during the integration phase or during the frame transfer phase for dynamic recording. The multiexposure scheme during frame transfer has possible applications in digital photoelasticity. This technique is feasible when the height of the object to be recorded is smaller than the image section of the CCD. The optics is to be so adjusted that only the lower portion of the imager section equal to the height of the scene of interest is to be exposed. The remaining top portion of the sensor is to be masked for obtaining high contrast. Referring to Fig. 11.11, one can obtain a temporal sequence of images spatially shifted relative to each other in the storage array. The source of illumination is from an LED of type H-3000 (Stanely) emitting at 660 nm. A proper synchronisation of the camera with the dynamic event is crucial. An adjustable phase shift device achieves this so that the illuminating light pulses are released either during the frame transfer time or the integration period of the camera.

The well-known Cranz-Schardin high-speed framing technique has been extended to the digital domain. Bretthauer et al. [31] have developed an 8-image electronic Cranz-Schardin camera. This uses a miniature array of eight high efficiency LED's (H-3000, Stanely) coupled with eight standard low-cost CCD cameras in bakelite mode and stored in an eight channel

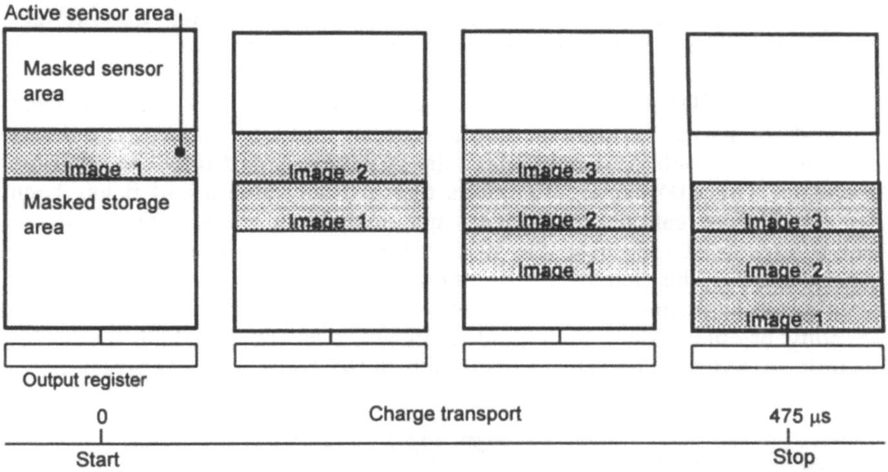

Fig. 11.11. Principle of grabbing multiple-images during the frame transfer operation of a FT CCD camera

image processor memory. The CCD cameras used by them are of the type having the full-frame interline transfer architecture (XC-77CE, Sony).

Synchronisation of the camera with the dynamic event is a crucial aspect. The cameras are set to run in frame integration mode to receive a picture over a whole frame. The trigger conditioner does the identification of the integration phase by extracting and separating the odd and even V-SYNC pulses from the video signal of the camera. The video output of the first camera is used to synchronise the other cameras.

The simplicity of the system lies in the easy design of the light pulse controller. Power V-MOS transistors providing square current pulses, which can be adjusted within the range of 0-10A, can control the LED's. The pulse width can be varied from 50 ns to 10 μs. Thus, the system provides a desirable range of exposure to study various phenomena by digital photoelasticity. It is reported that a high degree of timing accuracy could be maintained. To adjust the optical set up, the light pulse controller is set to cw mode providing a current ranging from 0-50 mA. Full camera resolution (756 × 581) could be used to record the scene and hence, the images obtained by the system have a higher spatial resolution. For special purposes, each camera can be operated in a multiexposure (10 times) mode during the integration phase as discussed previously [30]. Thus, up to 80 images could be recorded at a lower resolution.

Racca et al. [32] have also come up with a system, which uses multiple consumers grade CCD cameras. The system has a single entrance lens and optical path, so that it can be operated like a normal camera with a single light source. However, the intensity of the light source should be high, as the intensity of light transmitted is progressively decreased from one camera to another by beam splitters. They used a high intensity HLMP-8150

LED (Hewlett-Packard) coupled with two spherical mirrors for illumination. They have demonstrated a system with three CCD cameras (Philips type 56471) and the design can be extended up to 10 camera modules.

The use of TDI mode of operation of the CCD camera has also been explored for its use in high-speed photography [33–35]. In fact, the direct use of it in the context of photoelasticity is demonstrated only for this type of system. The TDI camera functions like a drum camera and it has been used for both streak photoelasticity and delayed microflash applications. Asundi and Sajan [33] used the EDC-1000 camera (Electrim Corporation, Princeton, NJ). The camera can be operated either in the conventional full-frame mode or in the TDI mode. The camera is aligned while operating in the conventional video mode and switched over to the TDI mode for dynamic recording. The camera resolution is 192 (H) × 165 (V) pixels.

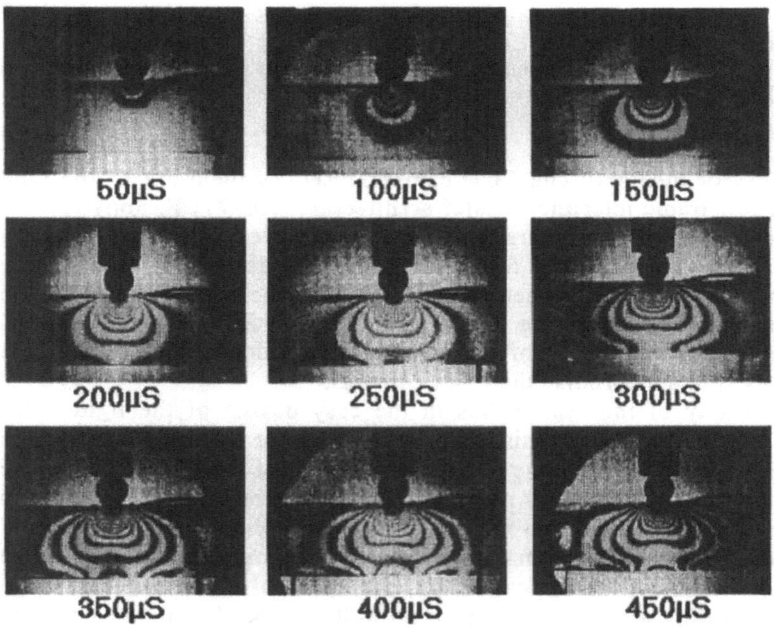

Fig. 11.12. Dynamic photoelastic fringe patterns recorded by a TDI camera (Courtesy of Prof. A. Asundi, Nanyang Technological University, Singapore)

Typical fringe patterns recorded, taken at different time intervals, using the TDI camera are shown in Fig. 11.12. The fringe patterns correspond to the impact loading of a specimen made of a birefringent material (PSM-4, Photoelastic Inc.) of size 230 mm × 45 mm × 6 mm, supported on two sharp edges separated by a distance of 15 cm. Loading is provided by a free falling ball from a height of 1 m. The ball weighing 27 g is guided by a tube and impacts at the midpoint between the supporting edges. TDI camera is set to the image saving mode just before releasing the ball for impact. No synchronisation of the event with shuttering is needed for image recording. The image is frozen during the short flash duration somewhere along the length of the image strip. The photoelastic fringe pattern is cut out from the saved image using image analysis software.

For image illumination a 10 mW laser diode with 100% depth of modulation facility has been used by Asundi and Sajan [33]. The 100% depth of modulation feature ensures that the light is completely cut off when the signal is off. The time delay circuit is assembled using widely available ICs [33]. The experiment is repeated with different time delays to obtain the fringe patterns shown in Fig. 11.12. Considering the several advantages of LED, Asundi and Sajan recommended the use of LED's in conjunction with the TDI camera [36, 37], instead of the laser diode initially used by them.

The dynamic recording methods discussed so far concerned only with freezing the ultra high-speed dynamic fringe patterns. This provides only whole field fringe patterns and one has to invoke fringe-thinning methodologies for data reduction. Fringe data over the field could be obtained by using the Fourier Transform approach proposed by Quan et al. [38]. This requires the design of quartz wedges to produce carrier fringes of required spatial frequency. Patterson and Wang developed a new apparatus to record four phase-stepped image simultaneously [39]. For an incident circularly polarised light on the model, the reflected/transmitted light is split into four beams of equal intensity by cube beam splitters (Melles Griot Inc., USA). Using four dedicated CCD (BE-201A, Hitachi) cameras and four sets of output quarter-wave plates and analysers, four phase-stepped images are recorded simultaneously. The basic optical scheme for the four-step phase shifting method was discussed in Chap. 5. The intensity equations are suitably modified to include the effect of the beam splitters. To minimise the ease of data interpretation, the role of stray light is not accounted for in the intensity equations. The usefulness of the equipment is demonstrated for static applications using a 250 W quartz-tungsten halogen source.

The equipment has rich potential for its use in truly extending digital photoelasticity for dynamic studies. For ultra high-speed dynamic recording, instead of the normal CCD camera, one can think of using a TDI camera. However, the choice of the suitable light source will be a challenge, which future research should pay attention to.

11.7 Application to Composites

Composite materials have established themselves as strong candidates for important structural applications. For their efficient utilisation, a complete stress analysis and reliable failure criteria are essential. Analytical methods tend to become difficult for composites due to material anisotropy, and multiplicity of failure modes. Consequently, experimental methods of stress analysis and failure analysis are very useful for composites [40]. The use of photoelasticity for stress analysis of composites is discussed in this section.

Dally and Alfirevich [41] employed isotropic birefringent coatings bonded to the composite. In this case, the strain field is determined from the coating using a reflection polariscope together with conventional methods of photoelastic analysis established for isotropic materials. Pipes and Dally [42] conducted detailed analysis near free edges of composite materials and pointed out the limitations of the method due to the mismatch of Poisson's ratios of the coating and the substrate.

Pih and Knight [43] initiated the use of a transparent birefringent model of the composite with anisotropic elastic and optical properties for photoelastic stress analysis. One of the important requirements for this technique is the development of appropriate model materials. Several investigators [43–49] have addressed this. Unlike isotropic 2-D photoelasticity, here, one requires three material constants to relate the optical response to stress information. The details of this are discussed next.

11.7.1 Photo-Orthotropic Elasticity Theories [50]

11.7.1.1 Stress-Optic Law

It is well established in photo-orthotropic literature that the fringe order obtained in a transparent composite is a function of three material constants F_L, F_T and F_{LT}. The stress-optic law is

$$N = \left[\left\{\frac{\sigma_L}{F_L} - \frac{\sigma_T}{F_T}\right\}^2 + \left\{\frac{2\tau_{LT}}{F_{LT}}\right\}^2\right]^{\frac{1}{2}} t \tag{11.21}$$

Let the principal stresses be (p, q) and the principal stress p be at an angle α to the L-axis, then Eq. (11.21) becomes

$$N = \left[\left\{\cos^2\alpha + \frac{q}{p}\sin^2\alpha - \left[\sin^2\alpha + \frac{q}{p}\cos^2\alpha\right]\frac{F_L}{F_T}\right\}^2 + \left\{1 - \frac{q}{p}\right\}^2\left\{\sin 2\alpha \frac{F_L}{F_{LT}}\right\}^2\right]^{\frac{1}{2}} t \frac{p}{F_L} \tag{11.22}$$

For a uniaxial stress-field (i.e., $q = 0$), using the definition $F_\alpha = tp/N$, it is possible to write

$$\frac{1}{F_\alpha}=\left[\left\{\frac{\cos^2\alpha}{F_L}-\frac{\sin^2\alpha}{F_T}\right\}^2+\frac{\sin^2 2\alpha}{F_{LT}^2}\right]^{\frac{1}{2}} \tag{11.23}$$

Equation (11.23) indicates that the material stress fringe value is a function of the fibre-orientation angle. Extending the analogy of Mohr's circle, Sampson [48] obtained the optical isoclinic angle (θ) as

$$\tan 2\theta=\frac{(2\tau_{LT}/F_{LT})}{\left\{\frac{\sigma_L}{F_L}-\frac{\sigma_T}{F_T}\right\}} \tag{11.24}$$

In the case of a transparent composite laminate, the state of stress at each point along a light beam through the model thickness is different. The optical response observed in a polariscope is due to the integral effect of the stress field through the thickness of the model. Cernosek [51], using the principle of integrated photoelasticity analysed the orthotropic composite and observed that for the case of an orthotropic composite, the optical effect could be represented by a retarder with the following retardation

$$2\delta_o=2n[\delta_f^2+\delta_m^2+2\delta_f\delta_m\cos 2\Omega]^{\frac{1}{2}} \tag{11.25}$$

where, $2\delta_m$ is the relative phase retardation gained in the matrix surrounding a single fibre, $2\delta_f$ is the relative phase retardation gained in single fibre, n is the number of fibres traversed by light, $\Omega = \theta_m - \theta_f$, θ_f is the angle of polarization in the fibre with respect to L-axis, θ_m is the angle of polarization in the matrix with respect to L-axis. The orientation of the retarder is given by Eq. (11.24).

11.7.1.2 Strain-Optic Law

The reduced influence of mechanical anisotropy on the strain-fringe value, compared to its effect on the stress-fringe value, has been recognised by Horridge [52], Hayashi [53], Grakh and Mozhenskaya [54] and Netrebko [55]. Pipes and Rose [56] developed a strain-optic law involving only one photoelastic constant. However, their results are applicable only to unidirectional composites having a very low fibre content. Following the approach of Sampson, Prabhakaran [57] developed a strain-optic law based on the concept of Mohr's circle of birefringence and introduced three material strain-fringe values, $F_{\varepsilon L}$, $F_{\varepsilon T}$ and $F_{\varepsilon LT}$. These are related to the observed fringe order in the model material by the relation

$$N = \left[\left\{ \frac{\varepsilon_L}{F_{\varepsilon L}} - \frac{\varepsilon_T}{F_{\varepsilon T}} \right\}^2 + \left\{ \frac{\gamma_{LT}}{F_{\varepsilon LT}} \right\}^2 \right]^{\frac{1}{2}} t \tag{11.26}$$

Agarwal and Chaturvedi [58] derived an exact strain-optic law for orthotropic materials using Pockel's [59] theory of crystalline photoelasticity. They also concluded that three material strain fringe values are required for the total characterisation of the composite model material. Their approach is similar to the method followed by Knight and Pih [60] for stress-optic law.

11.7.2 Calibration of Photo-Orthotropic Composites

Dally and Prabhakaran [49] have reported equations to evaluate F_L, F_T and F_{LT} as functions of the elastic and optical properties of the constituents of the composites. However, an experimental determination is desirable.

Using tension specimens with fibres oriented at 0° and 90°, it is easy to determine F_L and F_T. On the other hand, determination of F_{LT} is not easy since it requires loading the specimen in pure shear. One can adopt anyone of the following two approaches depending on the accuracy desired.

1. An approximate method to find F_{LT} is by finding $F_{\pi/4}$ initially, by conducting a 45° off-axis coupon test and then using this value in the following relation to get F_{LT}.

$$\frac{F_{LT}}{F_{\pi/4}} = \left[1 + \frac{1}{4} \left\{ \frac{F_{LT}}{F_L} \right\}^2 \left\{ 1 - \frac{F_L}{F_T} \right\}^2 \right]^{\frac{1}{2}} \tag{11.27}$$

   The error in equating F_{LT} to $F_{\pi/4}$ depends upon the ratios, F_{LT}/F_L and F_L/F_T. This error is small (usually less than 3%).

2. An exact method for finding F_{LT} has been reported by Chaturvedi [61]. It is seen from Eq. (11.23) that F_α is directly related to F_{LT} when

$$\frac{F_T}{F_L} = \tan^2 \alpha_o \tag{11.28}$$

   The values of α_o can be determined by knowing F_L and F_T of the composite. If the unidirectional composite is subjected to α_o-deg coupon test, then

$$F_{LT} = F_{\alpha_o} \sin 2\alpha_o \tag{11.29}$$

Strain fringe values can be determined by using the inter-relationships between stress-fringe and strain-fringe values [62].

In the case of isotropic materials, accurate determination of F_σ is very critical, since this is the only parameter which relates theoretical and experimental results. Following the same approach, it is essential that F_L, F_T and F_{LT} are determined with extreme care. In the determination of F_{LT}, the first method is widely used. However, the method proposed by Chaturvedi [61] is better and is recommended for cases where higher accuracies are demanded.

11.7.3 Influence of Residual Birefringence

Curing of all types of epoxy or polyester resin is accompanied by a change of volume. Due to this shrinkage, residual birefringence is introduced. This is unavoidable while fabricating a transparent composite for conducting photo-orthotropic experiments. It has been found experimentally [63] that the principal direction of residual birefringence is identical with the direction of the fibres.

In the literature, accounting for residual birefringence is basically done in two different ways. In one approach, the final fringe order and optical isoclinic angle are expressed as functions of residual birefringence. Cernosek using integrated photoelasticity and Mittal [64] based on Toupin's [65] theory of deformable dielectrics obtained identical expressions for fringe order and isoclinic angle incorporating the effect of residual birefringence. The total fringe order in the presence of residual birefringence is given by

$$N_{\mathrm{T}} = \left[\left\{ \frac{N_{\mathrm{r}}}{t} + \frac{\sigma_L}{F_L} - \frac{\sigma_T}{F_T} \right\}^2 + \left\{ \frac{2\tau_{LT}}{F_{LT}} \right\}^2 \right]^{\frac{1}{2}} t \tag{11.30}$$

and the isoclinic angle θ_{T} in the presence of residual stress is given by

$$\tan 2\theta_{\mathrm{T}} = \frac{(2\tau_{LT}/F_{LT})}{\left\{ \frac{N_{\mathrm{r}}}{t} + \frac{\sigma_L}{F_L} - \frac{\sigma_T}{F_T} \right\}} \tag{11.31}$$

The other approach is to extract the load-induced birefringence and the isoclinic angle. Cernosek [51] has shown that orthotropic composite can be modelled as a retarder without any rotation. Due to this, extraction of initial birefringence is very similar to isotropic cases [66]. This has been shown using tensorial nature of birefringence by Knight and Pih [60]. Load induced birefringence given by

$$N = [N_T^2 + N_r^2 - 2N_T N_r \cos 2(\theta_T - \theta_r)] \tag{11.32}$$

and

$$\tan 2\theta = \frac{N_T \sin 2\theta_T - N_r \sin 2\theta_r}{N_T \cos 2\theta_T - N_r \cos 2\theta_r} \tag{11.33}$$

Where N_T and θ_T are the total observed fringe order and isoclinic angle respectively. N_r and θ_r are respectively residual fringe order and residual isoclinic angle. θ_T and θ_r should be referred to L-axis of the composite. While deriving Eqs. (11.30) and (11.31), Cernosek and Mittal assumed θ_r as 0. Pipes and Dalley have experimentally confirmed this.

When a theoretical reconstruction of fringes is required, one can use Eqs. (11.30) and (11.31). On the other hand, Eqs. (11.32) and (11.33) are appropriate for experimental calculations.

11.7.4 Separation of Stresses in Photo-Orthotropic Elasticity

The complete optical characterisation of photo-orthotropic model materials requires three calibration constants. In general, the directions of principal stresses and strains are not identical. Hence, the stress-optic laws and the strain-optic laws are different for these model materials. The isochromatic fringes observed in such materials are neither representatives of contours of principal stress differences nor the principal strain differences. Further, the optical isoclinics seen neither represent the principal stress directions nor the principal strain directions.

In view of the above, separation of principal stresses or strains is not a simple task even in a two-dimensional problem. Various investigators have proposed several methods. Sampson [48] proposed that the components of birefringence according to the Mohr's circle of birefringence along with the differential equations of equilibrium could be used in a shear difference method. Another method to separate the principal stresses has been addressed by Agarwal and Jha [67]. Prabhakaran [68] has summarised various techniques for separating principal stresses.

11.7.5 Application of Digital Photoelasticity to Composites

From the foregoing discussions, it is clear that data interpretation in photo-orthotropic elasticity is quite complex. However, for data acquisition, the various techniques discussed for 2-D transmission photoelastic analysis could be directly extended for photo-orthotropic elasticity. This is so because, the integrated photoelastic analysis of the transparent composite by Cernosek has revealed that the transparent composite could be modelled as a simple retarder [51]. Voloshin and Burger reported the use of HFP for orthotropic materials [69]. Fringe contours were obtained by HFP for the problems of a concentrated load on a semi-infinite plate and disc under diametral compression. Patterson [70] extended the use of phase-shifting to photo-orthotropic elasticity.

11.7 Closure

Some of the recent initiatives in extending digital photoelasticity to acquire data in integrated photoelasticity and scattered light photoelasticity have been discussed in this chapter. Newer methodologies are continuously reported [71, 72]. The importance of developing a device independent software for digital photoelasticity has also been highlighted. The improvements in DIP hardware have opened up possibilities of extending digital photoelasticity to dynamic studies. The current research has achieved digital recording of dynamic events but the data analysis has to be improved further. In future, it is expected that this lacuna will be fulfilled.

The number of research groups working on digital photoelasticity have increased considerably over the years. This indicates the growing importance of the technique. Newer ideas are also steadily pouring in the field of digital photoelasticity. Use of "Wavelet Transform" for isochromatic fringe ordering has been recently reported [73].

Exercises

11.1 Develop a software to evaluate characteristic parameters theoretically. Apply it for the problem of a circular disc in oblique incidence

11.2 Develop phase shifting approach to calculate the experimental parameters in scattered light photoelasticity.

11.3 What are the various modes of CCD operation? How these are used for recording a dynamic event?

11.4 List the challenging issues that are to be addressed for developing an effective data acquisition system in digital dynamic photoelastic analysis.

11.5 Discuss the role of residual birefringence in photo-orthotropic elasticity. How does this affect data acquisition in digital photo-orthotropic elasticity.

11.6 Convert the softwares included in the CD to be device independent.

References

1. Srinath LS, Ramesh K, Ramamurti V (1988) Determination of characteristic parameters in three-dimensional photoelasticity. Opt Engng 27(3):222–230 Also In: Billings BH (ed) (1992) Applications of polarised light. SPIE Milestone Series, MS-57, SPIE press, pp 589–594

2. Srinath LS, Keshavan SY (1978) A fast iterative procedure to determine photoelastic characteristic parameters. Mech Res Commun 5(3):159–165
3. Aben HK, Saar Aj (1972) Experimental investigation of optical systems described by a unitary matrix. Opt Spektrosk 32(1):196–201 (Russian)
4. Desailly R, Lagarde A (1977) Rectilinear and circular analysis of a plate slice optically isolated in a three-dimensional photoelastic model. Mech Res Commun 4(2):99–107
5. Mangal SK, Ramesh K (1999) Determination of characteristic parameters in integrated photoelasticity by phase-shifting technique. Optics and Lasers in Engng (in press)
6. Quiroga JA, González-Cano A (1997) Phase measuring algorithm for extraction of isochromatics of photoelastic fringe patterns. Appl Optics 36(32):8397–8402
7. Berezhu SYu Berezhnyi IV, Takashi M Optical tomographical technique as a part of a hybrid method for elastic analysis. In: Allison (ed) Advances in design, testing and analysis. Proc of XI Int Conf on Experimental Mechanics, pp 489–494
8. Aben HK (1979) Integrated photoelasticity. McGraw-Hill, New York
9. Amarnath S, Simha KRY (1998) Tomographic data acquisition in axisymmetric specimens. Exp Tech 22:21–25
10. Dupré JC, Lagarde A (1997) Photoelastic analysis of a three-dimensional specimen by optical slicing and digital image processing. Exp Mech 37:393–397
11. Zenia a, Dupré JC, Lagarde A (1998) Theoretical and experimental approaches of a three-dimensional photoelastic medium. In: Allison (ed) Advances in design, testing and analysis. Proc of XI Int Conference on Experimental Mechanics, pp 501–506
12. Holst GC (1996) CCD arrays, cameras and displays. SPIE Optical Engineering press, Bellingham, Washington
13. Barbe DF (1976) Time delay and integration image sensors. In: Jespers PG, Van de Wiele F, White MH (eds) Solid state imaging, pp 659–671
14. Chamberlain SG, Jenkins PT (1990) Capturing images at 1000 feet per minute with TDI. Photonics Spectra 161–164
15. Farrier MG, Dyck RH (1980) A large area TDI image sensor for low light level imaging. IEEE transactions on Electron Devices, ED 27(8):1688–1693
16. Oliver D, Anderson S, Zigon B, McCord J, Gumas S (1993) Tricks of the graphics gurus. SAMS publishing, Indiana, USA
17. Ramesh K, Govindarajan R (1997) Towards a device independent digital image processing software for photoelastic analysis 33:115–120
18. Edgerton HE (1970) Electronic flash. McGraw-Hill, New York
19. Früngel F (1980) High speed pulse technology. Vol IV Academic Press, New York
20. Hadland J (1974) A review of the current state of the art. Proc of XI Int Cong High Speed Photography, 124, Chapman and Hall, London
21. Dubovik A (1981) The photographic recording of high speed processes. John Wiley and Sons, New York
22. Field CE (1982) High speed photography: techniques and applications. Opt Engng 21(4):709–717
23. Cranz C, Schardin H (1929) Kinematographic auf ru hendum film und mit exterm, hoher bildfrquenz, Z Phys 56:147–183
24. Christie DG (1955) A multiple spark camera for dynamic stress analysis. J Phot Sci 3:153–159
25. Wells AA, Post D (1957) The dynamic stress distribution surrounding a running crack: a photoelastic analysis. Proc SESA XVI:69–92
26. Dally JW, Sanford RJ (1982) Multiple ruby laser system for high speed photography. Opt Engng 21:704–708

27. Winkler S, Cranz-Schardin (1987) High speed camera recent developments. In: Lagarde A (ed) Static and dynamic photoelasticity and caustics – recent developments. Springer-Verlag, pp 510–522
28. Stenzal A (1959) Elektronisch gesteuerte cranz-Schardin-Funkenzeitlupe. Proc IV Int Cong High Speed Photo, Darmstadt:136–138
29. Ramesh K, Agarwal BD, Bajpai KK, Yadav J, Shukla A (1993) High speed photographic facility at IIT Kanpur. In: NK Gupta (ed) Plasticity and impact mechanics. Wiley Eastern Ltd, New Delhi, pp 507–518
30. Hiller WJ, Kowalewski TA (1989) Application of the frame transfer charge coupled devices for high speed imaging. Opt Engng 28:197–200
31. Bretthauer B, Meier GEA, Stasicki B (1991) An electronic Cranz-Schardin camera. Review of Scientific Instruments 62:364–368
32. Racca RG, Stephenson O, Clements RM (1992) High speed video analysis system using multiple shuttered charge coupled device imagers and digital storage. Opt Engng 31:1369–1374
33. Asundi A, Sajan MR (1994) Digital dynamic photoelasticiity. Optics and Lasers in Engng 20:135
34. Asundi A, Sajan MR (1994) Low cost digital polariscope for dynamic photoelasticity. Opt Engng 33:3052
35. Asundi A, Sajan MR (1996) Digital dynamic visualization system. In: Mruthyanjaya TS (ed) Advances in Mechanical Engineering. Narosa Publishing House, New Delhi, pp 1331–1348
36. Asundi A, Sajan MR (1995) Multiple LED camera for dynamic photoelasticity. Appl Optics 34:2236
37. Ng TW, Asundi A (1997) Light emitting diodes for illumination in photoelasticity. Exp Tech 21:28–30
38. Quan C, Bryanston-Cross PJ, Judge TR (1993) Photoelasticity stress analysis using carrier and FFT techniques. Optics and Lasers in Engng 18:79–108
39. Patterson EA, Wang ZF (1998) Simultaneous observation of phase-stepped images for automated photoelasticity. J Strain analysis for Engng Des 33:1–15
40. Whitney M, James, Daniel, Issac M, Pipes, Byron R (1984) Experimental mechanics of fiber reinforced materials. Revised Ed, Prentice Hall, Inc, N Jersey
41. Dally JW, Alfirevich I (1963) Application of birefringent coatings to glass-fiber--reinforced plastics. Exp Mech 3:97–102
42. Pih H, Knight CE (1969) Photoelastic analysis of anisotropic fiber reinforced composites. J Comp Materials 3(1):94–107
43. Pipes RB, Dally JW (1972) On the birefringent coating method of stress analysis for fiber reinforced laminates. Exp Mech 12:272–277
44. Agarwal BD, Chaturvedi SK (1970) Improved birefringent composites and assessment of photoelastic theories. Fiber Science and Technology 11:399–412
45. Tiwari N (1990) Development of transparent composite laminates and evaluation of mode I stress intensity factor using photo-orthotropic elasticity. MTech Thesis, Dept of Mech Engg, IIT, Kanpur
46. Prabhakaran R (1980) Fabrication of birefringent anisotropic model materials. Exp Mech 20:320–321
47. Agarwal BD, Tiwari N, Ramesh K (1991) Development and characterization of transparent composite laminates. Second annual progress report on application of image processing technique to the determination of SIF in orthotropic composite materials. Dept of Mech Engg, IIT, Kanpur, pp19–45
48. Sampson RC (1970) A stress-optic law for photoelastic analysis of orthotropic composites. Exp Mech 10:210–215
49. Dally JW, Prabhakaran R (1971) Photo-orthotropic elasticity. Exp Mech 11:346–356

50. Ramesh K, Tiwari N (1993) A brief review of photo-orthotropic elasticity theories. Sādhanā 18:985–997
51. Cernosek, J (1975) On the photoelastic response of composites. Exp Mech 15:354–357
52. Horridge GA (1955) A polarised light study of glass fiber laminates. Brit J Appl Phys 6:314–319
53. Hayashi T (1962) Photoelastic method of experimentation for stress analysis in orthotropic structures. Proc of the Fourth Intl Symposium on Space Tech and Science, Tokyo:156–169
54. Grakh II, Mozhenskaya AF (1971) A type of mechanically anisotropic optically sensitive bodies. Mekhanika Polimerov 5:835–839
55. Netrebko VP (1971) Optical polarization method of studying the stress state of anisotropic bodies. Mekhanika Tverdogo Tela 1:94–100
56. Pipes RB, Rose JL (1974) Strain-optic law for a certain class of birefringent composite. Exp Mech 14:355–360
57. Prabhakaran R (1975) A Strain-optic law for orthotropic model materials. AIAA J 13:723–728
58. Agarwal BD, Chaturvedi SK (1982) Exact and approximate strain optic laws for orthotropic photoelastic materials. Fiber Science and Technology 33:146–150
59. Pockels F (1906), Lehrbuch der Kristalloptik. Teubner BG, Leipzig, Germany
60. Knight CE, Pih H (1976) Orthotropic stress-optic law for plane stress photoelasticity of composite materials. Fibre Science and Technology 9:297–313
61. Chaturvedi SK (1978) Some experimental stress analysis techniques for fibrous composites. PhD Thesis, Dept of Mech Engg, IIT, Kanpur
62. Doyle JF (1978) A unified approach to optical constitutive relations. Exp Mech 18:416–420
63. Pipes RB, Dalley JW (1973) On the fiber-reinforced birefringent composite materials. Exp Mech 13:348–349
64. Mittal RK (1975) On the residual birefringence in anisotropic photoelastic materials. Strain 11:55–57
65. Toupin RA (1963) A dynamic theory of elastic dielectrics. Int J Engng Science 1(1):101
66. Friedrich Guenter, (1977) Superposition of photoelastic data. Exp Mech 17:33–36
67. Agarwal BD, Jha PK (1983) A stress separation technique for photo-orthotropic elasticity. Composites Tech Review 53:84–88
68. Prabhakaran R (1982) Photo-orthotropic elasticity: a new technique for stress analysis of composites. Opt Engng 21(4):679–688
69. Voloshin AS and Burger CP (1984) Half-fringe photoelasticity for orthotropic materials. Fibre Science and Technology 21:341–351
70. Patterson EA (1992) Application of phase stepping in photoelastciity to composite materials. Proc of Conference on Experimental Techniques and Design in Composite Materials, Cagliari:335–348
71. Tomlinson RA, Patterson EA (1998) Evaluating Characteristic parameters in integrated photoelasticity. In: Allison IM (ed) Advances in design, testing and analysis. Proc XII Int Conf on Experimental Mechanics, pp 495–500
72. Aben H, Ainola L, Anton J (1999) Half-fringe phase-stepping with separation of the principal stress directions. Proc Estonian Acad Sci Engng 5(3):198–211
73. Morimoto Y and Imamoto Y (1995) Determination of photoelastic fringe order by wavelet transform. Proc of the Japan Society for photoelasticity 15(2):29–34

Index

achromatic 223, 260
adhesive 59, 359
amplitude of light 149, 151
analog-to-digital 173
analyzer 11
angle
– of incidence 4, 5
– of reflection 4
– of refraction 4
annular hatch 355
apochromatic 263
applets 387
arithmetic operations 75
average intensity 86, 87
azimuth 7, 37, 143

Babinet-Soleil compensator 22, 64
band pass filter 86, 87
beam splitter 144, 159, 393, 395
binary image 75, 89, 92
birefringence 1, 4
birefringent coatings 66
– correction factors 52–55
– materials 58
– Poisson's ratio mismatch effects 56, 57
– reinforcing effects 53–55
bit 74
blur 94, 383, 388, 389
body forces 39, 40
boundary conditions 39, 315, 319, 327
B-rep 349
B-splines 349
byte 74

CAD 349–351
CAM 350
camera
– CCD 72, 98–103, 383, 384, 391–394
– Cranz-Schardin 390–392
– image converter 391
– rotating mirror 389
– solid-state 71, 72
– TDI 384, 394, 395
– vidicon 71
calibration
– of photoelastic coatings 62–64
– of photoelastic materials 10, 24–28
– of photo-orthotropic composites 398, 399
carrier fringes 188, 189, 395
casting techniques 59, 363
cathode ray tube 73, 391
CATIA 349, 350
CCD camera 72, 98–103, 383, 384, 391–394
CCD modes
– frame interline transfer 393
– frame transfer 383, 392, 393
– interline transfer 100
– TDI 384, 394, 395
CD 80
CFA 228, 229
characteristic directions 37
characteristic parameters 39
– determination 370–378
characteristic retardation 37
characteristic rotation 37
charge conversion 100
CIE 223
circular frequency 2
circular polariscope 18–20, 149, 150
CMY colour model 224
colour
– fundamentals 223
– hue 223–225
– models 224, 225
– primaries 223, 228–230
– saturation 223–225
– secondaries 228
colour models
– CMY 224
– HSI 224–226, 234
– RGB 223–226, 234–240
– YIQ 224
coloured filtered array 228, 229

compatibility conditions 39, 40
compensation techniques 23
compiler 80
COMPLEMENT 75
composite materials 395
composite video 229
contact stresses 268–273
– Hertzian 272
– non-Hertzian 272, 273
contrast stretching 89
convolution 94
co-ordinates
– frame-buffer 69, 266, 267
– model 266, 267
– pixel 266, 267
correlation factor 381
cracks
– criteria for direction of propagation 297
– stress field evaluation 284–290, 294–296
Cranz-Schardin camera
– conventional 390, 391
– digital 392, 393
– operating principle 390
critical temperature 37
CSG 349, 350

dark current 102, 230
data analysis 135
differential equations of equilibrium 39
digital dynamic photoelasticity 388–395
digital dynamic recording 391–395
digital fringe multiplication 109
digital image 69
digital image processing 77
digital-to-analog 174
digitisation 71, 103
discretization 281, 316, 323, 324
dispersion 256
display
– colour 224
– flat panel 104
– monochrome 73
Dos 206, 210, 387
dynamic range 84, 89, 101, 173
dynamic thresholding 92–94
edge detection 94–97
EIA 170 70
eigen value 35, 166, 167
ellipticity 42, 48, 143, 370
enhancement
– colour 225
– contrast stretching 89
– histogram equalisation 90
– image averaging 75
– image negative 89, 90
– max filter 98
– median filter 88
– min filter 98
– spatial filtering 86, 88
epoxy resin
– CIBA-GEIGY 5170 361
– CY230 292
– SG 95 363, 365
– SOMOS 6100 362
erosion 98
extensometer 63, 303
extraordinary ray 8

FDM 355–358
FEM
– formulation 319–322
– modelling 323–325
field equations
– compatibility 40
– equilibrium 39
– stress-strain 50, 52
field lens 390
fill factor 100, 101
filter
– band pass 86, 87
– convolution 94
– high pass 86, 87
– linear 88
– low pass 87, 88
– max 98
– median 88
– min 98
– non-convolution 95
– non-linear 88
– spatial 76, 86, 88
finite difference formulation 309–312
Fourier transform 89, 188–193
fractional retardation 153, 155–159, 161, 175, 180, 184, 185, 189, 191
frame buffer card 74
frame integration 393
frame interline transfer 393
frame rate 30, 391, 392
frame transfer 383, 392, 393
frequency 3, 14
– circular 2
fringe
– isochromatics 12
– isoclinics 12

– isopachics 326
fringe contours
– plotting of 29, 161, 225, 324–327
– pseudo contours 336, 337
fringe gradient 22, 32
fringe multiplication
– digital 109
– optical 107
fringe ordering 29–33, 135, 386
fringe sharpening 107, 108
fringe thinning
– binary-based 112
– global 115–126
– mask-based 113–115
fringe tracing 133, 132
fused deposition modelling 347

gamma 104, 111
global thresholding 92–94
gradient direction 136, 170
grey level
– quantization 69, 382
– shades 68
– transformation 89

half fringe photoelasticity 110
half-wave plate 9, 370
harmonic wave 2
hatch
– annular 355
– retracted hatch 353
– STaggered hatch 353
– Tri-Hatch 352–354, 361
– zigzag 355
HFP 110, 111, 137, 138, 400
histogram equalisation 90, 121, 122, 130
hole method 342
holography 144, 145, 325
HSI colour model 224–226, 234
hue 223–225

I-DEAS 349, 350, 366
illumination 89, 383, 384
image aquisition 80
image averaging 75
image differentiation 94–96
image enhancement 88
image negative 89, 90
image preprocessing 121, 122
image segmentation 82, 91, 105
image storage 74, 385
image storage format
– BMP 385
– FLAT 385
– GIF 385
– TARGA 385, 386
– TIFF 385
image subtraction 109, 137
intensity of light transmitted
– circular polariscope 149, 150
– plane polariscope 146–148
– reflection polariscope 247
– with white light as source 230–234
intensity variation 84–87, 218
isochromatics 12
isochromatic-isoclinic interaction 145, 157, 181
isoclinic-isochromatic interaction 158, 260
isoclinics 12
isopachics 326, 329
isotropic point 30–32

Java 387
Jones calculus
– circular polariscope 19, 149, 371
– plane polariscope 17, 146, 148, 373
– reflection polariscope 247
– Tardy's method 23
Jones matrix
– polarizer 17
– quarter-wave plate 17
– retarder 17

laser
– Argon-ion 351
– He-Cd 351
– He-Ne 183, 222
– Ruby 391
laser diode 395
least-squares method
– linear 27
– non-linear 270, 281, 333
LED 392–395
light
– monochromatic 3
– tricolour 255
– white 3
light ellipse 7, 37, 42, 370
light vector 2, 3
linearity 101
linker 80
load stepping 142, 191
loading fixture 291, 334
longitudinal wave 388
low pass filter 87, 88

macro 79, 121
magnification 213, 267
mask 76, 94–96
material fringe value
– strain 50
– stress 10, 26, 50
max filter 98
maximum shear stress 28
meaningful discretization 323, 327
median filter 88
Microsoft-C 80, 160, 168
MIL 385
min filter 98
mixed image 109
mixed-mode loading 276–278, 296
model making 348, 363
Mohr's circle
– of birefringence 397, 399, 400
– of stress 35, 303, 305, 340
Moiré 145, 325, 366
monochrome image 69, 385
morphology
– closing 98
– dilation 98
– erosion 98
MOS capacitor 98
multi-parameter method 281

neighbours of a pixel 74
noise
– dark current 102
– quantization 218
– shot 103, 230
– spatial discretization 218
NTSC 70, 73, 104
NURBS 349

oblique incidence 304, 339
OET 261, 263
optic axis 6
optical fringe multiplication 107
optically enhanced tiling 212
optically equivalent model 37–39
order of fringe 29
ordinary ray 5, 6
overdeterministic methods
– contact stress 270–272
– material stress fringe value 26–28
– stress field in fracture mechanics 281–287
– stress separation 314–316, 333

PAL 70, 73, 104
pan 81
phase difference
– relative 7–9, 13
phase shifting technique
– circular polariscope 155–180, 244–248
– colour domain 244–248
– monochrome domain 155–180
– plane polariscope 153–155
phase unwrapping 167, 184, 192
photoelastic materials 41, 58, 388
– calibration 24–28
– properties of 41, 48
photoelastic strain gauges 49
PHOTOSOFT_H 30–33, 46, 196
PIP 1024B 79
pixel 69, 74
pixel jitter 103
Poisson's ratio 56
polariscope
– circular 18–20
– plane 10–12
– reflection 48, 246
– scattered light 380
polarising axes 9
polarization 3
polarization stepping
– colour domain 253, 254
– monochrome domain 185–186
polarised light
– circularly 3, 4
– elliptically 3, 4
– linearly 3, 4
– states of 3, 4
polaroid 4
Prewitt 95–97
primary variable 319
principal strain 50
principal stress 9
Pro/ENGINEER 349, 350
progressive scan 383
properties of photoelastic materials 41
pseudo colour 82

quantization 69
quarter-wave plate
– error 20, 233, 234, 251
– Jones matrix 17
– mismatch 174, 177–180

Raleigh wave 388
rapid prototyping
– fused deposition modelling 355–356

– selective laser sintering 347
– solid ground curing 354–355
– stereolithography 351–354
rapid tooling 348–363, 367
real time 391
reference beam 145
reflection polariscope 48
refraction effects 304, 333
refractive index 4, 6, 9, 108
relative phase 6, 11, 397
residual birefringence 26
residual stress 59, 65, 365
retardation matrix 16
retracted hatch 353
RGB colour model 223–226, 234–240
RGBP 235, 240, 254
rotation matrix 15, 16, 38
RP 348–363
RT 348, 363–365

saddle point 30–32, 86, 87, 125
scattered light intensity 380
SCF 57, 66, 303
scroll 81
secondary principal stresses 304–305, 339
secondary variable 319
seed point
– primary 204, 205, 209–211
– secondary 204
segmentation 91–97
separation of stresses
– finite difference 309–314
– finite element 318–332
– hybrid techniques 316–317
– integration of compatibility condition 309–312, 318–332
– integration of stress difference equation 312–314
– least squares method 314–316, 333
– oblique incidence method 304–307, 332–341
– shear-difference method 307, 308, 340, 341
– using only isochromatic data 317, 318, 332–337
shear-difference method
– in three-dimension 340, 341
– in two-dimension 307–308
SIF
– definition 265, 274
– determination 278–283
sign convention 15, 24
silicone mould 363, 366
singular point 30–34
sink 30, 32
slicing of complex models 337
Snell's law 4–6
Sobel 95–97
software
– annul.c 119
– boundary.c 200
– compare.c 127
– curupip.h 80
– eigen.c 167
– excelplo.c 205
– femfrn.c 275, 326
– frnmul.c 109
– fsigma.c 28, 43
– noise.c 202
– obliq2d.c 306
– oring.c 182
– orthopip.h 118, 126
– pipdef.h 80, 81
– plus1.c 127
– pseudcol.c 82
– reconstr.c 29, 43
– shdiff.c 308
– shown.c 207
– snap512.c 81
– snapzoom.c 80
– thinpip.h 113, 114
– threspip.h 93
– tilepip.c 132
– tracepip.c 134
– unwrap.c 204, 205
– x3.c 127
– 2dplane2.c 182
– 2dtile.c 170
solid ground curing 354, 355
source 30, 32
spatial filter 86, 88
spatial resolution 71
speckle 144
spectral content analysis 249–251
spectral limits 234
spectral response 230, 231
spectrum 191
STaggered hatch 353
stair-step 351
STAR-WEAVE 353
stereolithography 351–354
STL format 350
storage devices 74
strain-optic law 50, 397
stray light 151

streak photoelasticity 394
streamline fillets 303
stress concentration factor 57
stress freezing 26, 36, 192
stress function
– Westergaard 273–276
stress intensity factor
– definition 265, 274
– determination 278–283
– multi-parameter method 281–283
– overdeterministic method 281–283
– two-parameter method 280
stress separation in
– 2-D photoelasticity 304–337
– 3-D photoelasticity 337–342
– reflection photoelasticity 342
– photo orthotropic elasticity 400
stress-optic coefficient 9, 10
stress-optic law 9, 10, 396
stress separator gauge 343
stress-strain relation 50, 52
strip coating 342
subslice 338
sub-subslice 338

Tardy's method 22–24, 111
TFP 222, 234–242, 252, 254
three fringe photoelasticity 222
thresholding
– dynamic 92–94
– global 92–94
tiling
– fringe skeleton 131–133
– OET 212
– phase unwrapping 210–211
– retardation calculation 169–172
time averaging 80, 81
time-edge effect 47, 327
tomography
– scalar 379
– tensorial 378–382
translation 267, 268
Tresca yield criterion 60
tricolour
– light source 222, 255
– photoelastic method 254–259
Tri-Hatch 352–354, 361
two-parameter method 280

ultraviolet light 351, 354
undefined zones of
– characteristic direction 377
– characteristic rotation 377
– isoclinics 157, 185
unwrapping 168, 169

variance 382
vector
– electric 2
– light 3
– magnetic 2
velocity of
– light 2
– Raleigh wave 388
– longitudinal wave 388
video
– component 224
– composite 229
video camera 71, 104
video format 70–72
video standards
– CCIR 70
– EIA 170 70
– NTSC 70,73,104
– PAL 70, 73, 104
– SECAM 70, 73, 104
vidicon camera 71
von Mises yield criterion 337–339

wave front 3
wave plates 12
wave propagation 41
wavelet transform 401
WEAVE 352
well capacity 384
white light 3
Windows 387
World Wide Web 387

Xenon flash lamp 389, 392

yield criteria
– Tresca 60, 303, 339–342
– von Mises 337
YIQ colour model 224
Young's modulus 358, 359

zero crossing 115
zeroth order fringe 33
zigzag hatch 355
zoom 79, 81

MIX
Papier aus verantwortungsvollen Quellen
Paper from responsible sources
FSC® C105338

If you have any concerns about our products,
you can contact us on
ProductSafety@springernature.com

In case Publisher is established outside the EU,
the EU authorized representative is:
Springer Nature Customer Service Center GmbH
Europaplatz 3, 69115 Heidelberg, Germany

Printed by Libri Plureos GmbH
in Hamburg, Germany